The Human Body in Equipment Design

The Human Body in Equipment Design

ALBERT DAMON
LECTURER ON ANTHROPOLOGY · HARVARD UNIVERSITY

HOWARD W. STOUDT
ASSOCIATE IN PHYSICAL ANTHROPOLOGY · HARVARD SCHOOL OF PUBLIC HEALTH

ROSS A. McFARLAND
GUGGENHEIM PROFESSOR OF AEROSPACE HEALTH AND SAFETY · HARVARD SCHOOL OF PUBLIC HEALTH

HARVARD UNIVERSITY PRESS · Cambridge · Massachusetts   1966

Material from Sections 10.3.1 and 10.3.2 of *Human Engineering Guide to Equipment Design,* edited by Clifford T. Morgan and others, copyright 1963, McGraw-Hill Book Company; and material from pages 82–86 and Figures 10, 13, 15, 16, and 18 from *Basic Motion Timestudy* by G. B. Bailey and R. Presgrave, copyright 1958, McGraw-Hill Book Company, are used by permission of the publisher.

Preface

Of the many practical applications of physical anthropology, which include medicine, forensic science, personnel selection, physical education, and engineering, the most immediate is the design of equipment for human use. It seems elementary that such equipment should suit human dimensions and biomechanical capabilities, but scientific progress in this direction has been slow.

There is a tradition at Harvard University for the study of physical anthropology in relation to equipment design. All of the authors of this book received some of their original inspiration and training from the late Earnest Hooton, Professor of Anthropology at Harvard University from 1913 to 1954. He was interested not only in human evolution and the more traditional phases of anthropology but also in man's physical adjustment to our industrialized civilization. For example, he was one of the first to recognize the applicability of anthropometry to the design of clothing, seats, and various types of equipment. It was his students who carried out much of the research which has been utilized by the military services and industrial concerns in the United States.

The first sustained, systematic attempt to delineate the field, to develop principles and accumulate data, and above all to apply them, took place during the Second World War. The most effective use of physical anthropology was in the fields of clothing, oxygen masks, and cockpit design. A publication summarizing these studies, entitled *Human Body Size in Military Aircraft and Personal Equipment,* was issued as Army Air Forces Technical Report No. 5501, Wright Field, Dayton, Ohio, 1946.

In the late 1940's investigations were undertaken at the Harvard School of Public Health to improve safety in the use of all types of equipment, civilian and military. Evaluations were made of representative trucks, buses, and automobiles from the point of view of human engineering and biotechnology, one of the most important factors being the extent to which the driver was effectively integrated with his equipment. An anthropometric survey was made of over 300 bus and truck drivers, and by means of percentile distributions, such measurements as sitting eye heights, arm reaches, and leg lengths were related to interior dimensions of the vehicles. One of the monographs by the present authors, resulting from these studies, was *Human Body Size and Capabilities in the Design and Operation of Vehicular Equipment,* Harvard School of Public Health, Boston, 1953.

Despite these efforts and the success of the few applied physical anthropologists currently active in military and industrial laboratories, the discipline remains an academic stepchild, untaught by university anthropologists and largely unavailable to their students as well as to students of psychology, engineering, architecture, and industrial design.

The lack of a comprehensive text or handbook like those available in psychophysiology and the other ergonomic sciences has for some time been a handicap to those professional workers involved in research, practice, and teaching within and among the disciplines concerned. The present volume is offered to help meet this need and to improve the design of all types of equipment for human use. A section of the Joint Services *Human Engineering Guide to Equipment Design,*

McGraw-Hill, New York, 1963 (Chapter 11, "Anthropometry"), was prepared by the present authors under contract from the Aerospace Medical Laboratories, Wright-Patterson Air Force Base, Ohio. Although this work forms the nucleus of the present volume, a great deal of additional material has been included from a wide variety of sources, such as the anthropometric phase of the U.S. National Health Examination Survey in which the authors participated.

We hope this book will prove useful to biological scientists—anthropologists, psychologists, and physicians; to architects, engineers, and designers of various kinds of equipment; and particularly to teachers and students in all these fields.

Albert Damon
Howard W. Stoudt
Ross A. McFarland

Harvard University
November 26, 1965

Acknowledgments

Our greatest debt is to those whose work we cite, particularly the authors of United States government publications. Part of this book originated in an Air Force contract monitored by H. T. E. Hertzberg of the Anthropology Branch, Aero Medical Laboratories, Wright-Patterson Air Force Base, Ohio. He and his colleagues—Walter Grether and Kenneth Kennedy at Wright-Patterson, Edmund Churchill and John McConville at Antioch College, Yellow Springs, Ohio—have aided us in many ways. Russell W. Newman of the U.S. Army Research Institute of Environmental Medicine and Robert H. White of the U.S. Army Natick Laboratories, Natick, Massachusetts, have graciously allowed access to unpublished data.

Sources of research support were the Guggenheim Center for Aerospace Health and Safety and the Department of Physiology, Harvard School of Public Health; the Program Project entitled "Occupational and Environmental Health," granted by the Bureau of State Services, U.S. Public Health Service, to the Division of Environmental Health Sciences and Engineering, Harvard School of Public Health; and the American Heart Association. The book was written in part while Albert Damon held an Established Investigatorship of the American Heart Association.

Additional sources of support have been the Commission on Military Accidents of the Armed Forces Epidemiological Board through funds from the Office of the Surgeon General, Department of the Army; the Office of Naval Research, Department of the Navy; the U.S. Public Health Service (National Health Examination Survey); the U.S. Veterans Administration (Normative Aging Study, Boston Outpatient Clinic); the General Motors Technical Center; the National Association of Motor Bus Operators; and the American Trucking Associations.

The following persons, organizations, and journals have kindly granted permission to reprint tables or figures, documented where they appear in the text: *Aerospace Medicine,* David E. Bass, Paul W. Braunstein, F. Gaynor Evans, Heywood-Wakefield Company, *Human Factors, Journal of the American Medical Association, Journal of Applied Physiology, Journal of Bone and Joint Surgery,* National Academy of Sciences, *New England Journal of Medicine,* W. B. Saunders Company, Charles C Thomas, Publisher, D. Van Nostrand Company.

We are especially grateful to the McGraw-Hill Book Company for permission to publish certain figures (in Chapter II) and portions of text (in Chapter IV) which we first prepared for the *Human Engineering Guide to Equipment Design,* edited by Clifford T. Morgan and others, copyright 1963.

Finally, we thank those concerned with the actual production of this book: Claire Wasserboehr for superb typing, John C. G. Loring for help with the index, and Virginia Wharton and Nancy Northam of the Harvard University Press for wise and patient editorial counsel.

Contents

Tables

Figures

The Human Body in Equipment Design

Introduction — I

1. Purpose

This book is intended as a guide for the designer of equipment involving human body size and mechanical capabilities. In summarizing the application of anthropometric data to design, it may serve as a text for physical anthropologists as well. Specific recommendations will be presented for many major biomechanical features of man-machine integration, together with data and methods applicable to the solution of other problems. Although the two aspects of applied physical anthropology, fitting men into spaces and fitting gear onto men, differ only in degree, tending in fact to coincide in the design of closely fitting space envelopes like the full pressure aerospace suit, fitting personal equipment to the man is a distinct field of study. Such equipment, which includes clothing, masks, goggles, and the like, will be considered here only as it affects the operator's dimensions and mechanical capabilities.

No sharp line can be drawn between civilian and military equipment since many civilian products, such as transport aircraft, automotive vehicles, furniture, and hospital supplies are used unchanged by the services for combat or "housekeeping" purposes. This book therefore presents anthropometric data on both civilian and military groups.

In addition to the customary anthropometric measurements, certain basic data on the human body—tissue and chemical composition, resistance to force, body circumferences, surface areas, and centers of gravity—are included because they find application in many engineering contexts, such as acceleration, blast and radiation effects, and dummy construction, and may eventually prove useful in ways we cannot now foresee.

A FEW DEFINITIONS

Anthropology, which means "the science of man," deals with man in his biological and social aspects. *Physical anthropology,* or human biology, is the study of the somatic characteristics of individuals or groups of men. Some of these bodily traits are quantitative, like height and weight, whereas others are qualitative, like skin color, hair form, and blood group. *Anthropometry* refers to human body measurement, with which the equipment designer is mainly concerned. Such measurement includes body dimensions and the strength, speed, and range of motions. Anthropometric data can be applied to various fields, including equipment design—as in this book—medicine, physical education, and personnel selection. The terms *applied anthropometry* or *applied physical anthropology* will be used here interchangeably to denote this "human factor" in equipment design.

Human engineering (human factors engineering, ergonomics, biotechnology) is not a single scientific discipline but a synthesis which integrates the biological sciences—psychology, anthropology, physiology, and medicine—with engineering. Human engineering means "fitting the machine to the man, and keeping him functioning with efficiency, with safety, and without discomfort in any environment" (Hertzberg, 1955). The human engineer's problems are consequently quite diverse. Among the most complex are the workspaces in military aircraft, with a profusion of "displays" to be interpreted and controls to be operated under the environmental stress of extremes of temperature, noise, vibration, and acceleration, and frequently under severe psychological stress as well. Much simpler are such problems as the design of a tool or control handgrip, or the height of a table surface.

Background of Human Engineering

The need for close coordination of men and machines became acute during World War II when the complex-ity of some types of modern military equipment began to outstrip the abilities of the men trained to operate them (Damon and Randall, 1944). The machines, superb examples of mechanical design, when properly handled could in theory perform just as the designers intended. In practice, however, the abilities necessary for efficient operation often exceeded those of the average or even the gifted operator. Man had become the weak link in the man-machine complex. With intensive training, he might be brought to peak efficiency, but such training is lengthy, expensive, and often impractical. Even then, some types of equipment could still not be operated at maximum efficiency, while others contributed many times more than the man to the total error of the man-machine system (Chapanis, 1951).

A major reorientation in the concept of equipment design was necessary to improve system efficiency. Since man cannot be redesigned, his dimensions, capabilities, and limitations must form the basis for machine design. This realization marked the first application of "human engineering" principles on a broad scale.

Although the need for human engineering first became apparent in the military services, the concept of designing in terms of human capabilities and limitations applies to civilian equipment as well. Military equipment must not fall short or fail in combat, and though the consequences of defective design are not so catastrophic in civil life, they are nevertheless significant.

It has been suggested that the increasing use of automation may virtually eliminate the need for human engineering by eliminating the human operator. In fact, the opposite is occurring. Some operators are eliminated, but the tasks of those who remain are more complex than ever, thereby enhancing the need for human engineering.

As already noted, there are many "human factors" in design, of which anthropometry is but one and hardly the most important. Most discussions of human engineering properly stress psychophysiology (Fogel, 1963; Human Engineering Guide to Equipment Design, 1963). Much basic work remains to be done in these disciplines, with application still farther in the future. On the other hand, anthropometry affords a practical, inexpensive method for improving man-machine performance immediately and to a tangible degree.

2. General Considerations

ANTHROPOMETRY IN THE MAN-MACHINE SYSTEM

The human body, in structure and mechanical function, occupies a vital though often neglected place in man-machine design. Failure to provide a few inches, which may be critical for the operator, can jeopardize the performance or even the existence of both man and machine. With forethought, these critical inches can usually be provided without compromising design. Reliable anthropometric data and procedures provide the most powerful tool available today for the optimal sizing of many mass-produced items, from oxygen masks to airplane cockpits and truck cabs.

GOALS OF THE DESIGN ENGINEER IN ACCOMMODATING HUMAN DIMENSIONS

(1) All men should be able to operate all machines. This being an ideal goal, compromise on 98 or 95 per cent or even 90 per cent of the intended users—but no less—may be necessary. Universal operability is desirable because: operating conditions often require crew interchangeability; the supply of qualified operators for complex machines is severely limited, and it is poor policy to limit this supply still further by restrictions on body size; and equipment which imposes size limitations complicates the selection and training of operators.

Universal or virtually universal operability of equipment is generally feasible because: few limitations on human size are imposed by the gross dimensions of a machine—the fault usually lies with design details (Fig. 1); the range of human variation is small, relative to machine dimensions, even in close quarters like airplanes or small automobiles; and the human size range, or at least 90 to 98 per cent of it, is in general readily accommodated by adjustable devices.

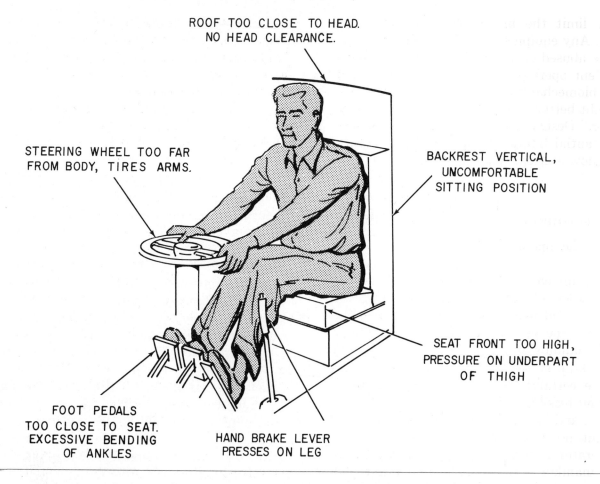

ROOF TOO CLOSE TO HEAD.
NO HEAD CLEARANCE.

STEERING WHEEL TOO FAR
FROM BODY, TIRES ARMS.

BACKREST VERTICAL,
UNCOMFORTABLE
SITTING POSITION

FOOT PEDALS
TOO CLOSE TO SEAT.
EXCESSIVE BENDING
OF ANKLES

HAND BRAKE LEVER
PRESSES ON LEG

SEAT FRONT TOO HIGH,
PRESSURE ON UNDERPART
OF THIGH

1. Design details—not over-all dimensions—usually cause the misfits.

(2) Do not limit the machine's performance by human failure. Any equipment, however cleverly engineered, may be abused or destroyed by an uncomfortable or inefficient operator. If this results from some anatomical or biomechanical oversight by the designer, the failure might better be termed "design error" than "operator error." Design error may in fact be responsible for a substantial fraction of aircraft and automobile accidents now ascribed to human failure.

GENERAL ANTHROPOMETRIC PRINCIPLES IN DESIGN

(1) Consider the operator. Once this is done, all else follows.

(2) Consider him early. Ideally, the man-machine system should take advantage of the most efficient functions of man and machine. This means that the human body must be kept in mind from the earliest discussion and planning stages of design. The drawing board and mock-up stages may be too late, and the prototype stage certainly is too late. At that point, even needed and feasible redesign may cost too much in time, money, and lost production to be considered. Use forethought, not afterthought (Fig. 2).

(3) The operator is functional. He is not a static template or a manikin in shirtsleeves and trousers. He may have to wear bulky gear and perform complex movements or assume unusual positions (Fig. 3). He requires an adequate visual field, inside and outside his workspace. He must be kept comfortable, safe, and efficient.

(4) Functional men vary in size and physical capacities. Age, sex, race, and occupation affect body dimensions and capabilities. The group of operators for whom an item is intended may differ significantly from those groups whose measurements are available. And within groups, individuals may vary widely. The significance of physical variations, both between groups and within groups, can be assessed only by direct test.

(5) Allow ample margins of safety for both men and equipment. Both may be subjected to unusual demands. Psychomotor adjustments possible in the laboratory or even in normal operation may fail under stress. Some operators at all times, and most under sufficient stress, will fall short of their potential performance. Therefore, always exceed minimum spatial and mechanical allowances rather than shave tolerances too closely.

(6) Evaluate human accommodation and performance in complete, functional equipment. Men of varying size, especially those toward each extreme of the range, should test machines under conditions simulating actual operation as closely as possible. (For specific instructions, see section 5 of this chapter.) The operator should wear complete sets of typical gear, and

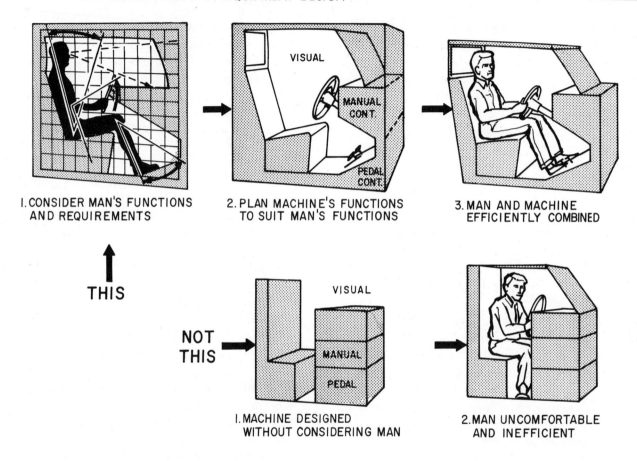

1. CONSIDER MAN'S FUNCTIONS
AND REQUIREMENTS

2. PLAN MACHINE'S FUNCTIONS
TO SUIT MAN'S FUNCTIONS

3. MAN AND MACHINE
EFFICIENTLY COMBINED

THIS

NOT
THIS

1. MACHINE DESIGNED
WITHOUT CONSIDERING MAN

2. MAN UNCOMFORTABLE
AND INEFFICIENT

2. Consider the operator early in design.

3. The operator is functional.

all machine accessories should be included in drawings or mock-ups, in wooden facsimile if not otherwise available (Fig. 4). Persons experienced in the use of an item of equipment, and physically representative of the ultimate user, should be consulted from the outset and throughout the design process. They should serve at formal mock-up and acceptance inspections. Prevention being better than cure, there is no substitute for close collaboration between the designer and his customer from the earliest stages of planning. Once the item is produced, man-machine performance should be continuously monitored in the field, both by specially assigned observers and by extending routine "unsatisfactory reports" to include design as well as mechanical defects.

(7) Design details, not over-all dimensions, usually cause the trouble. Minor rearrangements will often suffice.

3. The Use and Abuse of Statistics

The purpose of statistics in applying anthropometry to equipment design is to provide a quantitative guide to the evaluation of design and to the selection of operators and test subjects.

The need for statistics is imposed by human variability, as discussed in the following section. Simple statistical procedures will then be outlined on pages 16 to 29.

4. Test the operator's accommodation wearing complete gear in complete machines.

5. Human responses to identical external conditions vary.

HUMAN VARIATION; OR, WHY THE BIOLOGIST AND ENGINEER SPEAK DIFFERENT LANGUAGES

One way in which man differs from other elements in equipment design is his innate variation. A chemical compound has one melting point and one coefficient of conductivity. Metals, plastics, or textiles can likewise be described in terms of relatively fixed properties, with behavior precisely specified under given external conditions, as of heat, cold, vibration, mechanical stress, and the like. Man is not so uniform.

External Variation

Human body size and capabilities are of course subject to environmental modification. For example, overeating, starvation, physical activity, and disease can all influence body weight. Various body tissues may be altered by heat—desert troops maintain total weight but replace fat with muscle (Baker, 1955); or by barometric pressure (internal gases, as in sinuses, ear canals, intestines, or blood stream, expand at altitudes). Unlike inert material, however, human responses to identical external conditions will vary. "One man's meat is another man's poison" (Fig. 5).

Innate Variation

Also unlike inert material, man's inborn variation greatly exceeds that due to external conditions. Men vary, due to heredity,

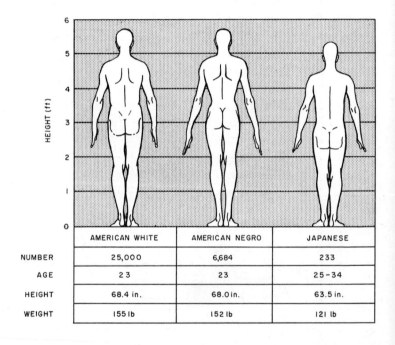

	AMERICAN WHITE	AMERICAN NEGRO	JAPANESE
NUMBER	25,000	6,684	233
AGE	23	23	25-34
HEIGHT	68.4 in.	68.0 in.	63.5 in.
WEIGHT	155 lb	152 lb	121 lb

6. Physique and approximate proportions of American white and Negro soldiers and of Japanese farmers (data from Newman and White, 1951; Ishii, 1957).

(1) From group to group. White persons, Negroes, and Orientals differ significantly in average physique, despite some overlapping (Fig. 6). Differences in body size and proportions, as indicated by inappropriate clothing-size tariffs, between American and Southeastern European troops led to a NATO anthropometric survey of Turkey, Greece, and Italy (Hertzberg *et al.,* 1963). Differences in age and sex are other obvious causes for physical variation. Less obvious but sometimes critical factors include occupation and long-term changes in physique. Even among American white men aged 18 to 40, for example, truck drivers differ from research workers (Fig. 7), and military personnel from the general population. Aircraft pilots differ from soldiers and aerial gunners. Bomber and fighter pilots differ from one another in some aspects of body form, as do navigators, bombardiers, and tank crews (Damon, 1955).

Within a country, physical measurements may change over a period of time. American soldiers of World War II averaged 0.7 inch taller and 13 pounds heavier than those of the first World War, 25 years earlier (Fig. 8). This trend, which has been noted in Japan as well as in many Western countries, seems to be continuing (Karpinos, 1961; Newman, 1963).

Thus, human variation is such that the group which will use the equipment must be precisely specified and,

	TRUCK DRIVERS	RESEARCH WORKERS
NUMBER	269	20
AGE	37	33
HEIGHT	68.5 in	70.6 in
WEIGHT	167 lb	167 lb

7. Truck drivers (Damon and McFarland, 1955) differ physically from research workers (Garn and Gertler, 1950).

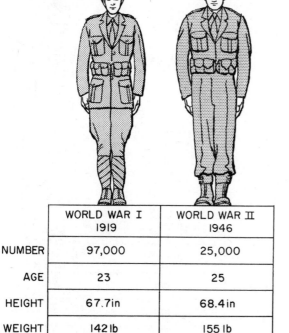

	WORLD WAR I 1919	WORLD WAR II 1946
NUMBER	97,000	25,000
AGE	23	25
HEIGHT	67.7 in	68.4 in
WEIGHT	142 lb	155 lb

8. Long-term size increase in American soldiers (data from Davenport and Love, 1921; Newman and White, 1951).

if possible, measured directly. One cannot assume that it resembles closely enough for design purposes other groups for whom anthropometric data are available.

(2) Within groups. Within any sizeable group, body dimensions are distributed along a bell-shaped "normal" curve (Fig. 9). Most persons are included in the mid-range of the distribution, with progressively fewer toward either extreme. The human biologist therefore deals with ranges rather than the sharp points to which the design engineer is accustomed and which he would prefer. There is no one standard man or height or weight, but only approximations to ranges. Hence, statistics are required to utilize anthropometric data for design purposes.

To summarize: men vary from group to group and within groups, in respect to dimensions and capabilities; the human engineer thus works not with points but with ranges, with tolerance limits broader than for inert material; his recommendations are approximations, valid for a limited set of conditions (age, race, sex, occupation), rather than precise formulas, universally applicable; he specifies a range of human dimensions to be accommodated, since a single equipment dimension may not fit his entire group; and adjustability becomes a keynote of design for human use (Fig. 10).

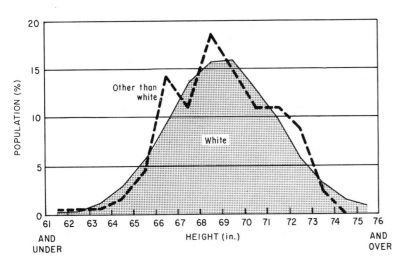

9. The "normal" bell-shaped curve. Stature among 5,911 white and 173 nonwhite Navy recruits (after U.S. Navy, 1949).

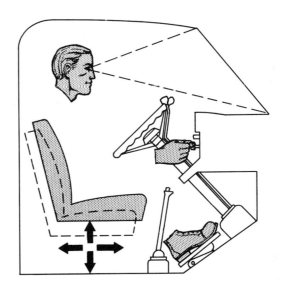

10. Adjustability to critical points should enable all or nearly all men to operate all machines.

ANTHROPOMETRIC DATA—
RELIABILITY AND LIMITATIONS

To be useful in equipment design, anthropometric data should satisfy certain criteria. Sample selection will be discussed more fully in section 6 of this chapter.

(1) The groups measured should be representative of the equipment user. Ideally, every person expected to use an item of equipment should be measured, but only rarely, as for a small group of test subjects or astronauts, can this ideal be realized. In general, samples must be selected for measurement. Such samples should reflect the characteristics of the intended users in respect to the major determinants of physique already noted—namely, age, sex, racial or ethnic ancestry, and occupation. Other relevant characteristics are geography (Pacific Northwest men average 0.6 inch taller and 3 pounds heavier than New Englanders—Brues, 1946; Newman and Munro, 1955) and socioeconomic status, college men being larger than noncollege men. The use of volunteers may make a sample unrepresentative ("biased"), since persons with poor or extreme physiques tend to avoid the anthropometrist.

If 100 per cent of operators cannot be measured, sample selection becomes critical, requiring formal statistical sampling techniques (Cochran, 1953). The most useful in the present context are random, stratified, and cluster sampling. Random in this sense does not mean haphazard, but means assigning consecutive numbers to subjects—alphabetically or in order of appearance—and then selecting according to published tables of random numbers. The aim in stratified sampling is to approximate the using population in respect to its major characteristics ("strata")—here, the determinants of physique—and then sampling randomly within each stratum. In cluster sampling, one selects several groups which *in toto* represent the population of interest, then samples randomly within each cluster. A military camp or processing center might be one such cluster, and samples from several regional centers could approximate a military population of equipment users.

Formal sampling is sometimes possible for industrial groups, but the foregoing paragraph may sound unreal to investigators in a military setting. Where ideal sampling is not feasible, the alternative is to measure a group matched to the intended operators in physique. The closer the matching dimension or dimensions are to those dimensions critical for the particular equipment concerned, the better. Although one usually knows little more than height and weight for a group beforehand, these two dimensions provide a fair general description of physique. Height correlates reasonably well with body lengths and weight with body breadths, depths, and girths.

Sample selection is discussed at such length here and in section 6 of this chapter because one cannot assume, without test, that anthropometric data from one group apply to another. For example, military personnel, highly selected in respect to age, health, and height-weight standards, cannot pretend to represent the general population. Even within a single military service, specialists are physically distinct, as already noted—fighter pilots differ from bomber pilots (Fig. 11), tank crews from military police, divers and submariners from shore patrols or honor guards.

(2) Samples should be large enough to yield reliable results reproducible from one sample to another. For most anthropometric purposes, 50 to 100 persons should be the minimum sample size, but the more the better. In general, larger samples are more vital for dimensions with a wide range, like weight, than for relatively range-restricted or small dimensions like those of the head, hand, and foot. See also section 6 of this chapter.

(3) Measuring techniques should be specified and standardized. This provides the only valid basis for comparing groups and for locating test subjects as percentiles of the user group (see "Percentiles," below). It makes a great difference whether measurements are made with or without clothing, in the erect or in the normal, slumped sitting position, or with the chest in inspiration or expiration. Most of the measurements presented here were taken by the techniques described by Hertzberg *et al.* (1963), with exceptions noted in the appropriate tables.

(4) Anthropometric data represent a compromise

	FIGHTER	BOMBER
NUMBER	210	1184
AGE	27	29
HEIGHT	68.8 in.	69.4 in.
WEIGHT	159 lb	166 lb

11. Fighter and bomber pilots differ in size (data from Hertzberg *et al.*, 1954).

among the foregoing criteria. One cannot take, on all drivers or pilots, let alone the civilian population, all the measurements that may conceivably become important for some item of equipment. The engineer must make educated guesses from available data and must sometimes measure small series himself, when a "trick" dimension is involved for which no data are at hand. The engineer should critically appraise the relevance of existing data to his own problem. For example, the amount of difference between bomber and fighter pilots, or Pacific Northwest Coast men and New Englanders, may not matter for certain dimensions or purposes. All women in military service (nurses, WAFS, WACS, WAVES, SPARS, Marines) are fairly comparable with one another.

This book presents the anthropometric data found most useful to date, specifies the number of subjects, and indicates the limitations, if any, of each study. Information missing from the tables reflects the scarcity of data currently available for major segments of the military services, notably sailors and marines and, as regards certain dimensions, for the civilian population as well. It is hoped that these gaps will eventually be filled.

PERCENTILES AND HOW TO USE THEM

The bell-shaped, "normal" distribution curve can be described in terms of its peak, or average value; its dispersion about this peak; and its range. The most useful statistics for the equipment designer are percentiles.

Definition: Percentiles are values corresponding to each man, if 100 men are lined up from least to greatest in any given respect. Percentiles are easily calculated for samples of any size.

Examples: (1) The 1st percentile in height is that height exceeded by 99 per cent of the group. (2) The 50th percentile, or median, is one kind of average, corresponding to the peak of the bell-shaped curve. Another statistical average, the mean,* is very close to the median for large, normally distributed samples. Half the members of a group have values below the 50th percentile, and half above. (3) The 95th percentile is that value which exceeds 94 per cent of the population and is exceeded by only 5 per cent.

Uses: Percentiles can serve the design engineer in several ways. The 50th percentile approximates the average value of a dimension for any group. Percentiles permit a more realistic concept of the range of dimensions to be accommodated than does the spread from least to greatest value encountered in a group. The extreme values represent chance occurrences which should be disregarded in designing equipment. Removing 1 per cent at both ends of the range will eliminate most of these "freak" values and leave a range covering 98 per cent of the population. For some dimensions and

---

*There are three principal averages, or measures of central tendency: the mean, or sum of measured values / number of individual measurements; the median, or 50th percentile; and the mode, or most commonly occurring value. The mode is of little importance in equipment design.

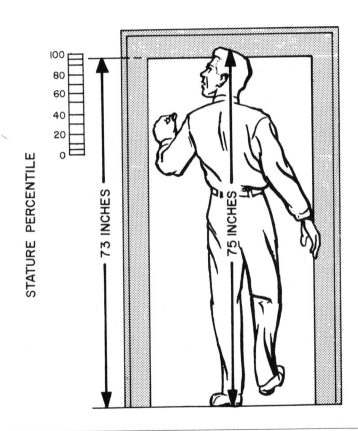

STATURE PERCENTILE

100
80
60
40
20
0

73 INCHES

75 INCHES

12. A door opening 73 inches high will inconvenience the 5 per cent of this group taller than 73 inches.

equipment, this range from 1st to 99th percentiles may be easily accommodated. For others the 5th-95th percentiles, or 90 per cent of the group, should be provided for. In general the designer should try to accommodate at least 90 per cent of his population, and 98 per cent or more if possible (see "Fallacy of the average man," page 20).

Percentiles afford a basis for estimating the proportion of a group accommodated or inconvenienced by any specific design.

Examples: (1) If a door is 73 inches high, this value approximates the 95th percentile of stature for the military and civilian populations—72 inches, plus 1 inch for shoes (Table 8).* Therefore, 5 per cent of the population will be inconvenienced (Fig. 12).

(2) A control placed 32.5 inches from the back of an operator's chair is at about the 55th percentile of functional arm reach (Table 25). Therefore, 55 per cent of the group—those with shorter arms—will reach it with difficulty, while only the larger 45 per cent are accommodated.

(3) To provide for 95 per cent of the population in respect to weight, a truck seat should be stressed for 215 pounds per nude civilian driver, and 192 pounds per nude Army driver (Table 7). A seat supporting only 150 pounds per nude driver would be inadequate for some 70 per cent of civilians and 55 per cent of soldiers.

(4) A seat adjustable over a 5-inch vertical range will

*The tables containing percentiles of body dimensions begin in chap. II with Table 7, Weight.

accommodate over 90 per cent of all civilian drivers, since 4.8 inches separate the 5th and 95th percentiles of sitting height (Table 14). The corresponding value for military drivers is 4 inches.

Percentiles permit the selection and accurate use of test subjects. A critical body dimension or physical ability of a test subject can be readily located as a percentile of the relevant population—provided, of course, that the population in question has been measured.

Percentiles aid in the selection of operators. If equipment imposes any limitation on the size of operators, misfits can be avoided by eliminating those whose critical dimension exceeds or falls short of the established cut-off point. To be sure, the cut-off point can be established in measurement units without using percentiles (say 70 inches tall and 180 pounds in weight), but percentiles indicate the proportion of potential operators rejected. If this proportion is too large, redesign is in order.

Example: During World War II, the Air Force considered raising gunners' height and weight limits to 73 inches and 180 pounds. However, even at the existing limits of 70 inches and 170 pounds, a significant percentage of gunners (30 per cent for one turret, 40 per cent for another) had trouble operating turrets. Larger men could not even enter the turrets. The proposal was accordingly rejected in favor of redesigning turrets.

How to Find Percentiles

Graphically, from normal probability paper: Anthropometric data may be presented in percentile form, but often are not. Even when they are, the engineer may need to find percentiles other than those given, such as beyond the 5th or 95th, or between the 60th and 70th percentiles. The use of normal probability paper, available from stationers or drafting suppliers, locates percentiles in a normally distributed group,

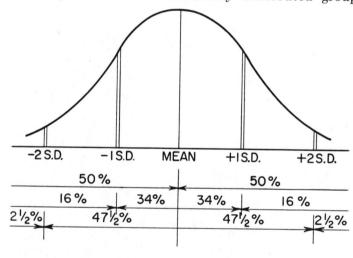

13. Normal curve showing percentages of the distribution covered by multiples of the standard deviation (after Emanuel et al., 1959; from Mainland, 1963).

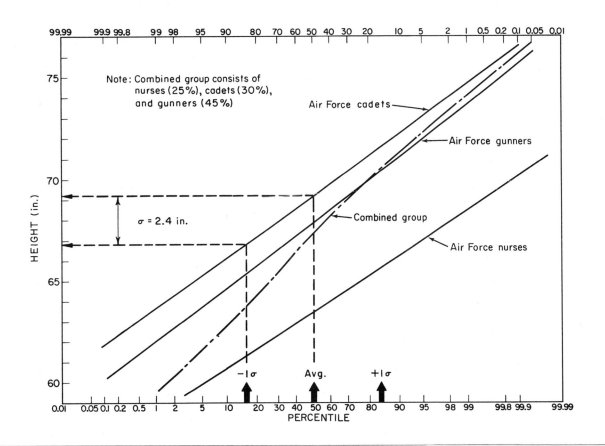

Note: Combined group consists of nurses (25%), cadets (30%), and gunners (45%)

σ = 2.4 in.

Air Force cadets

Air Force gunners

Combined group

Air Force nurses

−1σ   Avg.   +1σ

PERCENTILE

HEIGHT (in.)

14. Use of normal probability graph paper for plotting percentile distributions (after Roebuck, 1957a; data from Randall et al., 1946).

that is, one following the bell-shaped curve (Figs. 9 and 13), as do most human dimensions and capabilities. Any desired percentile can be read from the chart, provided that any two percentiles or any two of the following values are known: any single percentile; an average, whether mean or median (50th percentile); or the standard deviation (S.D.). Two points are needed because a normal distribution plots on normal probability paper as a straight line, which is defined by two points (Fig. 14).

The standard deviation (S.D.): This is a measure—in scale units like inches or pounds—of dispersion, variation, or scatter about an average. Its formula is

$$\text{S.D.} = \sqrt{\frac{\Sigma(x^2)}{N}}$$

where Σ = "sum of,"

x = deviation of each individual value from the arithmetic mean for the group,

N = total number of individuals in the group.

In a normally distributed group, the mean (or 50th percentile)

±1 S.D. includes 68 per cent,
±2 S.D. includes 95 per cent,
and ±3 S.D. includes 99.7 per cent or virtually the entire group (Fig. 13).

The S.D. can be easily determined from a normal distribution plotted on probability paper, without lengthy computation. The S.D. is the value corresponding to the 84th minus the 50th percentile.

Finding percentiles arithmetically, from the standard deviation: Factors for computing percentiles from the known standard deviation provide an alternative to the graphic technique (Table 1).

FALLACY OF THE "AVERAGE MAN"

Designing to fit the "average man" is a serious error which has resulted in many defects in the past. By definition, 50 per cent of a group may suffer from a design which accommodates the 50th percentile.

Examples: (1) Weight. Provision for larger men will also suit the smaller but not vice versa. Supports just adequate for men of average weight may collapse under the heavier 50 per cent (Fig. 15). (2) Reach. Here, accommodation for smaller men will also serve larger ones, but not the reverse. The shorter 50 per cent will be unable to reach the control just suited to the "average," or 50th percentile, operator (Fig. 16).

A further fallacy of the "average man" concept is that few persons are average in all respects or even in several. If one takes the middle third of a group in any dimension—say height—and then the middle third of that third in respect to an independent or uncorrelated

15. Fallacy of fitting the "average man": the average may mark the upper limit of accommodation. An elevator which will just support men of average weight (or men of various weights averaging out to the 50th percentile) will fail to support a heavier group.

Table 1. Factors for Computing Percentiles from
Standard Deviation.
(Roebuck, 1957a)

Percentile		K_1	Central Per Cent Covered	K_2[a]
30	70	0.524	40	1.045
25	75	.674	50	1.349
20	80	.842	60	1.683
15	85	1.036	70	2.073
10	90	1.282	80	2.563
5	95	1.645	90	3.290
2.5	97.5	1.960	95	3.920
1.0	99.0	2.326	98	4.653
0.5	99.5	2.576	99	5.152

[a] $K_2 = 2K_1$

Examples:

1. To find the 95th percentile, when mean =
   35.1 inches and S.D. = 1.5 inches
   1.5 x 1.645 = 2.5 inches
   35.1 + 2.5 = 37.6 inches, the 95th
   percentile.

2. To find the adjustment needed for the
   middle 90 per cent of the same group
   1.5 x 3.29 = 4.9 inches, the range
   of adjustment.

dimension (see Correlation Tables, below) and then repeats the process for a third independent dimension —one reaches $\frac{1}{3} \times \frac{1}{3} \times \frac{1}{3}$, or $\frac{1}{27}$, less than 4 per cent of the initial population, who are "average" in only three traits. (Even fewer, about 1 per cent, will be average in four traits.) This has been found true for height, weight, and chest circumference, despite the fact that

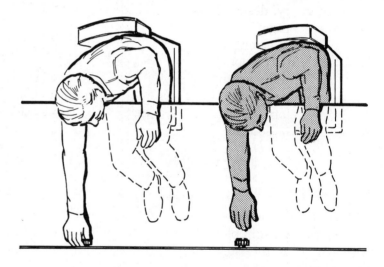

16. Fallacy of fitting the "average man": the average may mark the lower limit of accommodation. A control just reachable by the average man, or easily reached by those with longer arms, cannot be operated by the 50 per cent with shorter reaches.

these dimensions are not independent, but are correlated (Daniels and Churchill, 1952).

Not only is designing for the "average man" fallacious in theory, but in practice it has been found that designers and engineers may employ as "average" subjects persons who are far from the average of the intended operators.

In summary, design for at least a 90 per cent range, not "the average."

The details—such as which 90 per cent (high, low, or middle) to fit; the possibility of accommodating 98 per cent; and specific arrangements which will accomplish those goals—will differ for each item of equipment and each dimension. The data and principles in this book should suggest solutions.

CORRELATION TABLES

Frequently the designer is faced with a problem involving the relationship between body dimensions. He may want to know the range of arm reach for men of the middle 90 per cent in height or in sitting height. Or, given a severely restricted space in front of the knees, what will be the range of seated eye level for those short-legged men who alone could operate the equipment? Or, without data for a specialized or "trick" dimension—for example, height of lumbar (low-back) concavity above the seat—in a certain group, how

could such values be approximated from the commonly available height and weight data?

Fortunately, as already noted in section 3 of this chapter, the relationships between height and body lengths and between weight and body breadths, depths, and especially girths are close enough to allow rough predictions for groups of men, though not for individual men.

A "correlation table" or scatter diagram presents the number or percentage of subjects arranged in respect to Dimension A who fall at each value of Dimension B. A glance shows whether and to what extent Dimension B varies with Dimension A.

Examples: (1) 90 per cent of soldiers, those between 64.5 inches and 72 inches tall, the 5th and 95th percentiles, varied in sitting height roughly between 33 inches and 37 inches (Table 2). (2) Aviation cadets of the 90 per cent midrange in weight—5th to 95th percentiles, 128 to 184 pounds—varied in hip breadth, sitting (bitrochanteric breadth) from 12.6 to 15.6 inches (Table 3). (3) For civilian males in 1945, the 90 per cent midrange in "seat length" (buttock to back of knee)—5th to 95th percentiles, 17.5 to 21 inches—varied in "seat height" (floor to bottom of thigh) from 17.3 to 20.4 inches (Table 4).

For soldiers of known stature but unknown knee height, one enters the correlation table (Table 5) to find the probable distribution of their knee heights. One can do the same, in any group for which such tables

Table 2. Correlation Table - Stature and Sitting Height
Entries Represent Percentages of 24,404 Subjects.

(Newman and White, 1951)

U.S. Army Male Separatees (White)   Sitting Height

Stature	30.0-30.7	30.7-31.4	31.5-32.2	32.3-33.0	33.1-33.8	33.9-34.6	34.7-35.4	35.5-36.2	36.3-37.0	37.0-37.7	37.8-38.5	38.6-39.3	39.4-40.1	40.2-40.9	41.0-41.7	Total
58.7-59.4						.004										.004
59.5-60.2	.004	.008	.004	.012		.004										.032
60.2-60.9			.008	.012	.016	.004										.040
61.0-61.8		.004	.041	.057	.041	.012	.012	.004								.171
61.9-62.6		.008	.033	.160	.213	.061	.037	.008								.520
62.7-63.4		.004	.066	.213	.373	.254	.111	.004	.016	.004	.004					1.049
63.5-64.2		.008	.053	.262	.779	.856	.348	.053	.016	.016	.012					2.403
64.3-65.0		.004	.045	.320	1.049	1.365	.836	.262	.049	.016	.012					3.958
65.0-65.7		.004	.045	.229	.869	1.955	1.742	.660	.070	.016	.004	.008				5.602
65.8-66.5		.020	.025	.127	.783	2.323	2.836	1.746	.418	.012	.020	.008				8.318
66.6-67.3		.004	.016	.102	.516	2.135	3.704	2.901	1.029	.123	.004	.004				10.538
67.4-68.1		.008	.020	.033	.291	1.598	3.708	4.143	2.037	.422	.070	.020				12.350
68.1-68.8		.004	.016	.025	.197	.824	2.836	4.553	3.028	.971	.115	.004				12.573
68.9-69.6		.004	.008	.029	.066	.389	1.914	4.003	3.819	1.496	.201	.045				11.974
69.7-70.4			.012	.020	.049	.193	1.004	2.643	3.692	1.869	.504	.049				10.035
70.5-71.2				.008	.045	.078	.414	1.537	2.795	2.020	.717	.127	.008			7.749
71.3-72.0			.004	.008	.020	.061	.176	.758	1.655	1.754	.824	.291	.016			5.567
72.1-72.8					.004	.025	.066	.311	.844	1.192	.779	.275	.020			3.516
72.9-73.6					.008	.012	.008	.107	.283	.524	.512	.238	.037			1.729
73.7-74.4						.012	.016	.029	.131	.234	.299	.127	.033			.881
74.5-75.2						.004		.008	.029	.164	.217	.094	.053	.004		.573
75.3-76.0							.004			.037	.057	.082	.033	.004		.217
76.1-76.8							.004		.004	.012	.008	.033	.016	.004	.004	.085
76.9-77.6								.004			.012	.025	.033			.074
77.7-78.4								.004	.004			.004	.004	.004	.004	.024
78.4-79.1											.004			.004		.008
Total	.004	.080	.396	1.617	5.319	12.169	19.776	23.738	19.919	10.882	4.371	1.438	.253	.020	.008	99.990

Measurements are in Inches

Sitting Height: Mean = 35.771 S.D. = 1.337
Stature: Mean = 68.436 S.D. = 2.480

$r = .722$
$x = .389y + 9.150$
$y = 1.340x + 20.503$

are available, for *height* and the following: arm reach, shoulder-elbow, elbow-middle finger, sitting height, eye level from seat, shoulder height, buttock-knee, and knee heights; and for *weight* and shoulder breadth, chest breadth, depth and circumference, abdominal depth, seat and hip breadth, and breadth across elbows and knees. Correlation tables of the more important dimensions in the major military and civilian series are

Table 3. Correlation Table - Weight and Hip Breadth, Sitting (Bitrochanteric). Entries Represent Individual Subjects; total 2,972 men (Randall et al., 1946).

Bitrochanteric

Inches

Inches	110–115	116–121	122–127	128–133	134–139	140–145	146–151	152–157	158–163	164–169	170–175	176–181	182–187	188–193	194–199	200–205	206–211
11.8–12.5	4	3	5	3	1												
12.6–13.3	6	20	64	82	79	44	15	3	5								
13.4–14.1		3	20	68	186	276	257	182	81	41	7	4					
14.2–14.9			1	4	22	68	129	192	211	160	121	65	16	6	2		
15.0–15.6			1		4	4	3	25	42	68	75	82	60	36	21	4	1
15.7–16.4									1	5	8	10	9	14	7	6	2
16.5–17.2							1				1		1	2	1	1	
17.3–18.0																	
18.1–18.4															1		

Pounds / Weight

included in Hooton, 1945; Randall *et al.,* 1946; and Newman and White, 1951. Correlation tables are not yet available from the United States National Health Examination Survey (Stoudt *et al.,* 1965).

If one needs, for a special purpose, a "trick" dimen-sion not available for the group to be accommodated, one can approximate its distribution by using correla-tions. Two such examples which arose during World War II concerned height of the calf muscle above the ground, in connection with a combat boot (Fig. 17) and

Table 4. Correlation Table – Seat Length (Buttock-Popliteal) and Seat Height (Popliteal Height).
Entries Represent Individual Subjects; total 1,947 men (Hooton, 1945).

Seat Height	15.5–15.9	16.0–16.4	16.5–16.9	17.0–17.4	17.5–17.9	18.0–18.4	18.5–18.9	19.0–19.4	19.5–19.9	20.0–20.4	20.5–20.9	21.0–21.4	21.5–21.9	22.0–22.4	22.5–22.9	23.0–23.4
15.3–15.6	1															
15.7–16.0																
16.1–16.4	1		1		2											
16.5–16.8	1	3	3	1	4	3										
16.9–17.2	1	1	2	12	2	7	5	2								
17.3–17.6		2	5	10	19	25	15	5	1							
17.7–18.0	1		5	10	40	37	39	24	6							
18.1–18.4			3	23	55	83	52	45	19	9	1					
18.5–18.8			4	9	35	82	77	73	35	11						
18.9–19.2			3	8	26	54	92	86	55	30	7	1				
19.3–19.6				3	13	36	54	75	44	40	7	4	1			
19.7–20.0					3	16	29	47	41	26	11	7	1			
20.1–20.4					6	5	24	19	17	19	8	4				
20.5–20.8						2	9	9	18	13	6	4		1		
20.9–21.2							2	3	15	8	8	4		2		1
21.3–21.6									4	4	6	4		1		
21.7–22.0														1	1	1

Inches — Seat Length

Table 5. Correlation Table - Stature and Knee Height

Entries Represent Percentages of 24,415 Subjects.

(Newman and White, 1951)

U.S. Army Male Separatees (White)   Knee Height

Stature	15.0-15.7	15.8-16.5	16.5-17.2	17.3-18.0	18.1-18.8	18.9-19.6	19.7-20.4	20.5-21.2	21.3-22.0	22.1-22.8	22.9-23.6	23.6-24.3	24.4-25.1	25.2-25.9	26.0-26.7	26.8-27.5	Total
58.7-59.4						.004											.004
59.5-60.2					.008	.020											.028
60.2-60.9				.004	.012	.049	.012										.077
61.0-61.8			.004	.012	.020	.102	.041	.004	.012								.195
61.9-62.6					.066	.213	.176	.020	.012	.008	.004						.499
62.7-63.4		.004		.008	.025	.397	.516	.160	.033	.012	.004	.004					1.163
63.5-64.2	.004		.012	.008	.086	.516	1.040	.442	.070	.025	.008						2.211
64.3-65.0			.004	.008	.066	.487	1.892	1.241	.143	.025	.029						3.895
65.0-65.7			.004	.012	.020	.328	2.032	2.593	.623	.086	.029						5.727
65.8-66.5				.004	.033	.217	1.855	4.444	1.515	.176	.066						8.310
66.6-67.3			.012	.016	.008	.152	1.245	4.686	3.596	.414	.057	.008					10.194
67.4-68.1		.004	.004	.016	.008	.061	.659	4.227	5.431	1.311	.127	.020	.004	.004			11.876
68.1-68.8			.004	.020	.004	.029	.451	3.080	6.471	2.736	.197	.016	.008				13.016
68.9-69.6		.004		.012	.004	.008	.201	1.384	5.226	4.219	.532	.020	.008	.012			11.630
69.7-70.4				.008		.004	.061	.688	3.273	4.964	1.040	.078	.025	.008			10.149
70.5-71.2		.004			.004	.004	.029	.274	1.741	4.071	1.651	.152	.029	.012			7.971
71.3-72.0				.004			.016	.102	.614	2.396	1.823	.381	.033	.004			5.373
72.1-72.8							.020	.029	.274	1.094	1.556	.545	.041	.004		.004	3.567
72.9-73.6					.004		.016	.033	.049	.467	.868	.455	.070	.004			1.966
73.7-74.4							.004	.016	.025	.156	.438	.348	.082	.016			1.085
74.5-75.2							.004	.012	.008	.045	.172	.250	.123	.004	.004		.622
75.3-76.0									.012	.012	.037	.123	.041	.025			.250
76.1-76.8									.004		.004	.041	.037	.004			.090
76.9-77.6										.004		.004	.041	.012			.061
77.7-78.4										.004	.004		.012	.004			.024
78.4-79.1													.004				.004
Total	.004	.016	.044	.132	.368	2.591	10.270	23.435	29.132	22.225	8.646	2.445	.558	.113	.004	.004	99.987

Measurements are in Inches

Knee Height: Mean = 21.617   S.D. = 1.087
Stature: Mean = 68.471   S.D. = 2.510

$$r = .766$$
$$x = .332y - 1.115$$
$$y = 1.770x + 30.208$$

height of shoulder blade (scapula) above seat, for locating a sling type of back-support in troop-carrier aircraft. Both were body lengths and hence correlated with height. Rough approximations were obtained by measuring the special dimension on several subjects from each end of the military height range between the 5th and 95th percentiles, as well as several scattered throughout this range. Their distribution in respect to the trick measurement was noted, and it was assumed that just as the test subjects bracketed the height range, so they would the range of the trick dimension. Had special body breadths or depths been required, weight would have been the basic variable used. In general, the engineer will know only height and weight for the subjects in his experimental sample.

A much closer approximation could be obtained by setting up a correlation table between the trick dimension and the closest standard dimension already obtained on the intended users, in this case soldiers and airmen. Thus, one might use knee height for correlating with calf height, and shoulder height for shoulder-blade height, measuring knee and shoulder heights on experimental subjects selected (by the use of percentiles, see page 32) to represent the range in height, weight, and the standard knee and shoulder heights. The relationship between the trick and the standard dimension would obviously be closer than the relationship between the trick dimension and height.

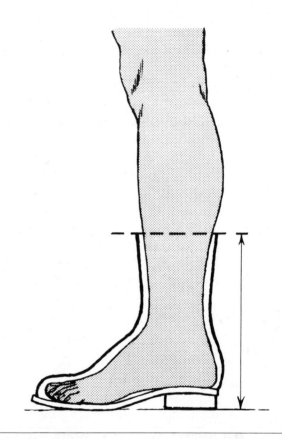

17. A "trick" dimension: calf muscle height, for combat boot design. Lacking direct measurement, its distribution can be approximated by using test subjects and correlation tables.

As a third approach, instead of correlations, the difference between the pair of dimensions on each test subject might be used to derive an average difference, which could then be added to or subtracted from the standard dimension to give the trick dimension. The nature of the problem, the accuracy required, and the measuring facilities available will indicate for any specific situation whether the correlations with height and weight are close enough, or whether additional measurements must be taken on experimental subjects.

Statistical Note on Correlation

The over-all strength of association between the two variables of the scatter diagram is given by the statistic r, the coefficient of correlation. When $r = 0$, the two variables are completely independent or uncorrelated; when $r = -1$, there is perfect negative correlation; when $r = +1$, there is perfect positive correlation. Tables 2 and 5 show r's of $+0.72$ between stature and sitting height, and $+0.77$ between stature and knee height, another component of stature. The r between height and weight was 0.46 in the same Army series of 24,374 men (Newman and White, 1951).

Prediction depends on r^2; thus, $0.46^2 = 0.21$, which means that only 21 per cent of the variability in height (or weight) will be accounted for (or depend on, or be reduced) by controlling the other. Even a correlation of 0.8, which is higher than between most pairs of body dimensions, will account for only 64 per cent of the variation in the dependent variable—too little for individual prediction, though helpful in group prediction. Thus, the psychological test battery ("stanine") for flyers during World War II correlated $+0.61$ with graduation from training. The use of cut-off points based on these tests reduced the number of failures, but 20 per cent of cadets so qualified failed, whereas some below the cut-off points succeeded.

The designer or engineer will find little use for the coefficient of correlation, although it is mentioned in chapter II, section 8 in connection with estimates of body density, for locating centers of gravity. For equipment design, the entries in specific cells of the correlation table are the important thing.

4. Effects of Clothing and Personal Equipment

Machines should be designed to suit the operator as he will use the machine under actual operating conditions. These include environments extreme in respect to heat, cold, altitude, pressure, and combat, all of which involve special protective devices worn by the operator. *Personal equipment must be considered in machine design.* The bulkiest cold-weather or high-altitude outfit, the parachute and life-raft, survival kit, body armor, and ammunition—all are added to the nude body to constitute the functional operator.

Failure to heed this principle has led to such design errors as escape hatches that did not permit passage of a flyer wearing a parachute (Fig. 18); gun-charging handles inoperable by a gloved hand (Fig. 19); and

18. Consequence of failure to consider personal equipment: aircraft escape hatch opening (dark arrow) too small for flyer wearing parachute (light arrows).

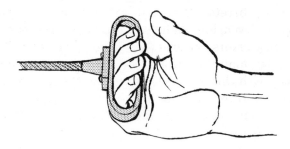

THE BARE HAND FITS

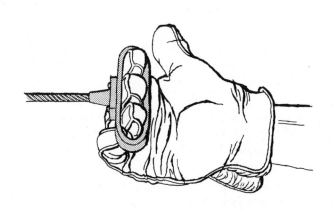

THE GLOVED HAND DOESN'T

19. Consequence of failure to consider personal equipment: gun-charging handle inoperable with the gloved hand.

aircraft gun turrets that provided insufficient space for gunners wearing helmets and oxygen masks. Equally gross errors can be cited from other fields. The designer and manufacturer must therefore acquaint themselves with the conditions and personal equipment which the operator of their machines will encounter. The facts are available, and actual items of equipment should be routinely on hand. It is likewise the responsibility of the purchaser to keep the designer up to date. Machines should be designed to fit the extremes of human size, lightly clad or encumbered with complete outfits of personal gear (Fig. 20). In most cases, this can be readily accomplished if the operator and his equipment are regarded as inseparable from the outset.

Space and weight added by personal gear to the operator are included, where available, in this book (chap. 2, especially Table 6).

5. What the Designer Should Do

The designer needs data on the human body in two contexts: evaluating the adequacy of current or preproduction equipment and specifying requirements for or actually planning new models. It is hoped that eventually all new models will be designed "from the man out." Until that utopian day arrives, the designer will usually be faced with an item of equipment and several

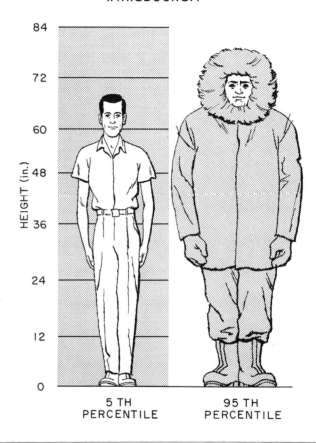

20. Machines should accommodate the smallest (5th percentile) operators in light clothing, as well as the largest (95th percentile) in the bulkiest gear.

questions: How well does it fit the intended operators? How serious are its shortcomings? What improvements should be made, which of them are the most urgent, which are feasible, and how can they best be effected? His procedure can be outlined as follows.

EVALUATION OF CURRENT OR PREPRODUCTION MODELS

(1) Obtain data on the physique of the intended operators. Although human beings vary widely in size and shape, their variation follows certain patterns. The designer's job is simplified by the fact that groups within the population vary less than the population as a whole. For example, Table 14 shows that the 5th to 95th percentile range of sitting height, and hence of required seat adjustment, is less for soldiers ("Army separatees," white and Negro—4.4 inches) or airmen ("Air Force Flying Personnel"—4.2 inches) than for the civilian population, men and women combined (6.0 inches). The limits of the 90 or 95 per cent range must be determined by actual measurement of samples from the group to be accommodated. It cannot be repeated too often that assumptions as to the similarity of any group of operators to other groups or to the "general population" must be tested.

This testing is not the designer's job. It is his responsibility, however, to seek such information. The data presented here will cover the major military and civilian groups for whom equipment is designed. If any group of intended users is likely to differ markedly from these series in respect to age, sex, race, or occupation, or if the desired data are not available for his own or comparable user groups, the designer should seek advice from any of the sources mentioned below in section 7 of this chapter.

(2) Select and measure a small group of representative test subjects. About ten subjects should be selected, with strong representation at both ends, around the 5th and 95th percentiles, of the height and weight distribution of the operators to be accommodated. This gives reasonable assurance that the range of other dimensions will be approximated as well. For the crudest approximations, five well chosen subjects may suffice. Height and weight unclothed or, if this is impossible, with socks and undergarments, are the only dimensions the designer must obtain. Experience has shown that he should obtain them himself, since surprisingly few persons know their own height and weight accurately.

Ideally, all of the test subjects should be measured in detail, as anthropologists do in their own laboratories. One can then determine more accurately what percentage of operators would or would not be accommodated by a given design feature. If, for example, a control is located too far away for some operators to reach, it would be more exact to estimate the percentage inconvenienced from arm reach measurements than from stature.

While designers should be encouraged to obtain such detailed data on their test subjects from or under the direction of anthropologists, using the instruments and techniques described by Randall and Baer (1947) and by Hertzberg *et al.* (1963), for some practical purposes height and weight will suffice. Perhaps a 20 per cent error is introduced by using height and weight rather than dimensions closer to the one involved. It makes little real difference whether 60, 50, or only 40 per cent of potential operators will be inconvenienced by a given design—the design is obviously poor. The gain in precision will only rarely justify the trouble of teaching engineers to become anthropometrists. The chief value of such training at present is to make the engineer aware of the human body. It is hoped that in the future the rudiments of anthropometry may become part of engineering education.

(3) Dress the test subjects in the widest range of standard personal gear that may be worn while operating the equipment. The machine should accommodate—and can, in most cases, by relatively simple devices—the smallest individuals (at the 1st or 5th percentile) in light clothing, as well as the largest wearing the bulkiest gear. The importance of considering all items of personal equipment, including items for combat, safety, and extremes of environment, has been stressed in section 4 of this chapter.

(4) Have the test subjects, wearing the full range of gear, operate the machine. If a mock-up is used, all items which will be in the finished machine should be present in their intended location, in wooden facsimile if not otherwise available. The subjects should not merely seek and hold a comfortable position, but should perform all necessary motions for the duration of an actual operation. Many difficulties are noted immediately, but others become apparent only with time—for example, a cramped space about the knee is tolerable for several minutes, but over the course of hours can become not merely uncomfortable but dangerous or even fatal. Immobility of the lower legs over such periods has caused thrombophlebitis, or blood clots in the calf veins, which can break off, lodge in the lungs, and cause death by a reflex cardiorespiratory arrest (Homans, 1954; Nareff, 1959).

(5) Note difficulties in respect to comfort, efficiency, vision, and safety due to human body size and capabilities. Check lists are the most efficient way to ensure that all major provisions for the operator, in four main categories, are covered. As regards human dimensions, *comfort* is divided, for convenience of analysis, into static and dynamic fit, the latter including not only placement of controls to be operated, but also the subject's fit during such operation (Fig. 21). Fit is further

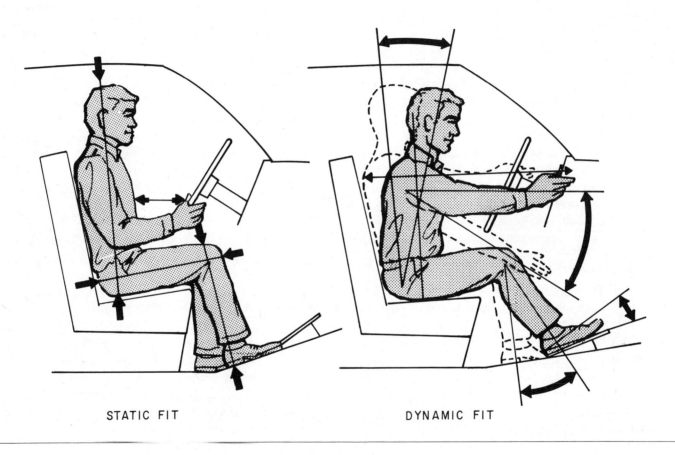

STATIC FIT   DYNAMIC FIT

21. Static fit—clearances. Dynamic fit—operations.

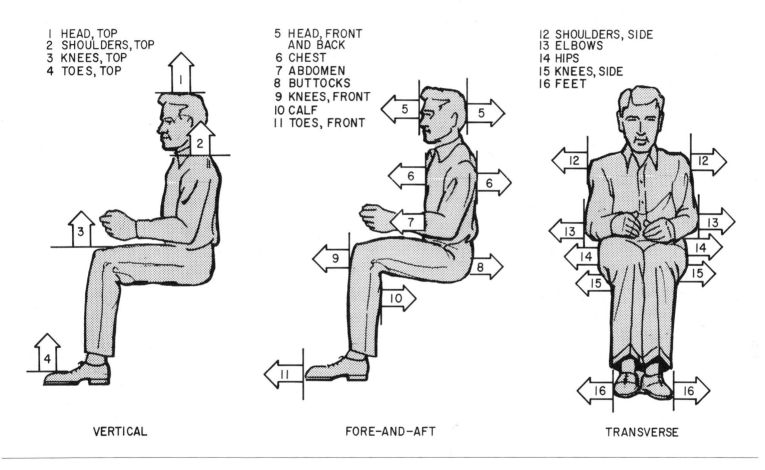

1 HEAD, TOP
2 SHOULDERS, TOP
3 KNEES, TOP
4 TOES, TOP

5 HEAD, FRONT AND BACK
6 CHEST
7 ABDOMEN
8 BUTTOCKS
9 KNEES, FRONT
10 CALF
11 TOES, FRONT

12 SHOULDERS, SIDE
13 ELBOWS
14 HIPS
15 KNEES, SIDE
16 FEET

VERTICAL

FORE-AND-AFT

TRANSVERSE

22. The three dimensions of workspace fit.

23. Visual fields from three drivers' eye positions—from top to bottom, the 5th, 50th, and 95th percentiles—in a 1954-model automobile (modified from Sutro and Kydd, 1955).

subdivided into fore-and-aft, vertical, and transverse clearances at different body levels (Fig. 22). *Efficiency* involves the forces, ranges of motion, and sequence in the pattern of movements required. *Vision,* as measured from the operator's eye location (which is itself determined by human body size and machine design), comprises those portions of the total field which can be seen, as well as the quality of vision afforded. For example, the location of certain structures may result in "blind spots" (Fig. 23), or the optical properties of transparent surfaces may change with varying eye positions. And finally, *safety* includes ease of access and exit by the operator or by a rescuer in case of accident; "packaging" or restraint during operations; and reduction of mechanical hazards to health or life (Fig. 24).

These considerations hold for new or unorthodox equipment as well as for conventional designs. Whatever the equipment or mission, human needs remain the same.

(6) Relate shortcomings in design to percentiles of the operator population and recommend redesign accordingly. Few items of design are wholly bad; a wholly bad feature would have been obvious from the outset and would never survive the mock-up stage. The same is true, with even more force, of an entire machine or workspace. Someone must have been accommodated in the course of its development, if only the engineer him-

24. Emergency escape hindered by obstructions at knee and foot.

self in shirtsleeves. The usual problem is to isolate those defects that seriously hamper the operator. As we have often mentioned, these are almost always specific, remediable, relatively minor items in the total design. By stating the percentage of operators inconvenienced, one can estimate which of the problems are serious and roughly how serious. Concentrating on specific items and specifying just how bad they are, in terms of percentage of operators discommoded, will effect more correction than will blanket criticism of a whole machine as "unsatisfactory."

Estimation of the percentages accommodated or inconvenienced is of course a rough approximation. If a test subject at a known percentile of height or weight cannot fit into a space, it is assumed that all those larger will likewise fail. If a subject cannot reach a control, it is likely that all those smaller in the relevant dimension will fail. The critical percentile can be estimated more closely by determining the percentile of the critical subject in the most pertinent dimension (such as knee height or arm reach), in addition to height and weight, and by having several subjects available around the critical percentile. Hence the desirability of subjects scattered throughout the midrange as well as at the extremes. With only one or two critical subjects and only height and weight percentiles known, the severity of inconvenience will serve as an additional guide to the percentage discommoded.

Other things being equal, an item which discommodes 80 per cent of operators calls more strongly for correction than one which discommodes 20 per cent. But other things are not always equal. Inconvenience itself is relative, varying from the awkward or annoying to the intolerable. For example, in one civilian truck model 20 per cent of drivers could not operate the clutch or the foot brake without lifting their knees against the steering wheel. In the same vehicle, 100 per cent of the drivers could not reach the forward hand (emergency) brake without twisting the body out of normal driving position. Despite the smaller percentage discommoded, the first defect took priority because of the frequency of use and critical role of the controls involved.

To replace the assumptions and extrapolations of this "mock-up" approach, experiments are needed to identify the points at which spatial cramping produces a performance decrement, and the type and amount of decrement so produced. Such research comprises one of the major areas of unfinished business in applied physical anthropology.

Human body size is of course only one among many criteria for evaluating equipment. Other considerations may outweigh the apparent importance of redesign based solely on body size considerations. In the final analysis, the cost of redesign, in money and lost production, will determine which if any initial defects are ever corrected, however serious they may be. Hence the importance of considering the operator from the outset.

PLANNING NEW DESIGNS

It is hoped that the recommendations in this book will guide the designer or engineer in his planning for new models which, when they reach the mock-up stage, may be tested by the procedure just outlined. But even in the blueprint, the designer should have some indication of the operator's size and shape. Jointed manikins in various scales have been used as templates for this purpose. If it is realized that these models rarely reflect personal equipment accurately, that they have only two dimensions, and—most important of all—that they cannot simulate the operator's natural position or movements, they may be used, but only as the roughest of guides. Some anthropologists regard them as more misleading than helpful for these reasons, but their view is probably extreme.

In planning a new design, the designer first determines the population to be accommodated, seeking reliable data on its dimensions and capabilities. If needed information is not readily available, he should consult anthropologists at one of the laboratories mentioned below. Military samples are fairly well covered in this book, but the civilian population has not been studied in as much detail, so that most questions will arise in connection with civilian equipment. Using the recommendations in this book, the designer should attempt to provide the structural strength, clearances, and placement and type of controls that will fit at least 90 per cent of his intended population. Again, while it is hoped that the major human requirements in most situations and for most groups in the population have been covered, it is impossible to anticipate all the problems that will arise.

6. Technical Note: Selection of Samples

The following comments are directed primarily to the anthropologist, but they are of importance to the designer also in determining the basis for the anthropologist's recommendations, in evaluating such recommendations, and in applying them properly to his own problems.

In general, the accuracy required by the machine or workspace designer is less than that needed by either the anthropologist or the designer of personal equipment, which must fit very closely. The aircraft, automotive, or mechanical design engineer will usually find that the rough-and-ready procedure just outlined, with five to ten subjects, will suffice.

SELECTION OF SAMPLES
By Edmund Churchill*

When the need arises to obtain anthropometric or similar data by measuring a sample of men, the twin problems of how large the sample should be and how to select its members must be faced.

Sample Size

The determination of the optimum sample size is important, since too small a sample will give results of inadequate accuracy while too large a sample will entail unprofitable expenditure of time and effort.

In order to select the best sample size, two pieces of information are usually needed: the required level of accuracy in the final results and some estimate of the variability of the characteristic which is to be measured.

The required level of accuracy must be determined in each instance by the nature of the problem at hand and the way the desired data are to fit into its solution. The determination of this accuracy is a task for the design engineer and for him alone. The problem of estimating the variability of the data will be discussed below.

The size of the sample required depends on the sta-

* Associate Professor of Mathematics, Antioch College, Yellow Springs, Ohio.

tistic or statistics which are to be computed. The procedure for computing the sample size also depends somewhat on whether the statistic of interest is based on measured or counted data.

Measured Data

If it is necessary to determine the average (mean) value, the standard deviation (S.D.), or some percentile within plus or minus E units, compute:

$$N = \frac{K(\text{S.D.})^2}{E^2}$$

where $K = 4$ for the mean;
$\quad\quad K = 2$ for the standard deviation;
$\quad\quad K = 7$ for percentiles from the 30th through 70th;
$\quad\quad K = 8$ for the 20th, 25th, 75th, and 80th percentiles;
$\quad\quad K = 12$ for the 10th and 90th percentiles;
and $\quad K = 18$ for the 5th and 95th percentiles.

Example: Suppose it is desired to determine the mean value of some body dimension within ±0.1 inch, and that the standard deviation of this dimension has been estimated as 0.6 inch. Here $K = 4$, S.D. $= 0.6$, $E = 0.1$; hence

$$N = \frac{4(0.6)^2}{(0.1)^2} = \frac{4 \times 0.36}{0.01} = 144.$$

Counted Data

If the characteristic under study is the proportion of men who fall into a specified category, and if it is desired to determine this proportion within the quantity plus or minus E, the correct sample size can be estimated as:

$$N = 1/E^2.$$

Example: Suppose a sample of men are to be interviewed to determine what proportion of them prefer a new type of helmet to the one they are currently using, and that this proportion needs to be estimated within about 10 per cent. In this case, $E = 0.10$, and $N = 1/(0.10)^2 = 1/0.01 = 100$.

If the proportion is either quite small or quite large, this formula will overestimate N. If the proportion (p) can be estimated in advance, the alternative formula:

$$N = 4p(1 - p)/E^2$$

can be used.

Example: If, in the previous example, it was known that the value of p was in the neighborhood of 80 per cent, then a sample of

$$N = 4(0.80)\ (0.20)/(0.10)^2 = 64$$

would be suggested. Note that in all cases the size of the sample depends on the square of the required accuracy; thus, to double the accuracy, the sample size must be made four times as large.

Estimating Variability

The problem of estimating the variability of the data in advance may be handled in a number of ways. Sometimes it is practical to begin a survey before the total number of men to be measured need be specified. In such a case the data from the first dozen or so men could be analyzed far enough to provide an estimated standard deviation, and this estimate used to establish the proper sample size.

For many types of body size data, another method of estimating the variability is available. Most body dimensions fall into one of a small number of groups of dimensions—lengths, major circumferences, miscellaneous surface measurements, etc.—which are characterized by more or less equal coefficients of variation (coefficient of variation = standard deviation divided by the mean or average value). If the dimension to be investigated falls into one of these groups, the coefficient of variation is thus, at least roughly, established. To convert the coefficient of variation to an estimate of the standard deviation requires only that the dimension be measured on a single (preferably, more or less typical) individual and the value for this one may be multiplied by the coefficient of variation (abbreviated to C.V.).

For the purpose of this type of estimating, it will be practical to use the following estimates of the coefficient of variation: (1) lengths (stature, sitting and eye height, arm and foot length, etc.) and major head measurements, C.V. = 4.5%; (2) major circumferences and breadths, C.V. = 6%; and (3) miscellaneous surface measurements, C.V. = 8%.

Example: A body length to be measured was found to have a value of 30 inches on a man of average height. The standard deviation of this length can be estimated as $30 \times 4.5\% = 30 \times 0.045 = 1.35$ inches.

Selection of Members of the Sample

The dangers of unrepresentative or of "biased" samples have been amply stressed above. Nevertheless, the convenience of using the men most readily available in place of a carefully selected sample often successfully tempts investigators to ignore these dangers. In addition, the problem of selecting truly random samples from a military population is usually insurmountable.

Perhaps the most satisfactory method of avoiding serious bias in the selection of a sample is to pick one which is representative, in one or more major respects, of the group for which the equipment is being designed. Thus, if the purpose of the sample survey is to provide data for use in the design of equipment for USAF flyers, the sample should be picked to correspond in height, in weight, or in both, to the statistics for these dimensions given in Tables 7 and 8.

The characteristic in which the sample is made to correspond to the population statistics should be selected on the basis of the closest possible relation to the characteristic to be measured on the sample. Thus if arm reach is to be studied, stature or arm length should be used as the basis of selecting the sample; if clearance breadths or depths are to be studied, weight could serve; and so on.

If the sample is to be picked on the basis of such a correspondence, it is important to perform the selection so that sample and population agree not only in terms of their average values, but in terms of the over-all distribution of the dimension. One way to obtain such agreement is to pick the sample on the basis of the percentile values. For example, to pick a sample of forty men to correspond to USAF fliers in terms of stature, one might pick two men whose statures are below the 5th percentile, two men with statures between the 5th and 10th percentiles, and so forth.

7. Sources of Further Information

When the designer needs more help, either to find specialized data or to solve problems beyond the scope of this book or the references cited (especially Hansen, Cornog, and Hertzberg, 1958), he can more easily consult an applied physical anthropologist than attempt to become one himself. The Ergonomics Society in Britain and the Human Factors Society in the United States publish membership lists and journals containing occasional papers on applied anthropometry, as do the *American Journal of Physical Anthropology* and *Human Biology*. The anthropologist will be acquainted with the relevant literature and, much more important, with its interpretation and application.

There are several anthropometric laboratories to which the engineer—or the biologist seeking more information—can turn. All of the Armed Forces have active programs and will gladly cooperate with manufacturers and scientists. Service laboratories include the Aerospace Medical Laboratory, Wright-Patterson Air Force Base, Dayton, Ohio; and the Quartermaster Research and Engineering Command and the Army Research Institute of Environmental Medicine, both at Natick, Massachusetts. Other sources of information include the Naval Electronics Laboratory, San Diego, California, and the Naval Procurement Office, Brooklyn, New York. The Royal Air Force has an anthropometric laboratory at Farnborough, Hampshire, England, and the R.C.A.F. at Toronto, Canada, as does the U.S. Federal Aviation Agency, at Oklahoma City, Oklahoma. Departments of anthropology, engineering, psychology, or industrial hygiene in several universities are being developed as training and research centers, and a few private firms undertake anthropometric research. The designer of military equipment should ask the specialized laboratory in the service procuring his products, whether his problem is best handled by service laboratory personnel, a university, or a private consulting firm in human engineering.

Anthropometry and Human Engineering —— II

1. Introduction

Two kinds of body dimensions, static and dynamic, are pertinent to the anthropometric problems of human engineering. Static dimensions, taken on the human body with the subjects in rigid, standardized positions, are easily obtained and readily applied to equipment design. Dynamic dimensions, those taken on the human body at work or in motion, are more complex and difficult to measure. Functional arm reach, for example, a dynamic dimension, is not a simple derivative of anatomical arm length, but is a resultant of such mechanical factors as shoulder height and breadth; length of the several segments of the arm and hand; and range of motion at the shoulder, elbow, wrist, and fingers. Functional arm reach therefore changes with each placement and motion of the body, arm, hand, or fingers.

2. Application of Anthropometric Data

Equipment and workspace dimensions derived from static body measurements are of three basic kinds, minimum, maximum, and adjustable dimensions, whereas those derived from dynamic body measurements relate mainly to arm and leg reach capabilities. If one wishes to describe a group for human engineering purposes, the ten most important dimensions to obtain are, in order: height; weight; sitting height; buttock-knee and buttock-popliteal lengths; breadths across elbows and hips, seated; knee and popliteal heights; and thigh clearance height. All are static dimensions.

Minimum dimensions establish clearances in work or resting spaces. One such dimension is the vertical seat-to-roof distance in vehicle cabs or airplane cockpits. The percentiles for erect sitting height (Table 14) can guide the designer in accommodating any desired percentage of the population with respect to clearance above the head. Whenever minimum dimensions are decreased, the larger members of the operating population will be cramped. Minimum dimensions can generally be increased, to give added clearance and to accommodate even larger percentages of the population.

Maximum dimensions, on the other hand, insure the

accommodation of the smaller members of a group. Seat height above floor is one such dimension. If maximum seat height is increased, the shorter operators previously accommodated will be unable to touch the floor with their feet, and will find the front edge of the seat cutting into their thighs.

Adjustable dimensions are required where the arrangements that accommodate operators at either end of the range do not accommodate those at the other. A door or passageway high and wide enough for large persons will also admit small ones; a control reachable by small operators is easily worked by larger ones. This is not true, however, for such a dimension as the range of fore-and-aft seat adjustment in vehicles. The driver with short legs must be seated close to the control area, while the driver with long legs needs to move his seat much farther to the rear. To take another example, vertical seat adjustment is required to bring the eye level of operators of varying sitting heights to the same point.

Dynamic body measurements, chiefly functional arm and leg reaches, determine the location of controls, tools, or other items which must be reached and operated. The arc described by the trunk and head when a passenger restrained by a seat belt is suddenly decelerated would be another dynamic measurement. Design dimensions based on such dynamic measurements are usually laborious to determine and vary with each change in body position.

In the following sections design recommendations cannot be given for every type of equipment dimension, for every distinct population or subgroup, or for any single population wearing every possible combination of clothing and personal equipment. Consequently, the design recommendations are limited to standard equipment or workspace dimensions appropriate for military or civilian populations, with a few known minimum or maximum clothing and personal equipment increments.

To determine the desired equipment dimensions: (1) select the most relevant body measurement, such as sitting height (Table 14) for seat-to-roof distance; the group of operators involved; and the percentage of the population to be accommodated—such as 95 or 99 per cent. The basic value or range of values derived from nude operators can be read directly from the pertinent body dimension table. (2) Determine the type of clothing worn, such as flying gear or civilian clothing, and add the increments from Table 6 to the "nude" values of step (1). (3) Add any other increments—or decrements, such as the slump from erect to normal sitting height—indicated in the text which accompanies Tables 7 to 45. (4) Include additional factors where specified, such as extra clearances, as a safety factor, type and angulation of seating surface, and visual requirements. The final figure will establish the desired equipment or workspace dimension.

To simplify the designer's task, the discussion of each major body dimension includes sizing recommendations for the corresponding equipment.

3. Factors Influencing Human Body Size

Equipment is designed for use by the general population or by specialized occupational groups differing significantly in body size from the general population and from one another. As noted in chapter I, group differences in body size result from a variety of factors, both biological and environmental. Factors like age and sex influence all body dimensions, whereas others—like the normal seated slump—may affect very few dimensions. This section discusses briefly the factors influencing body size as a whole. Certain more specific factors will be noted in connection with the measurements affected.

(1) Age. All body dimensions increase consistently, though sometimes irregularly, from birth to the late teens or early twenties. The precise year at which growth is complete varies with the individual and the dimension. For height, as for most other body lengths, full growth is attained for practical purposes by the age of 20 in males and 17 in females. At some time after middle age, stature decreases, but current evidence is inadequate to establish the onset of this decre-

ment, its extent, or its bearing on other body lengths (Damon and Stoudt, 1963). Certain head dimensions, nose length, and ear length increase slowly throughout life; chests deepen in old age as a result of kyphosis (increased curvature of the thoracic spine) and flaring of the upper rib cage. Weight and its correlated body breadths, depths, and girths on the whole increase significantly through middle age, decreasing again in old age. Weight is of course influenced by diet and physical activity as well as by age.

(2) Sex. Men are larger than women, at any given percentile, for most body dimensions. The extent of the difference varies considerably from one dimension to another. Women are consistently larger than men only in hip breadths and circumferences and in thigh circumference. Men's arms and legs are not only absolutely longer than women's, but are also longer relative to stature and to trunk height. Pregnancy markedly affects certain female body dimensions, mainly of the abdomen and pelvic region, but also of the breasts. Such changes begin to reach anthropometric significance by about the fourth month of pregnancy.

(3) Race. There are wide differences in body size among races, subraces, and national or ethnic groups. Size extremes are represented by the pygmies of Asia and Africa and by African Nilotic Negroes or some peoples of Northwest European ancestry. Although racial or national differences in body size are sometimes

small, they can be of practical significance for equipment design. British aircraft gun turrets were too small for American gunners early in World War II; unmodified American fighter planes are difficult for Asiatic pilots to operate; and Southeastern Europeans require different size tariffs for American clothing.

Americans, a national group of mixed racial origins, may be considered, comparatively speaking, a "large" people. As a group they are larger in body size than most other white national groups, most Negroes, and almost all Mongoloids.

Racial differences occur in body build as well as in gross size. Some groups are characteristically tall and slender (Nilotic Negroes, Scandinavians), others short and slender (Mediterraneans), and still others short and stocky, like the Japanese (Fig. 6).

(4) Body build. Although not strictly a factor determining body size, physical habitus or body build deserves mention at this point as an aspect of physique intermediate between total body size and specific dimensions. Body build varies among races, as just noted, and also among individuals of the same race. At the same height one person may be thin, another fat, and a third heavily muscled. Similarly a 150-pound man might be tall and thin, short and fat, short and muscular, or simply "medium" in physique.

Body proportions also differ among individuals. Men of the same height may have a varying proportion of their height in the legs and trunk, such as long legs and short trunk or the reverse. Other individual differences include the differential development of fat, muscularity, and linearity in some parts of the body, such as long, thin legs with short and fat or muscular arms. Within the general population such disproportions tend to be randomly distributed, though the concentration of certain types may vary in selected groups, as noted below.

(5) Occupation. Differences in body size and proportions among occupational groups are common. On the average, steel workers, truck drivers, and athletes are more muscular than bookkeepers, college professors, and artists. Military personnel differ from the general civilian population. Although numerous factors contribute to the physical differences among occupations, such as age, diet, health status, and physical activity, there also appear to be selective factors tending to make certain occupational groups more homogeneous physically than would be expected by chance. Some tasks require specific physical abilities, and workers having these abilities tend to be similar in physique. Weight-lifting is a classic if extreme case in point. In addition to physical abilities, jobs may require or attract distinctive personalities; and the possible association between physique and temperament is an area of current research activity. For a review of occupational anthropology, with additional data on bus and

truck drivers, see Damon and McFarland (1955).

(6) Diet. Diet has an important influence on many human body dimensions. Malnutrition or undernutrition during growth inhibits attainment of the maximum potential of all body dimensions. After maturity, diet most markedly affects measurements taken over fatty tissue, thereby influencing body circumferences and many breadths and depths more than the relatively stable body lengths. Undernutrition, dieting, or malnutrition can reduce body measurements, often very substantially, as in the case of abdomen depth or of waist or seat circumference. In starvation, such body dimensions decrease drastically, whereas stature and other body lengths decrease to a lesser, though still appreciable, degree (Keys *et al.,* 1950). As a result of too much food and too little exercise, body circumferences, breadths, and depths increase from the early twenties to the sixties. A 45-year-old group will be much larger in those dimensions than a 20-year-old group drawn from the same population. The larger body size of groups from higher social and economic strata reflects mainly better nutrition and possibly less growth-inhibiting disease.

(7) Health. Disease may alter body dimensions, but significant changes in body size (except weight) resulting from disease are rare in the general population and negligible in the armed forces. They will be ignored in this book.

(8) Physical activity and exercise. Exercise decreases measurements reflecting fat and, to a lesser extent, increases those reflecting muscle. The amount of change varies with the intensity and duration of exercise. Dimensions revert to or toward their original values after the course of exercise stops (Seltzer, 1946; Tanner, 1952).

(9) Posture and body position. Some human dimensions vary with posture or body position. For purposes of standardization and comparison, the anthropometrist usually requires rigid, erect positions which are rarely those assumed by people at work or at rest. "Normal" stature, sitting height, and eye height involve slump and are therefore significantly less than when measured with the body erect (1.75 inches less for sitting height). Standing height is smaller than prone or supine length, and is 0.2 to 0.8 inch less when the subject stands erect but unsupported, than when he stretches to full height against a wall. The smaller difference between the two methods can be expected under ideal laboratory conditions, while the larger difference of 0.8 inch may approximate that between "true" stature and the stature obtained in large-scale surveys (Damon, 1964). Buttock breadth and abdominal depth are larger in the seated than in the standing position. Most dynamic dimensions are altered by body movement; thus, maximum arm reach with free movement of the shoulder or trunk is much greater than

with the shoulder and trunk restrained.

(10) Voluntary changes. Some body dimensions can be altered at will by the individual. Abdominal depth may be decreased by drawing in the abdomen; chest circumference, breadth, and depth are smaller on expiration than on inspiration; stature is reduced by crouching. Such voluntary changes in body dimensions help in accommodation to restricted spaces, but these expedients are temporary and undesirable.

(11) Time of day. The time of day when measurements are made significantly affects some body dimensions, principally heights. A person will "lose" height after being up and about, owing to compression of the intervertebral discs. Body heights are therefore greatest immediately after arising and least before retiring. The difference averages about 0.5 inch in children (Malling-Hansen, 1886; Kelly *et al.,* 1943) and 0.95 inch in adult males (Backman, 1924). Weight, on the other hand, is generally least in the morning. Changes in weight result from eating, drinking, and the elimination of body wastes as feces, urine, and sweat. A reasonable value for such normal variation is one or two pounds, perhaps up to a maximum of 2 per cent of total body weight (Dempster, 1961).

(12) Long-term changes. Changes in human body size have been taking place from prehistoric times to the present. There has been a worldwide trend toward an increase in height and most other body dimensions, beginning in Europe over 100 years ago (Hansen, 1912; Bowles, 1932). These changes have been large enough in recent years to invalidate anthropometric surveys of the nineteenth and early twentieth centuries for present-day use in human engineering. Comparison between Harvard fathers and sons showed the sons to be significantly taller (by 1.3 inches) and heavier (by 10 pounds) than their fathers at the same age (Bowles, 1932). This trend appears to be continuing (Karpinos, 1961; Newman, 1963). Demonstration of continued growth by tall young college men (Newman, 1963) disposes of Morant's (1947) argument that the trend is merely an acceleration of growth, with adult stature unchanged. Although environmental amelioration, chiefly medical and nutritional, is responsible for most of the secular increase in body size, a genetic factor may also play a role. This factor is heterosis or hybrid vigor resulting from increased mobility, both geographic and social, and marriage outside of local inbred groups (Hulse, 1957).

(13) Clothing and personal equipment. Clothing and personal equipment influence human body dimensions, sometimes very considerably. Minimum increments are added by light indoor or warm-weather outdoor clothing, comprising cotton underwear, shirt, trousers, socks, shoes, and perhaps a jacket. Maximum increments result from bulky cold-weather protective clothing, which usually consists of the following: (a) For civil-

ians, heavy woolens, including long underwear, trousers, shirt, sweater, jacket, overcoat, socks, boots, hat or cap, and gloves. (b) For military ground forces, long underwear, trousers, shirt, combat suit, overcoat or parka with liner, cap, steel helmet and liner, socks, boots, and gloves, plus varying combinations of weapons, ammunition, and other equipment. (c) For flyers, long underwear, partial or full pressure suit, ventilation suit, antiexposure suit and liner, flying helmet, and boots.

Clothing and personal equipment can sometimes decrease rather than increase body measurements. Stature, for example, a static measurement, is increased significantly when shoes and headgear are worn, but a partial pressure suit when inflated reduces this dimension because the flyer cannot stand fully erect. Heavy gear carried on the back can decrease sitting height as well as stature by preventing full extension of the spine. Functional arm reach, a dynamic measurement, is decreased by bulky jackets or coats.

The anthropometric data in the following tables were taken on nude subjects unless otherwise indicated. For human engineering purposes, measurements on fully clothed and equipped subjects would be preferable, but the tremendous range of possible combinations precludes the presentation of anthropometric values for each combination. Table 6, though inade-

quate for many outfits, notably civilian attire, indicates the effect of some typical clothing on nude body dimensions. Trousers and undershorts, for example, have virtually no effect on sitting eye level height, increasing it by only 0.04 inch. At the other extreme, Air Force high-altitude clothing, comprising the partial or full-pressure suit, the ventilation suit, and the antiexposure suit, increases elbow-to-elbow breadth by as much as 11.0 inches.

In each of the following sections dealing with a single anthropometric dimension, design recommendations are presented first for the nude or very lightly clothed body. Increments to be added to the basic recommendations are then given for various clothing outfits. Since the increments for many types of clothing and personal equipment are unknown, the designer may have to interpolate or make his own measurements. This is no formidable task, since very few subjects (say four or five) will suffice. The subjects should be measured by the same technique, both nude and clothed. The difference is the clothing increment, to be added to the nude dimension of the population who will use the equipment.

Another method of determining clothing and personal equipment increments is the photographic technique of Hall *et al.* (1950) and Kobrick (1956, 1957). Kobrick included a numerical scale in photographs of clothed 5th and 95th percentile subjects in various

Table 6. Clothing Increments for Nude Body Measurements.
All Body Dimensions in Inches, Weight in Pounds.

	CIVILIANS	ARMY				AIR FORCE		
	Underwear, shirt, trousers, and tie (or dress, or blouse and skirt), jacket, shoes (Ashe, Bodenman, and Roberts, 1943, Pett and Ogilvie, 1957)	Column 2: Underwear, khakis or O.D.'s or fatigues, socks, shoes, helmet and liner (Roberts et al., 1945)	Column 3: Underwear, khakis or O.D.'s or fatigues, blouse or field jacket, socks, shoes, helmet and liner (Roberts et al., 1945)	Column 4: Underwear, khakis or O.D.'s or fatigues, blouse or field jacket, overcoat, socks, shoes, helmet and liner (Roberts et al., 1945)	Column 5: Underwear, khakis or O.D.'s or fatigues, combat suit, overcoat, socks, shoes, gloves, wool cap, helmet and liner (Emanuel, n.d.)	Column 6: T-1 partial pressure suit-inflated, ventilation suit-deflated, MD-1 antiexposure suit and MD 3A liner, long cotton underwear (Emanuel, n.d.) Column 7: T-5 partial pressure suit, uninflated, K-1 pressure helmet and boots (U.S. Air Force, 1953) Column 8: World War II heavy winter flying clothing includes jacket, trousers, helmet, boots and gloves (Damon, 1943)		
	1	2	3	4	5	6	7	8
Weight	4-6 [a]	9.4	11.8	18.6	22.9			20.0
Stature	1.0 [b, c]	2.65	2.65	2.65	2.75	-2.0	3.3	1.9
Abdomen depth	1.2	.94	1.18	1.95	2.54	5.0		1.4
Arm reach, anterior		.04	.08	.20	.37			0.4
Buttock-knee length	0.3	.20	.30	.54	.70	2.0		0.5
Chest breadth						2.5		0.6
Chest depth		.41	.96	1.80	1.54	4.5	0.8	1.4
Elbow breadth	1.0	.56	1.04	1.84	2.12	11.0		4.4
Eye height sitting	0.1							0.4
Foot breadth	0.2-0.3	.04	.08	.16	.22			1.2
Foot length	1.2-1.6	.20	.20	.20	.20			2.7
Hand breadth		1.60	1.60	1.60	1.60			0.4
Hand length					.30			0.4
Head breadth		2.8	2.8	2.8	2.8			0.4
Head length		3.5	3.5	3.5	3.5			0.4
Head height		1.35	1.35	1.35	1.45			0.2
Hip breadth		.56	.76	1.08	1.40		2.9	1.3
Hip breadth sitting	0.8	.56	.76	1.08	1.40	5.5		1.7
Knee breadth (both)		.48	.48	.72	1.68	9.5		2.5
Knee height, sitting	1.0 [b]	1.32	1.32	1.44	1.44		0.4	1.8
Shoulder breadth		.24	.88	1.52	1.16	6.0		0.7
Shoulder-elbow length		.14	.50	.94	.62			0.3
Shoulder height sitting		.16	.58	.92	.80			0.6
Sitting height	0.1 [c]	1.39	1.43	1.61	1.67		2.1	0.6

[a] for women, 3 to 4. [b] for women, 0.5 to 3.0. [c] add another 1.0 ± for headgear.

positions. Although photography permits the determination of body areas, useful in problems of temperature, radiation, blast, and the like, the compressibility of clothing cannot be determined, making some dimensions unrealistically large. Direct clothing measurements are preferable for equipment sizing problems.

4. Sources of Anthropometric Data

Military populations, particularly the United States Air Force and Army, are well represented in the following tables of anthropometric data. The general civilian population can also be described, for the first time, in respect to height, weight and the ten most important "human engineering" dimensions. These were taken in 1960–1962 by the National Health Examination Survey of the United States Public Health Service on a sample of 6,672 persons carefully selected to represent racially, geographically, and socioeconomically the nonmilitary and noninstitutionalized American population between the ages of 18 and 79. For measurements other than these twelve, the general civilian population has been less well described. Hooton's (1945) classic study on railway travelers includes eight measurements relevant to the design of railway coach seats. Because of the special measuring chair used, some of the measurements are not comparable to those taken with standard techniques; moreover, the data are now some 20 years old.

O'Brien and Shelton's (1941) detailed anthropometric study of women is excellent for its intended purpose, garment and pattern construction, but consists mostly of body heights, circumferences, and skin surface measurements, only a few of which relate to equipment design. The life insurance industry (Society of Actuaries, 1959) has provided data on heights and weights for economically selected groups of people measured by unstandardized techniques and wearing a variety of shoes and clothing. Other than such sources, the civilian population is represented chiefly by studies of the aged (Roberts, 1960; Damon and Stoudt, 1963) and by groups like truck and bus drivers and college students.

5. Static Human Body Dimensions and Design Recommendations

Each of the following sections deals with a single human body measurement related to some aspect of equipment or work space design. Discussion is held to a minimum, and only data needed by the designer are presented. Each human body measurement section contains the following subsections: (1) A definition, in words and diagram, of the body measurement and how it was taken. The arrows in the diagrams point centrally rather than peripherally to connote the impingement of the mechanical environment on the human dimensions. (2) The data, in tabular form for the 1st, 5th, 50th, 95th, and 99th percentiles; the standard deviation; and the age and number of subjects in each group. The source is specified. (3) Related equipment or workspace dimension—the specific design dimensions related to, and which should be based upon, that body measurement. (4) Design recommendations that will accommodate specified percentages of the population. (5) Conversion factors for obtaining female dimensions from those of males. (6) Correction factors for clothing, personal equipment, and other relevant variables.

NOTES ON THE ANTHROPOMETRIC DATA

Some features of Tables 7 to 44 must be understood for their proper interpretation:

(1) Measurements were made on nude subjects standing erect, unless otherwise noted.

(2) Any deviation from standard measuring technique, as defined in the discussion of each measurement, is noted under each table.

(3) Some groups, notably the military, have upper and lower height-weight limits for acceptance, thus eliminating extremely tall, short, stocky, and thin persons.

(4) The column headed "50th percentile" contains in some cases the average or arithmetic mean instead of the median. Ordinarily there is little or no practical difference between these two measures of central tendency. Weight is one exception, however; because of its skewed distribution, the mean exceeds the median, or 50th percentile, by about 2 pounds. Those groups whose 50th percentile is given are: U.S. Air Force Flying Personnel (total group only), Basic Trainees, Cadets, and Gunners; Army Separatees, white and Negro; Army drivers, white and Negro; Army Aviators, Navy Pilots, Aviation Cadets, Enlisted Men, Submarine Officers and Enlisted Men; Marine Corps; Selective Service Registrants; WAF Basic Trainees; WAC's and Army Nurses; Truck and Bus Drivers; Railroad Travelers (male and female); Spanish-American War Veterans, and Civilian Population, male and female. For all other groups, only means are available.

(5) Some of the percentiles presented in the following Tables 7 to 44 have been computed on normal probability paper (see page 18). This technique has been used where original frequency distributions were not available and percentiles could be obtained in no other way. The graphic technique is accurate only to the extent that the measurement is normally distributed (see page 12). Although many human body dimensions approximate the normal distribution closely enough to justify using the graphic technique, an occa-

sional group deviates because of biased or nonrandom selection of subjects or variation in measuring technique. Weight, as just noted, is the physical trait with a distribution departing farthest from normality. In Table 7 (Weight), percentiles which were obtained graphically rather than computed from the (unavailable) frequency distributions are indicated.

(6) For two groups, Army separatees and Army drivers, the data on whites and Negroes have been processed separately, and are so presented. For most other groups the data have been treated as a single population and are not identified by racial make-up.

(7) The percentile values for heights and weights for the Selective Service Registrants measured in 1957–1958 are age specific, *i.e.*, weighted according to the age distribution of the U.S. population within the two indicated age ranges.

(8) The anthropometric data included in the following tables are those available in June 1965. New anthropometric surveys, planned for all four branches of the U.S. Armed Services, will change some of the values presented here as well as the apparent relationships among the several military groups.

(9) The "Civilian Population" included in the following tables is a specially selected probability sample of the adult, civilian, noninstitutionalized population of the mainland United States measured by the U.S. National Health Examination Survey. The subjects for this survey were selected as follows: the entire population was first stratified by broad geographic region and by size of place of residence (rural areas, small cities, etc.). These strata were then subdivided into segments and households, and these units were then randomly sampled until a total of 7700 prospective subjects were obtained. It may be assumed that these subjects are reasonably representative of the entire U.S. population, as defined above, in terms of those variables which are known to influence body size, including age, sex, race, socioeconomic status, etc. The survey was conducted in 1960–1962.

Of the 7700 subjects originally selected, 6672 were actually measured in the field, a response rate of 86 per cent. Investigation of a subsample of respondents and nonrespondents showed that no major features of the U.S. adult population were seriously distorted by the absence of the nonrespondents. For a detailed description of the procedures employed in this study, see Stoudt *et al.* (1965).

NOTES ON THE DESIGN RECOMMENDATIONS

The design recommendations accompanying the following anthropometric data require some explanatory comments for their proper interpretation and application.

(1) The design recommendations for civilian men and

women are based on the anthropometric data from the National Health Examination Survey (see above) whenever possible. Where the needed measurements are not available on this population, other less representative civilian groups have, of necessity, had to be used. This may result in some minor inconsistencies in the design recommendations from measurement to measurement, but the magnitude of these differences will probably not exceed a few tenths of an inch. Where measurements were completely lacking on female civilians, WAFS, flying nurses, and Women's Auxiliary Service pilots of World War II were used to determine the male-female differences.

(2) The design recommendations for Air Force flyers are based on the 1950 Air Force Survey. This group is the best described anthropometrically for human engineering and other design purposes. As compared to the population as a whole, Air Force flyers are younger, taller, larger in most body heights and lengths, lighter in weight, and considerably leaner. They are also smaller in those body breadths, depths, and circumferences to which fatty tissue, which increases with age, contributes substantially. There is little doubt, however, that present Air Force flyers are slightly larger in body size than this series measured in 1950, because of the demonstrated secular increase in body size (Newman, 1963).

(3) The design recommendations for Army troops in the following sections are based primarily on the large series of Army separatees measured in 1946, but modified considerably by data from more recent surveys, primarily that of Army drivers in 1960, which indicate that present Army personnel are becoming both taller and heavier.

(4) "Minimum" clearance dimensions are based on the 95th and 99th percentiles in body size. "Maximum" design dimensions, such as reach distances, which cannot exceed the abilities of small persons, are based on the 1st and 5th percentiles in body size. "Adjustable" design dimensions, which must accommodate large and small persons and those in between, are based on the middle 90 per cent (5th to 95th percentiles) and 98 per cent (1st to 99th percentiles) of the group.

(5) Additional allowances have been incorporated into most design recommendations as a safety factor to insure adequate fit. These allowances are as follows: (a) 0.2 inch has been added to the 95th and 99th percentiles of body and trunk measurements used in determining minimum clearance dimensions. (b) 0.1 inch has been added to the 95th and 99th percentiles of head, hand, and foot measurements used in determining minimum clearance dimensions. (c) 1.5 inches have been added to the 95th and 99th percentiles of shoulder height, standing and sitting, to correct for the difference between acromial height (the edge of the shoulder, where the measurement was taken) and true functional

shoulder height. (Note: the three foregoing allowances are minimal, and additional space should be provided if possible.) (d) 0.2 inch has been subtracted from the 1st and 5th percentiles of popliteal height in determining seat height, and the same amount subtracted from the 1st and 5th percentiles of buttock-popliteal length in determining seat length. Shortening and lowering the seat by 0.2 inch decreases the pressure of the seat front against the popliteal region. (e) 3.0 inches have been subtracted from all percentiles of elbow height used to determine the height of standing work surfaces. (f) 2.0 inches have been subtracted from the 1st and 5th percentiles of knuckle height in determining the base-to-handle height of portable equipment. The 2-inch allowance permits hand-held equipment to clear the ground. (g) No corrections have been applied to the percentiles of eye height, standing or sitting, used to determine the eye level reference point; nor to the body measurements involved in the reach distances of the arm, hand, and leg; nor to the interpupillary distance used in spacing binocular eyepieces.

Weight

Definition: Nude body weight, in pounds, taken on a reliable scale which is frequently calibrated (Tables 7 and 9).

> *Relevant Equipment Feature:* Structural strength.
> *Equipment Involved* (partial list only): Body sup-

ports, especially those subject to accelerative forces. Workspaces, resting spaces, means of transport. Floors, platforms, benches, seats. Beds, couches, hammocks, sofas, litters; ladders, stairs, chutes, slides (especially in aircraft). Aircraft, spacecraft, manned rockets, land vehicles, ships and submarines; elevators, escalators; parachutes; bridges; sleds. Structural members, such as braces, supports, struts, springs, harness, supporting and restraining belts and fabric, rigging.

Design Recommendations for Structural Strength

For civilian men, a structure stressed for 239 pounds will support all but the heaviest 1 per cent of the population and 215 pounds will support all but the heaviest 5 per cent. For civilian women 234 pounds will support all but the heaviest 1 per cent and 197 pounds all but the heaviest 5 per cent. For Air Force flyers the 99th percentile is 216 pounds and the 95th percentile 201 pounds. For Army soldiers the corresponding figures are 233 and 205 pounds.

These figures represent actual percentiles of nude weight, without any safety margin.

Clothing: Add 5 pounds for men's light clothing, 3.5 pounds for women's light clothing, and up to 23 pounds or more for arctic military clothing (Table 6).

Other factors: G* forces, as in acceleration and decel-

---

*G is a unit of reactive force equal in magnitude but opposite in sign to g (the acceleration due to gravity, 32 feet per second^2).

Table 7. Weight (Pounds).[a]

Group	Date of Survey	Refer-ence*	Number of Subjects	Age (Mean or Range)	Percentile 1	5	50	95	99	S.D.
Males										
Civilian population[b]	1960-62	27	3,091	18-79	110	124	164	215	239	-
			411	18-24	113	122	155	212	229	-
			675	25-34	112	127	167	221	246	-
			703	35-44	119	132	169	217	242	-
			547	45-54	114	129	169	217	239	-
			418	55-64	110	121	163	211	228	-
			265	65-74	97	115	159	205	223	-
			72	75-79		105	144	196		-
Selective Service registrants	1943-44	22	110,251	18-19	-	(106)	141	(176)	-	21.1
			97,795	20-24	-	(109)	146	(183)	-	22.4
			69,905	25-29	-	(110)	151	(192)	-	24.8
			68,792	30-34	-	(110)	153	(195)	-	25.8
			39,194	35-37	-	(111)	154	(197)	-	26.1
	1957-58	23	37,603	17-19	97	109	145	187	209	25.0
			235,294	20-25	106	118	156	202	225	26.2
Railroad travelers[c]	1944	20	1,959	38	-	126	161	212	-	-
Truck and bus drivers	1951	26	305	36	-	129	164	213	-	-
Airline pilots	1946	25	7,238	31	-	(134)	168	(201)	-	20.3
Pulp industry workers[c]	1952-58	36	3,099	38	-	(124)	164	(204)	-	24.5
College students misc. regions	1928-30	11	23,122	16-21	-	(112)	142	(172)	-	18.1
Harvard students	pre-1930	2	480	18	-	(122)	150	(178)	-	17.2
	1938-42	18	258	19	-	(132)	159	(187)	-	16.6
Univ. Kansas students	1948-52	14	1,487	18-22	-	(118)	156	(195)	-	23.5
Univ. Chicago students	1941	4	259	18	-	(115)	148	(180)	-	19.7
Spanish-American War veterans	1959	8	130	81	112	119	151	192	204	23.19

Table 7 (continued).

Group	Date of Survey	Refer-ence*	Number of Subjects	Age (Mean or Range)	Percentile 1	5	50	95	99	S.D.
Air Force flying personnel	1950	19	4,052	27	123	133	162	201	216	20.9
Multi-engine pilots			1,184	28	123	-	166	-	217	20.5
Fighter pilots			210	26	123	-	159	-	225	20.7
Student pilots			508	23	123	-	159	-	199	17.4
Bombardiers			445	30	126	-	169	-	211	20.6
Navigators			1,011	27	125	-	165	-	214	20.6
Observers			140	30	113	-	166	-	217	22.4
Flight engineers			520	30	124	-	166	-	222	23.3
Gunners			277	26	121	-	158	-	214	21.3
Radio operators			117	26	115	-	157	-	199	19.0
Air Force basic trainees	1952	9	3,332	18	109	118	145	186	208	21.0
Air Force cadets	1942	32	2,960	23	-	129	153	184	-	-
Air Force gunners	1942	32	584	23	-	120	147	173	-	-
Army inductees	1954	5	163	< 20	-	(111)	159	(206)	-	29.4
			234	> 20	-	(122)	162	(202)	-	23.9
	1957-58	23	25,424	17-19	102	114	150	192	213	24.4
			100,560	20-25	110	122	159	201	223	24.2
Army separatees, white	1946	28,37	24,506	23	114	124	153	192	215	20.6
Army separatees, Negro	1946	37	6,684	23	-	(120)	152	(183)	-	19.2
Army aviators	1959	43	500	30	124	136	167	200	213	18.9
Army drivers, white	1960	7	431	24	119	127	158	210	244	24.8
Army drivers, Negro	1960	7	79	27	125	128	164	223	252	30.0
Navy recruits	1947	40	5,004	18	-	(110)	140	(171)	-	18.5
	1952	16	2,173	17-25	-	(119)	152	(185)	-	20.6

Continued on next page.

Table 7 (continued).

Group	Date of Survey	Reference*	Number of Subjects	Age (Mean or Range)	Percentile					S.D.
					1	5	50	95	99	
Navy enlisted men	1955	41	124	18-30	-	132	160	197	-	19.9
Navy, mixed sample	1946	24	141	23	-	(126)	162	(197)	-	21.5
Navy pilots	1958	17	1,190	-	128	138	167	199	214	18.3
Navy aviation cadets	1955	41	472	18-25	-	135	166	196	-	-
Navy submarine enlisted men[d]	1957	42	863	19	-	132	159	192	-	-
Navy submarine officers[d]	1957	42	121	24	-	140	173	206	-	19.7
Marine Corps recruits	1947	40	1,074	18	-	(112)	143	(174)	-	18.7
Marine Corps	1949	38	302	26	130	139	170	212	228	22.4
Females										
Civilian population[b]	1960-62	27	3,581	18-79	91	102	135	197	234	-
			534	18-24	89	97	124	168	216	-
			746	25-34	90	100	128	189	237	-
			784	35-44	98	107	135	202	236	-
			705	45-54	93	104	141	203	238	-
			443	55-64	93	110	144	209	242	-
			299	65-74	90	104	143	194	212	-
			70	75-79		93	135	191		-
Railroad travelers[e]	1944	20	1,908	35	-	100	129	175	-	-
Clothing survey subjects	1939-40	29	10,042	34	91	100	129	184	213	26.0
Working women	1932	1	100	36	-	(110)	136	(163)	-	16.0
White women, New York City	1951-56	6	139	45	95	108	140	200	229	27.2
Negro women, New York City	1951-56	6	103	36	85	104	143	193	210	34.5
College students, misc. regions	1928-30	12	17,127	16-21	-	(94)	121	(149)	-	17.1
College students, East	pre-1930	2	572	17	-	(101)	125	(149)	-	15.2
College students, Midwest	1939	13	937	17-21	-	(99)	126	(154)	-	16.9
Univ. Kansas students	1953-57	15	3,992	17-21	-	(97)	125	(153)	-	16.8

Table 7 (continued).

Group	Date of Survey	Refer-ence*	Number of Subjects	Age (Mean or Range)	Percentile					S.D.
					1	5	50	95	99	
WAF basic trainees	1952	10	851	19	95	102	122	148	162	14.5
Air Force pilots (WASPS)	1944	32	446	18-35	102	106	129	155	169	-
Air Force flying nurses	1943	32	150	-	104	107	122	135	143	-
WACs and Army nurses	1946	33,37	8,530	26	97	105	129	170	192	20.0
WAC officers	1946	31	468	31	-	(105)	132	(158)	-	16.1
WAC enlisted women	1946	31	4,299	26	-	(97)	130	(163)	-	20.6
Army nurses	1946	31	3,487	26	-	(95)	129	(162)	-	20.2

[a]Percentiles in brackets were computed from the mean and standard deviation. Because of the non-normal distribution of weight, these values may often be a few pounds lower than the true figures and should be used with caution.

[b]Two pounds have been subtracted from partly clothed weight to approximate nude weight.

[c]Six pounds have been subtracted from fully clothed weight to approximate nude weight.

[d]Three pounds have been subtracted from partly clothed weight to approximate nude weight.

[e]Four pounds have been subtracted from fully clothed weight to approximate nude weight.

*See references 1-43 on pp. 325-327.

eration, multiply the above weights (page 264). Allow a considerable margin of safety for outsize individuals, unusual personal gear such as parachutes and body armor, and extra G forces.

Stature

Definition: Vertical distance from floor to the top of the head; subject stands erect, looking straight ahead (Fig. 25 and Tables 8 and 9).

   Relevant Workspace Dimension: Head to Foot, vertical and horizontal.

   Equipment Involved: Rooms, passageways, vehicles, elevators, booths. For the erect position, overhead structures such as ceilings, roofs, doorways, archways, lintels, beams, rafters, ridgepoles, and struts, as well as obstructions or projections like pipes, lighting fixtures, and knobs or handles. For the prone or supine position —beds, cots, hammocks, sleeping bags, litters, coffins, escape chambers or capsules, platforms.

Design Recommendations for
Head to Foot, Vertical

For civilian men 74.8 inches will accommodate all but the largest 1 per cent and 73.0 inches will accommodate all but the largest 5 per cent. For civilian women the 99th percentile is 69.0 inches and the 95th percentile, 67.3 inches. For Air Force flyers the comparable figures

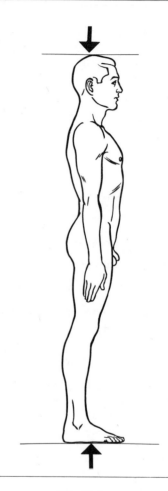

25. Stature.

Table 8. Stature (Inches).

Group	Date of Survey	Reference*	Number of Subjects	Age (Mean or Range)	Percentile					S.D.
					1	5	50	95	99	
Males										
Civilian population	1960-62	27	3,091	18-79	61.7	63.6	68.3	72.8	74.6	-
			411	18-24	62.6	64.3	68.6	73.1	74.8	-
			675	25-34	62.6	64.4	69.0	73.8	76.0	-
			703	35-44	62.3	64.2	68.6	72.5	74.1	-
			547	45-54	62.3	64.0	68.3	72.7	74.0	-
			418	55-64	61.2	62.9	67.6	72.2	73.5	-
			265	65-74	60.8	62.7	66.8	70.9	72.0	-
			72	75-79		61.3	66.2	70.5		-
Selective Service registrants	1943-44	22	110,251	18-19	62.0	63.8	68.0	72.3	74.1	2.61
			97,795	20-24	62.1	63.9	68.2	72.4	74.2	2.60
			69,905	25-29	61.9	63.7	68.1	72.4	74.2	2.63
			68,792	30-34	61.7	63.5	67.8	72.1	73.9	2.66
			39,194	35-37	61.3	63.2	67.6	72.0	73.8	2.64
	1957-58	23	37,603	17-19	61.4	63.3	67.9	72.5	74.4	2.79
			235,294	20-25	62.0	63.9	68.5	73.2	75.1	2.83
Railroad travelers[a]	1944	20	1,959	38	61.5	63.5	68.0	72.8	74.6	-
Truck and bus drivers	1951	26	306	36	63.0	64.6	68.4	72.5	74.1	-
Airline pilots	1946	25	7,238	31	64.4	66.0	70.0	73.9	75.6	2.4
Pulp industry workers[a]	1952-58	36	3,099	38	64.4	66.1	70.3	74.4	76.2	2.46
College students, misc. regions	1928-30	11	23,122	16-21	62.5	64.4	68.7	73.1	74.9	2.68
Harvard students	pre-1930	2	480	18	64.5	66.1	69.9	73.8	75.4	2.38
	1938-42	18	258	19	65.0	66.5	70.2	74.0	75.5	2.3
Univ. Kansas students	1948-52	14	1,487	18-22	64.2	65.9	70.0	74.1	75.8	2.49
Univ. Chicago students	1941	4	259	18	63.2	65.0	69.1	73.3	75.0	2.6
Spanish-American War veterans	1959	8	119	81	61.6	63.3	66.1	69.9	70.3	2.09
Air Force flying personnel	1950	19	4,062	27	63.5	65.2	69.1	73.1	74.9	2.44
Multi-engine pilots	1950	19	1,184	28	64.4	65.9	69.4	73.3	74.9	2.31
Fighter pilots	1950	19	210	26	63.8	65.2	68.8	72.6	74.2	2.24
Student pilots	1950	19	508	23	63.6	65.2	69.2	73.1	74.7	2.45
Bombardiers	1950	19	445	30	63.5	65.2	69.1	73.0	74.5	2.32
Navigators	1950	19	1,011	27	63.5	65.2	69.2	73.3	75.0	2.46
Observers	1950	19	140	30	63.8	65.4	69.1	72.8	74.2	2.44
Flight engineers	1950	19	520	30	63.1	64.8	69.0	73.2	75.0	2.51
Gunners	1950	19	277	26	62.4	64.2	68.3	72.2	73.7	2.43
Radio operators	1950	19	117	26	63.0	64.6	68.3	71.8	73.2	2.37

Continued on next page.

Table 8 (continued).

Group	Date of Survey	Refer-ence*	Number of Subjects	Age (Mean or Range)	Percentile 1	5	50	95	99	S.D.
Air Force basic trainees	1952	9	3,331	18	62.5	64.2	68.6	72.7	74.7	2.61
Air Force cadets	1942	32	2,961	23	63.8	65.4	69.2	73.1	74.7	-
Air Force gunners	1942	32	584	23	62.0	63.4	67.9	71.7	73.5	-
Army inductees	1954	5	163	‹20	62.4	64.3	68.7	73.1	74.9	-
			235	›20	62.7	64.6	69.0	73.4	75.2	2.65
	1957-58	23	25,424	17-19	62.1	63.9	68.4	72.9	74.7	2.69
			100,560	20-25	62.7	64.5	69.0	73.4	75.2	2.65
Army separatees, white	1946	28,37	24,508	23	62.7	64.3	68.5	72.6	74.5	2.52
Army separatees, Negro	1946	37	6,684	23	62.3	64.0	68.0	72.2	74.0	2.58
Army aviators	1959	43	500	30	64.4	65.8	69.4	73.3	74.8	2.25
Army drivers, white	1960	7	431	24	62.5	64.5	69.2	74.0	75.8	2.71
Army drivers, Negro	1960	7	79	27	61.6	64.4	69.5	74.2	76.2	2.95
Navy recruits	1947	39	5,010	18	62.8	64.5	68.5	72.6	74.2	2.5
	1952	16	2,173	17-25	62.9	64.6	68.6	72.7	74.4	2.48
Navy enlisted men	1955	41	124	18-30	63.2	64.8	69.5	73.5	75.5	2.48
Navy pilots	1958	17	1,190	-	64.9	66.5	70.3	74.3	75.8	2.36
Navy, mixed sample	1946	24	141	23	64.1	65.7	69.7	73.5	75.1	2.34
Navy aviation cadets	1955	41	472	18-25	65.1	66.6	70.1	73.8	75.2	-
Navy submarine enlisted men[a]	1957	42	861	19	62.3	64.0	68.9	73.1	75.1	-
Navy submarine officers[a]	1957	42	121	24	62.7	64.6	69.8	73.7	75.4	2.8
Marine Corps recruits	1947	39	1,074	18	63.0	64.6	68.6	72.5	74.1	2.4
Marine Corps	1949	38	302	26	64.4	66.1	69.7	73.5	74.5	2.18
Females										
Civilian population	1960-62	27	3,581	18-79	57.1	59.0	62.9	67.1	68.8	-
			534	18-24	58.4	60.0	63.9	67.9	69.3	-
			746	25-34	58.1	59.7	63.7	67.3	69.0	-
			784	35-44	57.6	59.6	63.4	67.2	69.0	-
			705	45-54	57.3	59.1	62.8	67.2	68.7	-
			443	55-64	56.0	58.4	62.3	66.6	68.7	-
			299	65-74	55.8	57.5	61.6	65.5	67.0	-
			70	75-79		55.3	61.8	64.9		-

Table 8 (continued).

Group	Date of Survey	Reference*	Number of Subjects	Age (Mean or Range)	Percentile					S.D.
					1	5	50	95	99	
Railroad travelers[b]	1944	20	1,908	35	57.1	58.8	62.9	67.1	68.8	-
Clothing survey subjects	1939-40	29	10,042	34	57.4	59.1	63.2	67.2	68.8	2.48
Working women	1932	1	100	36	58.1	59.7	63.6	67.5	69.2	2.43
White women, New York City	1948-56	6	367	34	56.5	58.4	62.8	67.3	69.2	2.69
Negro women, New York City	1948-56	6	259	30	57.3	59.0	63.1	67.3	69.0	2.53
College students, misc. regions	1928-30	12	17,127	16-21	58.5	60.0	63.8	67.6	69.2	2.33
College students, East	pre-1930	2	571	17	59.8	61.2	64.8	68.3	69.7	2.15
College students, Midwest	1939	13	937	17-21	58.8	60.5	64.4	68.4	70.0	2.36
Univ. Kansas students	1953-57	15	3,922	17-21	60.2	61.5	65.2	68.9	70.2	2.25
WAF basic trainees	1952	10	851	19	59.3	60.3	64.0	68.2	69.9	2.34
Air Force pilots (WASPS)	1944	32	447	18-35	60.8	61.7	64.9	68.3	70.0	-
Air Force flying nurses	1943	32	152	-	59.0	60.2	63.5	67.7	69.3	-
WACs and Army nurses	1946	33,37	8,549	26	58.4	59.9	63.9	68.0	69.7	2.42
WAC officers	1946	31	466	31	59.2	61.0	64.5	68.9	70.6	2.4
WAC enlisted women	1946	31	4,300	26	58.3	60.0	63.9	68.0	69.6	2.4
Army nurses	1946	31	3,488	26	58.7	60.4	64.1	68.3	70.0	2.4

[a]One inch has been subtracted from height with shoes.
[b]Two inches have been subtracted from height with shoes.
*See references 1-43 on pp. 325-327.

are 75.1 and 73.3 inches, and for Army personnel, 75.7 and 73.9 inches.

The above values represent nude height plus 0.2 inch, an arbitrary figure made purposely small to emphasize that the design recommendations refer to the nude body envelope. Each equipment designer must substitute his own population percentage to be accommodated, clothing and personal gear increments, additional clearance desired (as in mobile equipment, to prevent bumped heads), and other special features.

Clothing: Add 1.0 inch for men's shoes, 1.3 inches for military boots, up to 3.0 inches for women's shoes,

Table 9. Mean Heights and Weights of White Americans,
Birth to 19 Years.*

| | Males | | | | Females | | | |
| Age | Height (In.) | | Weight (Lb) | | Height (In.) | | Weight (Lb) | |
	Mean	S.D.	Mean	S.D.	Mean	S.D.	Mean	S.D.
Birth	20.0	1.0	7.6	1.3	19.7	1.0	7.5	1.1
½ Month	20.7	1.0	8.5	1.3	20.4	1.0	8.2	1.1
1½	21.5	1.0	9.9	1.6	21.3	1.0	9.4	1.2
2½	23.0	1.0	12.1	1.7	22.5	1.0	11.2	1.3
3½	24.0	1.0	13.9	1.9	23.6	1.0	12.8	1.5
4½	25.2	1.0	15.4	1.8	24.6	1.0	14.2	1.6
5½	26.0	1.0	16.8	2.0	25.4	1.0	15.6	1.8
6½	26.6	1.0	18.0	2.1	26.1	1.0	17.0	1.9
7½	27.4	1.0	19.2	2.3	26.6	1.0	17.9	2.0
8½	27.9	1.1	20.2	2.3	27.2	1.0	18.6	2.1
9½	28.4	1.1	20.9	2.4	27.8	1.0	19.4	2.1
10½	28.8	1.1	21.7	2.5	28.2	1.0	20.1	2.2
11½	29.4	1.1	22.4	2.5	28.7	1.0	20.8	2.3
1 Year	29.7	1.1	23	3	29.3	1.0	21	3
1½	32.3	1.1	26	3	31.9	1.1	24	3
2	34.5	1.2	28	3	34.1	1.2	27	3
2½	36.3	1.3	30	3	36.0	1.4	29	3
3	37.8	1.3	32	3	37.5	1.4	31	4
3½	39.3	1.4	34	3	39.1	1.5	34	4
4	40.8	1.9	37	5	40.6	1.6	36	5
4½	42.1	1.9	39	5	41.7	1.7	38	5
5	43.7	2.0	42	5	43.8	1.7	41	5
5½	44.7	2.0	44	5	44.5	1.8	43	5
6	46.1	2.1	47	6	45.7	1.9	45	5
7	48.2	2.2	54	7	47.9	2.0	50	7
8	50.4	2.3	60	8	50.3	2.2	58	11
9	52.8	2.4	66	8	52.1	2.3	64	11
10	54.5	2.5	73	10	54.6	2.5	72	14
11	56.8	2.6	82	11	57.1	2.6	82	18
12	58.3	2.9	87	12	59.6	2.7	93	18
13	60.7	3.2	99	13	61.4	2.6	102	18
14	63.6	3.2	113	15	62.8	2.5	112	19
15	66.3	3.1	128	16	63.4	2.4	117	20
16	67.7	2.8	137	16	63.9	2.2	120	21
17	68.3	2.6	143	19	64.1	2.2	122	19
18	68.5	2.6	149	20	64.1	2.3	123	17
19	68.6	2.6	153	21	64.1	2.3	124	17

*See reference 35 on p. 326.

roughly 1.0 inch for civilian headgear, 1.4 inches for steel helmets, and up to 2.6 inches for flying helmets.

Other factors: The average slump from the erect to the normal standing position is about 0.75 inch. At the end of the day, erect stature decreases by approximately 0.95 inch among adults, because of compression of the intervertebral discs in the erect position. Stature measured with the subject stretching against a wall, the preferred measuring technique, exceeds that taken with an anthropometer on the free-standing subject. The difference averages between 0.2 inch under laboratory conditions to 0.8 in large-scale surveys. A good summary estimate would be 0.4 inch (Damon, 1964).

Design Recommendations for
Head to Foot, Horizontal
Add 1.0 inch to the above recommendations for Head to Foot, Vertical.

Eye Height
(*Internal Canthus Height*)

Definition: Vertical distance from the floor to the inner corner of the eye; subject stands erect, looking straight ahead (Fig. 26 and Table 10).

Relevant Equipment or Workspace Dimension: Eye

26. Eye height.

Level, or the vertical distance from the floor to the eye position which affords the best visual field to the standing operator. The visual field includes areas inside and outside the workspace; in the outside area, overhead vision must not be neglected. In some cases, there are no constraints on the placement of the eye reference point, and the present discussion is irrelevant. But where there are obstructions to vision, or if certain objects in the work space or outside areas can be seen only from certain eye positions, it may be necessary to locate the eye reference point within a restricted vertical range. Eye placement is a cardinal feature of design, to which the other features must conform. Adjustability to a given eye level may be achieved by varying floor height or by permitting the operator to slump by supporting his arms. Provision for arm support requires additional forward space and equipment.

Equipment Involved: Streetcars, certain delivery trucks, scanning or lookout stations (as in some World War II tanks, aircraft gun turrets, and submarine periscope stations).

Design Recommendations for Eye Level

For men, 64.7 inches will accommodate the "average" or 50th percentile Air Force flyer. A range from 60.8 inches (5th percentile) to 68.6 inches (95th percentile) will accommodate the middle 90 per cent; a range from 59.2 inches (1st percentile) to 70.3 inches (99th percent-

Table 10. Eye Height (Inches).

Group	Date of Survey	Refer-ence*	Number of Subjects	Age (Mean)	Percentile					S.D.
					1	5	50	95	99	
Males										
Air Force flying personnel	1950	19	4,063	27	59.2	60.8	64.7	68.6	70.3	2.38
Army drivers, white	1960	7	431	24	58.3	60.0	64.8	69.1	70.9	2.65
Army drivers, Negro	1960	7	79	27	57.5	60.1	64.6	69.7	71.9	2.96

*See references 1-43 on pp. 325-327.

ile) will accommodate the middle 98 per cent.

There are no data for other groups, whose eye levels may be approximated by subtracting 5 inches, the average male head height, from stature. For women, subtract 4.5 inches from the male figures.

Clothing: Add 1.0 inch for men's shoes, 1.3 inches for military boots, up to 3.0 inches for women's shoes.

Other factors: The average slump of 0.75 inch in stature from the erect to the normal standing position; the decrease in erect stature of about 0.95 inch at the end of the day. The eye is located 0.7 inch behind glabella (the leading point on the forehead) and 7.0 inches forward of the occiput (the back of the head).

Shoulder Height
(*Acromial Height*)

Definition: Vertical distance from the floor to the most lateral point on the superior surface of the acromion process of the scapula; subject stands erect. Note: for human engineering purposes shoulder height should be measured to the highest point on the shoulder, wherever found (Fig. 27 and Table 11).

Relevant Equipment or Workspace Dimension: Shoulder Height Distance available from floor to top of shoulder.

27. Shoulder height.

Design Recommendations for Shoulder Height

For men, 63.6 inches will accommodate all but the largest 1 per cent of Air Force flyers and 61.9 inches, all but the largest 5 per cent. There are no data on any groups outside the Air Force. These figures consist of the nude percentiles, plus 1.5 inches for vertical distance between the acromion and the highest shoulder point, plus 0.2 inch for the minimal clearance factor. For women, subtract 4.0 inches from the above figure.

Clothing: Add 1.0 inch for men's shoes, 1.3 inches for military boots, up to 3.0 inches for women's shoes, 0.2 inch for light clothing, and 0.9 inch or more for heavy clothing.

Other factors: The average slump of 0.75 inch from the erect to the normal standing position; mobility at the shoulder joint. Increments added by hunching or shrugging have not yet been determined.

Elbow Height
(*Radiale Height*)

Definition: Vertical distance from the floor to radiale, the depression at the elbow between the bones of the upper arm (humerus) and forearm (radius); subject stands erect, arms hanging naturally at sides (Fig. 28 and Table 12).

Relevant Equipment or Workspace Dimension: Work- or Rest-Surface Height—Vertical distance be-

Table 11. Shoulder Height (Inches)

Group	Date of Survey	Refer- ence*	Number of Subjects	Age (Mean)	Percentile					S.D.
					1	5	50	95	99	
Males										
Air Force flying personnel	1950	19	4,063	27	51.2	52.8	56.6	60.2	61.9	2.28
Air Force basic trainees	1952	9	3,326	18	50.3	52.0	55.9	59.9	61.8	2.41
Females										
WAF basic trainees	1952	10	849	19	46.9	48.2	51.9	55.4	57.3	2.18

*See references 1-43 on pp. 325-327.

Table 12. Elbow Height (Inches).

Group	Date of Survey	Refer- ence*	Number of Subjects	Age (Mean)	Percentile					S.D.
					1	5	50	95	99	
Males										
Air Force flying personnel	1950	19	4,063	27	39.5	40.6	43.5	46.4	47.7	1.77

*See references 1-43 on pp. 325-327.

tween the floor and the top of tables, desks, and work-benches used in the standing position. The most comfortable and efficient location for the standing work surface was experimentally found to be 3 inches below elbow height (Ellis, 1951). Rest surfaces should be about 1 inch below elbow height (our "armchair" estimate).

Design Recommendations for
Work-Surface Height

For men, 40.5 inches will accommodate the "average" or 50th percentile Air Force flyer.

A range from 37.6 inches (5th percentile) to 43.4 inches (95th percentile) will accommodate the middle 90 per cent, and a range from 36.5 inches (1st percentile) to 44.7 inches (99th percentile) will accommodate the middle 98 per cent. There are no data for other groups. For women, subtract 3.5 inches from the above figures. This is a tentative estimate, no data being available on women.

Clothing: Add 1.0 inch for men's shoes, 1.3 inches for military boots, and up to 3.0 inches for women's shoes.

Other factors: The kind of task performed and the tools used may change the preferred work-or rest-surface height.

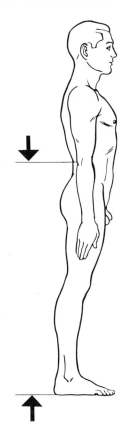

28. Elbow height.

Knuckle Height
(*Metacarpale III Height*)

Definition: Vertical distance from the floor to the largest knuckle of the middle finger (where finger meets palm), or proximal interphalangeal joint. The subject stands erect, palm flat against side of thigh (Fig. 29 and Table 13).

Relevant Equipment or Workspace Dimension: Base to Handle Height of Portable Equipment—Maximum permissible vertical distance between the base of an object and the underside of an attached carrying handle. Suitcases, boxes, and milk or ammunition cans would be items of equipment which should not exceed in height the figures given below—otherwise, they will not clear the ground when carried with the arms hanging down.

Design Recommendations for Base-to-Handle Height of Portable Equipment

For men, 24.7 inches will accommodate all but the smallest 1 per cent and 25.7 inches will accommodate all but the smallest 5 per cent of Air Force flyers. No other groups have been measured. These figures allow 2 inches of clearance above the ground. For women, subtract approximately 2.5 inches (tentative: no data are available on women).

Clothing: Add 1.0 inch for men's shoes, 1.3 inches for military boots, up to 3.0 inches for women's shoes.

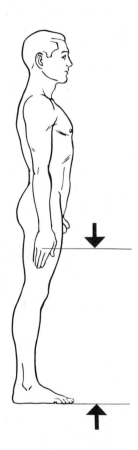

29. Knuckle height.

Table 13. Knuckle Height (Inches).

Group	Date of Survey	Refer-ence*	Number of Subjects	Age (Mean)	Percentile					S.D.
					1	5	50	95	99	
Males										
Air Force flying personnel	1950	19	4,059	27	26.7	27.7	30.0	32.4	33.5	1.45

*See references 1-43 on pp. 325-327.

Sitting Height

Definition: Vertical distance from the sitting surface to the top of the head; subject sits erect, looking straight ahead, with knees and ankles forming right angles (Fig. 30 and Table 14).

Relevant Equipment or Workspace Dimension: Head to Seat, Vertical—Distance between the seat surface and any overhead structure or object located at or above the top of the head.

Design Recommendations for Head to Seat, Vertical

For civilian men, 39.1 inches will accommodate all but the largest 1 per cent and 38.2 inches will accommodate all but the largest 5 per cent. For civilian women the corresponding figures are 36.8 inches and 35.9 inches. Air Force flyers have the same sitting height at the 99th and 95th percentiles as do civilian men. For

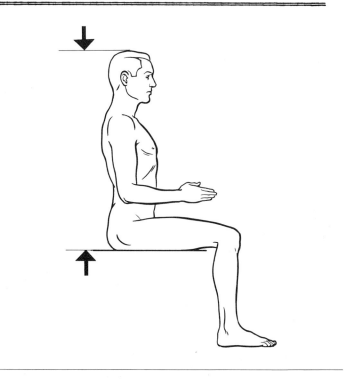

30. Sitting height.

Table 14. Sitting Height (Inches).

Group	Date of Survey	Refer-ence*	Number of Subjects	Age (Mean or Range)	Percentile					S.D.
					1	5	50	95	99	
Males										
Civilian population, erect	1960-62	27	3,091	18-79	31.9	33.2	35.7	38.0	38.9	–
			411	18-24	31.8	33.3	35.9	38.3	39.1	–
			675	25-34	32.5	33.9	36.1	38.4	39.0	–
			703	35-44	32.2	33.7	36.0	38.0	38.9	–
			547	45-54	32.8	33.5	35.7	38.0	38.9	–
			418	55-64	31.4	32.9	35.3	37.7	38.7	–
			265	65-74	31.3	32.5	34.8	36.9	37.7	–
			72	75-79		31.8	34.3	36.7		–
Civilian population, normal slump	1960-62	27	3,091	18-79	30.4	31.6	34.1	36.6	37.6	–
			411	18-24	30.5	31.9	34.2	36.7	37.8	–
			675	25-34	31.0	32.1	34.4	36.8	37.8	–
			703	35-44	30.8	32.0	34.3	36.7	37.7	–
			547	45-54	30.8	31.8	34.2	36.7	37.7	–
			418	55-64	30.2	31.3	33.9	36.0	36.9	–
			265	65-74	30.1	31.2	33.4	35.7	36.4	–
			72	75-79		29.8	33.3	35.8		–
Truck and bus drivers, erect	1951	26	310	36	33.5	34.3	36.3	38.2	39.0	–
Truck and bus drivers, normal slump	1951	26	313	36	31.8	32.6	34.7	36.6	37.5	–
Harvard students	pre-1930	2	479	18	31.2	32.7	36.3	39.8	41.4	2.24
	1940	4	174	18	32.8	33.9	36.5	39.2	40.2	1.57
Spanish-American War veterans	1959	8	119	81	32.5	33.0	34.7	37.0	37.2	1.21
Air Force flying personnel	1950	19	4,061	27	32.9	33.8	36.0	38.0	38.9	1.29
Air Force cadets	1942	32	2,959	23	33.7	34.5	36.4	38.5	39.4	–
Air Force gunners	1942	32	584	23	32.9	33.6	35.9	37.8	38.7	–
Army separatees, white	1946	28,37	24,352	23	32.5	33.5	35.8	38.0	39.0	1.34
Army separatees, Negro	1946	37	6,642	23	31.2	32.2	34.3	36.5	37.4	1.35

Table 14 (continued).

Group	Date of Survey	Refer- ence*	Number of Subjects	Age (Mean or Range)	1	5	50	95	99	S.D.
Army aviators	1959	43	500	30	32.5	33.5	35.6	37.7	38.7	1.27
Army drivers, white, erect	1960	7	431	24	32.5	33.6	35.9	38.1	39.0	1.31
Army drivers, white, normal slump	1960	7	431	24	30.7	31.9	34.1	36.7	37.9	1.39
Army drivers, Negro, erect	1960	7	79	27	31.1	32.5	34.8	37.5	37.8	1.52
Army drivers, Negro, normal slump	1960	7	79	27	29.7	30.6	33.2	36.1	36.2	1.48
Navy enlisted men	1955	41	124	-	33.4	34.3	36.2	38.3	39.1	1.16
Navy pilots	1958	17	1,190	-	32.5	33.5	36.0	38.4	39.9	1.45
Navy aviation cadets	1955	41	469	-	33.8	34.7	36.5	38.5	39.3	-
Navy mixed sample	1946	24	141	23	33.6	34.5	36.7	39.0	39.9	1.35
Females										
Civilian population, erect	1960-62	27	3,581	18-79	29.5	30.9	33.4	35.7	36.6	-
			534	18-24	30.4	31.4	33.7	35.9	36.7	-
			746	25-34	30.3	31.4	33.8	35.9	36.8	-
			784	35-44	30.3	31.5	33.7	35.8	36.8	-
			705	45-54	30.1	31.2	33.5	35.6	36.4	-
			443	55-64	30.0	30.7	33.0	35.4	36.4	-
			299	65-74	28.6	29.7	32.2	34.5	35.8	-
			70	75-79		28.1	32.1	34.8		-
Civilian population, normal slump	1960-62	27	3,581	18-79	28.2	29.6	32.3	34.7	35.7	-
			534	18-24	29.2	30.1	32.6	34.8	35.7	-
			746	25-34	28.9	30.1	32.6	34.9	35.9	-
			784	35-44	29.2	30.2	32.6	34.9	35.8	-
			705	45-54	28.7	29.7	32.3	34.6	35.5	-
			443	55-64	28.3	29.7	32.1	34.4	35.4	-
			299	65-74	27.0	28.7	31.2	33.9	34.9	-
			70	75-79		27.1	31.0	33.4		-
Normal working women	1932	1	100	36	30.9	31.7	33.7	35.7	36.5	1.15
College students, East	pre-1930	2	198	17	31.6	32.4	34.2	36.0	36.7	1.10
Univ. Tennessee students	1930	3	161	19	31.0	31.7	33.6	35.4	36.2	1.06
Air Force pilots (WASPS)	1944	32	446	18-35	31.8	32.4	34.1	35.8	36.3	-
Air Force flying nurses	1943	32	152	-	31.1	31.9	33.7	35.7	36.6	-

*See references 1-43 on pp. 325-327.

Army soldiers, 39.2 inches will accommodate all but the largest 1 per cent and 38.3 inches all but the largest 5 per cent.

The foregoing figures refer to the nude, erect posture, and contain a small clearance increment of 0.2 inch above the nude percentile.

Clothing: Add 0.2 to 0.3 inch for heavy clothing under buttocks, roughly 1.0 inch for civilian headgear, 1.4 inches for steel helmets, and up to 2.6 inches for flying helmets.

Other factors: The decrease (about 1.5 inches) in erect sitting height among the aged; the average slump from the erect to the normal sitting position of 1.6 inches for men and 1.1 inches for women; cushions, springs or other yielding seat surfaces; inclination of seat; the variety of civilian and military headgear; seat parachute or escape gear raising the buttocks above the seat.

Eye Height, Sitting
(*Internal Canthus Height, Sitting*)

Definition: Vertical distance from the sitting surface to the inner corner of the eye; subject sits erect, looking straight ahead (Fig. 31 and Table 15).

Relevant Equipment or Workspace Dimension: Seated Eye Level—Vertical distance between the seat surface and the eye position which affords the best

field of vision inside and if necessary outside the workspace.

Before using this section, read the previous comments on eye placement (p. 67). Adjustment to a given seated eye level is best afforded by seats which are vertically adjustable in at least four increments of no

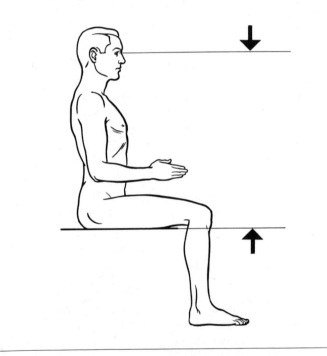

31. Eye height, sitting.

more than one inch each, since the middle 90 per cent of most groups is 4 to 5 inches (Table 14).

Design Recommendations for Seated Eye Level

For civilian men 31.2 inches fits the "average" or 50th percentile. A range from 29.3 inches (5th percentile) to 33.2 inches (95th percentile) will accommodate the middle 90 per cent and a range from 28.6 inches (1st percentile) to 33.9 inches (99th percentile) will accommodate the middle 98 per cent. (The average slump for this group, 1.6 inches, has been added to the values in Table 15.) For women subtract 2.0 inches from these values. For Air Force flyers 31.5 inches fits the 50th percentile, the 90 per cent midrange extends from 29.4 to 33.5 inches, and the 98 per cent midrange from 28.5 to 34.4 inches. These values are for nude, upright posture.

Clothing: Add 0.2 to 0.3 inch for heavy clothing under the buttocks.

Other factors: As for sitting height; *viz.,* the decreased sitting height among the aged (about 1.5 inches); the average normal slump, 1.6 inches for men

Table 15. Eye Height, Sitting (Inches).

Group	Date of Survey	Refer-ence*	Number of Subjects	Age (Mean or Range)	Percentile 1	5	50	95	99	S.D.
Males										
Truck and bus drivers[a]	1951	26	309	36	27.0	27.7	29.6	31.6	32.3	-
Air Force flying personnel	1950	19	4,061	27	28.5	29.4	31.5	33.5	34.4	1.27
Army aviators	1959	43	500	30	28.1	28.8	30.9	33.1	34.5	1.28
Navy pilots	1958	21	1,190	-	27.7	29.2	31.7	34.2	35.0	1.50
Females										
Air Force pilots (WASPS)	1944	32	443	18-35	27.9	28.5	30.0	31.6	32.4	-
Air Force flying nurses	1943	32	151	-	26.3	27.3	29.3	31.1	32.2	-

[a]Includes normal "slump."
*See references 1-43 on pp. 325-327.

and 1.1 inches for women, from erect to comfortable or usual sitting posture; yielding seat surfaces (cushions, springs); inclination of seat; equipment raising the buttocks above the seat (parachute, escape gear). The eye is located 0.7 inch behind the leading point on the forehead and 7.0 inches forward of the occiput (in Air Force flyers).

Shoulder Height, Sitting
(*Acromial Height, Sitting*)

Definition: Vertical distance from the sitting surface to the most lateral point on the superior surface of the acromion process of the scapula; subject sits erect (Fig. 32 and Table 16). Note: for human engineering purposes shoulder height should be measured to the highest point on the shoulder, wherever found.

Related Equipment or Workspace Dimension: Shoulder to Seat, Vertical—Distance between the seat surface and objects located at or above the shoulder.

Design Recommendations for Shoulder to Seat, Vertical

For civilian men, 27.6 inches will accommodate all but the largest 1 per cent and 26.9 inches will accommodate all but the largest 5 per cent. For women, subtract 2.0 inches from the male values. For Air Force flyers

corresponding figures are 27.5 inches and 26.8 inches; for Army soldiers 27.5 and 26.7 inches.

These figures consist of the nude percentile plus arbitrary increments of 1.5 inches for vertical distance from edge of shoulder to highest point, and 0.2 inch for minimum clearance.

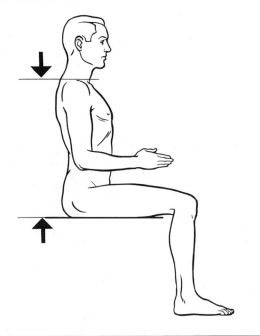

32. Shoulder height, sitting.

Table 16. Shoulder Height, Sitting (Inches).

Group	Date of Survey	Refer- ence*	Number of Subjects	Age (Mean or Range)	Percentile					S.D.
					1	5	50	95	99	
Males										
Truck and bus drivers[a]	1951	26	311	36	21.3	22.0	23.7	25.2	25.9	-
Air Force flying personnel	1950	19	4,057	27	20.6	21.3	23.3	25.1	25.8	1.14
Air Force cadets[a]	1942	32	2,957	23	21.6	22.2	23.8	25.3	25.9	-
Air Force gunners[a]	1942	32	583	23	20.9	21.5	23.3	24.8	25.6	-
Army separatees, white[a]	1946	28,37	24,715	23	20.1	21.0	23.0	25.0	25.8	1.17
Army separatees, Negro[a]	1946	37	4,847	23	19.4	20.2	22.1	23.9	24.7	1.19
Army drivers, white[b]	1960	7	431	24	22.6	24.1	25.9	27.7	28.4	1.15
Army drivers, Negro[b]	1960	7	79	27	21.9	23.3	25.2	27.1	27.6	1.18
Navy pilots	1958	21	1,190	-	21.6	22.5	24.6	27.1	29.3	1.47
Females										
Air Force pilots (WASPS)[b]	1944	32	445	18-35	21.8	22.4	23.8	25.2	25.9	-
Air Force flying nurses[b]	1943	32	152	-	20.4	21.1	23.1	24.8	25.9	-
WACs and Army nurses[a]	1946	33,41	8,502	26	18.6	19.4	21.2	23.0	23.7	1.09

[a]Trunk height, as measured to the upper edge of the suprasternal notch, generally averages 1/4 inch less than shoulder height.

[b]Measured to a point on the shoulder, midway between the angles made by the shoulder with the arm and neck.

*See references 1-43 on pp. 325-327.

Clothing: Light wear adds 0.2 inches, heavy clothing about 1.0 inch.

Other factors: The decreased shoulder height among the aged (about 1.25 inches); the average slump in shoulder height of about 1.5 inches from the erect to the normal sitting position; mobility at the shoulder, as in hunching or shrugging (amount unknown).

Elbow Rest Height, Sitting

Definition: Vertical distance from the sitting surface to the bottom of the right elbow; subject sits erect, upper arm vertical at side, forearm at right angle to upper arm (Fig. 33 and Table 17).

Relevant Equipment or Workspace Dimension: Elbow Rest Height, from Seat—Vertical distance between the seat surface and the top surface of an elbow rest.

Design Recommendations for Elbow Rest Height, from Seat

For men, civilian or military, 8.0 inches will accommodate many around the "average" or 50th percentile, and a range from about 5.5 to 11.0 inches will accommodate virtually everyone. For women, subtract about 0.25 to 0.50 inch from the above figures. These values are based on the nude percentiles minus 1.0 to 1.5 inches for slump.

Clothing: Makes no difference, that under the buttocks being balanced by that under the elbow.

Other factors: Sitting posture, erect or normal, and with elbows against or away from body; seat inclination; "give" of seat surface. The long arms of Negroes lower their elbow rest height.

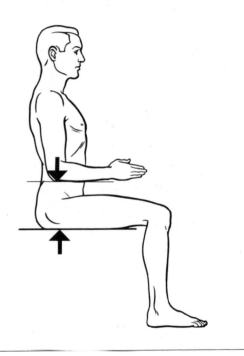

33. Elbow rest height, sitting.

Table 17. Elbow Rest Height, Sitting (Inches).

Group	Date of Survey	Refer- ence*	Number of Subjects	Age (Mean or Range)	Percentile					S.D.
					1	5	50	95	99	
Males										
Civilian population	1960-62	27	3,091	18-79	6.3	7.4	9.5	11.6	12.5	-
			411	18-24	6.3	7.6	9.6	11.9	12.8	-
			675	25-34	7.0	8.0	9.7	11.7	12.6	-
			703	35-44	6.5	7.8	9.7	11.8	12.6	-
			547	45-54	7.0	7.7	9.6	11.5	12.0	-
			418	55-64	6.0	7.2	9.3	11.4	12.2	-
			265	65-74	6.1	7.1	9.0	10.9	11.9	-
			72	75-79		6.5	8.6	10.6		-
Railroad travelers[a]	1944	20	1,959	38	7.5	8.1	9.6	11.1	11.7	-
Air Force flying personnel	1950	19	4,063	27	6.6	7.4	9.1	10.8	11.5	1.04
Army drivers, white	1960	7	431	24	6.9	8.0	9.8	11.6	12.2	1.08
Army drivers, Negro	1960	7	79	27	5.5	5.7	8.5	10.6	10.9	1.54
Females										
Civilian population	1960-62	27	3,581	18-79	6.1	7.1	9.2	11.0	11.9	-
			534	18-24	6.2	7.2	9.1	10.8	11.8	-
			746	25-34	6.1	7.4	9.3	11.1	11.9	-
			784	35-44	6.7	7.5	9.4	11.3	12.0	-
			705	45-54	6.4	7.3	9.3	11.0	12.1	-
			443	55-64	5.9	7.1	9.0	10.9	11.9	-
			299	65-74	5.4	6.4	8.5	10.2	11.3	-
			70	75-79		6.4	8.4	10.0		-
Railroad travelers[a]	1944	20	1,908	35	7.8	8.4	9.7	11.1	11.7	-

[a]Measured to a vertically adjustable elbow rest; its fixed lateral distance from the body sometimes results in slightly increased values. Subjects clothed.
*See references 1-43 on pp. 325-327.

Thigh Clearance Height, Sitting

Definition: Vertical distance from the sitting surface to the top of the thigh at its intersection with the abdomen; subject sits erect, knees and ankles at right angles (Fig. 34 and Table 18).

Relevant Equipment or Workspace Dimension: Thigh to Seat, Vertical—Distance between the seat surface and the underside of a table, desk, steering wheel, or other object located above the thighs.

*Design Recommendations for
Thigh to Seat, Vertical*

For civilian men and women alike, 7.9 inches will accommodate the 99th percentile and 7.1 inches the 95th percentile. For Air Force flyers 7.0 inches will accommodate all but the largest 1 per cent and 6.7 inches will accommodate all but the largest 5 per cent. Corresponding figures for Army soldiers are 7.6 and 7.1 inches. These recommendations consist of the nude percentiles plus an arbitrary 0.2 inch minimum clearance.

Clothing: Add 0.1 to 0.2 inch for light clothing, 1.4 inches or more for heavy clothing.

Other factors: Advancing the foot lowers thigh clearance height but requires more space forward of the operator. The effect of doubling the lower leg back under the thigh has not been determined, but this highly undesirable "solution" impairs blood flow to and from the leg.

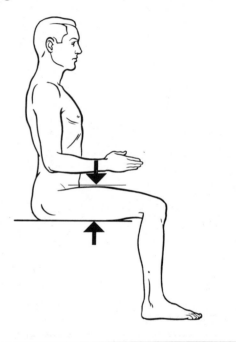

34. Thigh clearance height, sitting.

Table 18. Thigh Clearance Height, Sitting (Inches).

Group	Date of Survey	Refer- ence*	Number of Subjects	Age (Mean or Range)	Percentile					S.D.
					1	5	50	95	99	
Males										
Civilian population	1960-62	27	3,091	18-79	4.1	4.3	5.7	6.9	7.7	-
			411	18-24	4.1	4.3	5.7	6.9	7.7	-
			675	25-34	4.1	4.5	5.8	7.0	7.9	-
			703	35-44	4.1	4.6	5.8	7.0	7.8	-
			547	45-54	4.1	4.4	5.6	6.9	7.1	-
			418	55-64	4.0	4.2	5.5	6.8	7.4	-
			265	65-74	4.0	4.2	5.4	6.7	7.0	-
			72	75-79		4.1	5.2	6.6		-
Air Force flying personnel	1950	19	4,061	27	4.5	4.8	5.6	6.5	6.8	0.52
Army drivers, white	1960	7	431	24	4.7	5.0	5.9	6.9	7.4	0.56
Army drivers, Negro	1960	7	79	27	4.8	5.3	6.1	7.5	7.6	0.58
Females										
Civilian population	1960-62	27	3,581	18-79	3.8	4.1	5.4	6.9	7.7	-
			534	18-24	3.6	4.1	5.4	6.7	7.0	-
			746	25-34	4.0	4.2	5.4	6.9	7.7	-
			784	35-44	4.0	4.2	5.5	7.0	7.8	-
			705	45-54	4.0	4.2	5.5	6.9	7.7	-
			443	55-64	3.5	4.1	5.4	6.9	8.3	-
			299	65-74	3.4	4.1	5.3	6.6	7.0	-
			70	75-79		4.0	5.2	6.5		-

*See references 1-43 on pp. 325-327.

Knee Height, Sitting

Definition: Vertical distance from the floor to the uppermost point on the knee (not the patella or knee-cap); subject sits erect, knees and ankles at right angles (Fig. 35 and Table 19).

Relevant Equipment or Workspace Dimension: Knee Height—Distance between the floor, footrest, or foot control and the underside of a table, desk, wheel, dashboard, or other object located above the knees. A "minimum" or clearance dimension.

Design Recommendations for
Knee to Foot, Vertical

For civilian men, 24.3 inches will accommodate all but the largest 1 per cent and 23.6 inches will accommodate all but the largest 5 per cent. For civilian women allow 22.4 inches for the 99th percentile and 21.7 inches for the 95th percentile. Corresponding figures for Air Force flyers are 24.2 and 23.5 inches and for Army soldiers, 24.5 inches and 23.7 inches. These figures consist of the nude percentiles plus an arbitrary 0.2-inch minimum clearance.

Clothing: Add 1.0 inch for men's shoes and light clothing, 1.5 inches or more for military boots and heavy clothing, up to 3.0 inches for women's shoes and light clothing.

Other factors: Angulation of knee, ankle, or seat front. Pedal operation may raise the foot and hence the knee.

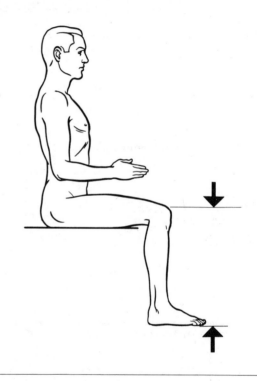

35. Knee height, sitting.

Table 19. Knee Height, Sitting (Inches).

Group	Date of Survey	Reference*	Number of Subjects	Age (Mean or Range)	Percentile					S.D.
					1	5	50	95	99	
Males										
Civilian population	1960-62	27	3,091	18-79	18.3	19.3	21.4	23.4	24.1	-
			411	18-24	18.3	19.4	21.5	23.4	23.9	-
			675	25-34	19.0	19.8	21.6	23.7	24.6	-
			703	35-44	18.4	19.4	21.5	23.4	24.4	-
			547	45-54	18.2	19.3	21.4	23.3	23.9	-
			418	55-64	18.1	19.1	21.1	23.1	24.0	-
			265	65-74	18.2	19.2	21.0	22.9	23.7	-
			72	75-79		19.0	20.7	22.7		-
Truck and bus drivers	1951	26	301	36	19.3	20.1	21.7	23.5	24.2	-
Spanish-American War veterans	1959	8	132	81	19.4	19.9	21.2	22.6	23.4	0.85
Air Force flying personnel	1950	19	4,060	27	19.5	20.1	21.7	23.3	24.0	0.99
Air Force cadets	1942	32	2,959	23	19.7	20.4	22.0	23.6	24.3	-
Air Force gunners	1942	32	583	23	19.2	19.8	21.5	23.0	23.7	-
Army separatees, white	1946	28,37	24,419	23	19.0	19.8	21.6	23.5	24.3	1.09
Army separatees, Negro	1946	37	6,623	23	19.6	20.3	22.2	24.0	24.7	1.14
Army drivers, white	1960	7	431	24	19.1	19.8	21.5	23.4	24.3	1.06
Army drivers, Negro	1960	7	79	27	18.6	20.2	22.1	24.1	24.7	1.18
Navy, mixed sample	1946	24	141	23	18.8	19.7	21.8	23.8	24.7	1.25
Females										
Civilian population	1960-62	27	3,581	18-79	17.1	17.9	19.6	21.5	22.4	-
			534	18-24	17.3	18.1	19.7	21.6	22.7	-
			746	25-34	17.2	18.0	19.7	21.6	22.5	-
			784	35-44	17.2	18.0	19.6	21.5	22.4	-
			705	45-54	17.1	17.6	19.5	21.6	22.5	-
			443	55-64	16.6	17.8	19.5	21.4	21.9	-
			299	65-74	17.1	17.8	19.2	21.0	22.0	-
			70	75-79		17.3	19.4	20.9		-
Air Force pilots (WASPS)	1944	32	445	18-35	18.3	18.7	20.1	21.5	22.2	-
Air Force flying nurses	1943	32	152	-	17.7	18.1	19.5	20.8	21.5	-
WACs and Army nurses	1946	33,37	8,412	26	16.6	17.2	18.8	20.3	21.1	0.95

*See references 1-43 on pp. 325-327.

Popliteal Height, Sitting

Definition: Vertical distance from the floor to the underside of the thigh immediately behind the knee; subject sits erect, knees and ankles at right angles, bottom of thighs and back of knees barely touching the sitting surface. In practice, taken from foot support to sitting surface with subject properly positioned (Fig. 36 and Table 20).

Relevant Equipment or Workspace Dimension: Seat Height—Vertical distance between the floor and the highest point on the front of the seat surface.

This is a "maximum" dimension, in that exceeding the recommended values will inconvenience the short-legged person. To the point of flexing (angling back) the lower leg on the thigh, a seat is better too low than too high. A low seat can always be raised by cushions, but if the seat is too high, one cannot always place objects under the feet. Lowering the seat 1 inch below popliteal height will increase comfort by relieving pressure, as will sliding the buttocks forward so that the popliteal region clears the seat edge. For greatest comfort, therefore, as in a riding rather than a working seat, ample forward leg and foot space must be provided. With fixed forward knee and foot space, too low a seat produces discomfort and could lead to thrombophlebitis (see page 33).

Design Recommendations for Seat Height

For civilian men, 14.7 inches will accommodate all but the smallest 1 per cent and 15.3 inches all but the smallest 5 per cent. For civilian women the comparable

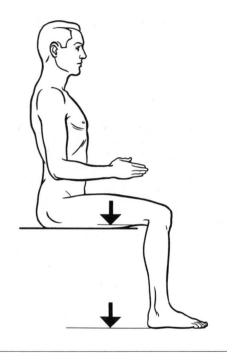

36. Popliteal height, sitting.

Table 20. Popliteal Height, Sitting (Inches).

Group	Date of Survey	Refer- ence*	Number of Subjects	Age (Mean or Range)	Percentile					S.D.
					1	5	50	95	99	
Males										
Civilian population	1960-62	27	3,091	18-79	14.9	15.5	17.3	19.3	20.0	-
			411	18-24	15.2	16.0	17.5	19.6	20.4	-
			675	25-34	15.1	16.0	17.5	19.7	20.6	-
			703	35-44	15.0	15.6	17.3	19.1	19.9	-
			547	45-54	14.7	15.5	17.2	19.1	19.9	-
			418	55-64	14.9	15.3	17.1	19.0	19.8	-
			265	65-74	14.2	15.1	17.1	18.9	19.8	-
			72	75-79		15.2	16.6	18.4		-
Railroad travelers[a]	1944	20	1,959	38	16.9	17.6	19.0	20.6	21.1	-
Spanish-American War veterans[b]	1959	8	131	81	15.4	15.7	17.2	18.6	19.2	0.83
Air Force flying personnel	1950	19	4,059	27	15.3	15.7	17.0	18.2	18.8	0.77
Army drivers, white[b]	1960	7	431	24	14.8	15.7	17.2	19.3	20.1	1.00
Army drivers, Negro[b]	1960	7	79	25	14.8	15.6	17.9	20.1	20.1	1.23
Females										
Civilian population	1960-62	27	5,381	18-79	13.1	14.0	15.7	17.5	18.0	-
			534	18-24	13.5	14.2	16.1	17.8	18.5	-
			746	25-34	13.2	14.1	15.8	17.5	18.2	-
			784	35-44	13.1	14.0	15.7	17.5	17.9	-
			705	45-54	13.1	13.8	15.5	17.5	18.3	-
			443	55-64	13.1	13.6	15.4	17.1	17.9	-
			299	65-74	13.0	13.9	15.3	17.0	17.9	-
			70	75-79		13.5	15.6	17.2		-
Railroad travelers[a]	1944	20	1,908	35	16.2	16.7	18.1	19.5	20.1	-

[a]Street shoes and a different measuring device result in values higher than those obtainable with standard anthropometric techniques.

[b]Measured to the seat surface.

*See references 1-43 on pp. 325-327.

figures are 12.9 inches and 13.8 inches. For Air Force flyers, 15.1 inches will be adequate for all but the smallest 1 per cent and 15.5 inches will fit all but the smallest 5 per cent; for Army soldiers the figures are 14.6 and 15.5 inches.

The above figures represent nude percentiles minus an arbitrary minimal clearance of 0.2 inch. As noted, 1.0 inch of clearance would be more comfortable for the smaller operators, but at the risk of discommoding the taller ones if forward foot space is inadequate.

Clothing: Add 1.0 inch for men's shoes, 1.3 inches for military boots, and up to 3.0 inches for women's shoes.

Other factors: Footrests and pedals may raise or lower the popliteal region, depending on their placement. A sloping seat front may relieve popliteal pressure without lowering the seat. As discussed above, ample forward leg and foot space will permit the buttocks to slide forward and the popliteal region to clear the seat front.

Buttock-Knee Length

Definition: Horizontal distance from the plane of the rearmost point on the buttocks to the front of the knee; subject sits erect, knees and ankles at right angles (Fig. 37 and Table 21).

Relevant Equipment or Workspace Dimension: But-tock-Knee, Horizontal—Distance between the seat back and objects located in front of the knees. A "minimum" or clearance dimension, usually static.

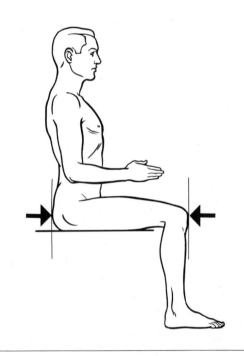

37. Buttock-knee length.

Table 21. Buttock-Knee Length (Inches).

Group	Date of Survey	Refer- ence*	Number of Subjects	Age (Mean or Range)	Percentile					S.D.
					1	5	50	95	99	
Males										
Civilian population	1960-62	27	3,091	18-79	20.3	21.3	23.3	25.2	26.3	-
			411	18-24	20.4	21.3	23.3	25.4	26.5	-
			675	25-34	20.8	21.6	23.6	25.7	26.8	-
			703	35-44	20.3	21.3	23.4	25.1	26.2	-
			547	45-54	20.4	21.3	23.4	25.2	26.1	-
			418	55-64	19.6	21.2	23.1	24.9	25.8	-
			265	65-74	20.1	21.0	23.0	24.8	25.9	-
			72	75-79		21.0	22.6	24.7		
Truck and bus drivers	1951	26	310	36	21.3	22.1	23.8	25.8	26.5	-
Spanish-American War veterans	1959	8	132	81	21.0	21.8	23.2	25.0	25.4	0.96
Air Force flying personnel	1950	19	4,060	27	21.2	21.9	23.6	25.4	26.2	1.06
Air Force cadets	1942	32	2,954	23	21.2	22.0	23.6	25.6	26.2	-
Air Force gunners	1942	32	582	23	20.5	21.1	23.1	24.7	25.6	-
Army separatees, white	1946	28,37	24,244	23	20.7	21.5	23.4	25.2	26.0	1.12
Army separatees, Negro	1946	37	6,594	23	21.1	21.9	23.8	25.8	26.6	1.17
Army aviators	1959	43	500	30	21.4	22.1	23.8	25.8	26.7	1.08
Army drivers, white	1960	7	431	24	20.9	21.7	23.7	25.6	26.5	1.14
Army drivers, Negro	1960	7	79	27	21.1	22.5	24.7	26.5	26.5	1.11
Navy enlisted men	1955	41	124	-	21.7	22.5	24.5	26.5	27.3	1.23
Navy, mixed sample	1946	24	141	23	20.6	21.4	23.4	25.0	25.8	1.18
Navy aviation cadets	1955	41	472	-	21.8	22.6	24.3	26.2	26.9	-
Females										
Civilian population	1960-62	27	3,581	18-79	19.5	20.4	22.4	24.6	25.7	-
			534	18-24	19.3	20.3	22.2	24.6	25.6	-
			746	25-34	20.0	20.5	22.4	24.6	25.6	-
			784	35-44	20.0	20.5	22.5	24.7	25.9	-
			705	45-54	19.4	20.3	22.4	24.6	25.5	-
			443	55-64	19.4	20.3	22.3	24.7	25.7	-
			299	65-74	19.4	20.2	22.2	24.6	25.9	-
			70	75-79		19.9	22.2	23.9		-
Air Force pilots (WASPS)	1944	32	447	18-35	20.4	21.1	22.6	24.2	25.0	-
Air Force flying nurses	1943	32	152	-	20.2	20.9	22.4	24.0	24.8	-

*See references 1-43 on pp. 325-327.

*Design Recommendations for Buttock
to Knee, Horizontal*

For civilian men, 26.5 inches will accommodate all but
the largest 1 per cent, and 25.4 inches will accommodate all but the largest 5 per cent. For civilian women
allow 25.9 inches for the 99th percentile and 24.8 inches
for the 95th percentile. For both Air Force flyers and
Army soldiers 26.4 inches will be adequate for all but
the largest 1 per cent, and 25.6 inches for all but the
largest 5 per cent. These figures represent nude percentiles plus the minimum clearance increment of 0.2
inch.

Clothing: Add 0.2 inch for light clothing, 0.7 inch
or more for heavy clothing, 2.9 inches for the partial
pressure suit.

Other factors: Shifting the buttocks forward on the
seat will require additional clearance. Seat or knee
angulation will change the dimensions, as will yielding
materials in the seat back.

Buttock-Popliteal Length

Definition: Horizontal distance from the plane of the
rearmost point on the buttocks to the back of the
lower leg at the knee; subject sits erect, knees and
ankles at right angles (Fig. 38 and Table 22).

Relevant Equipment or Workspace Dimension: Seat
Length—Distance from seat reference point (junction
of sitting surface and seat back) to front edge of seat.
A "maximum" dimension, in that too long a seat will
severely discommode short-legged operators. A seat is
better too short than too long.

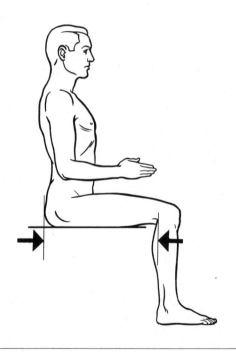

38. Buttock-popliteal length.

Table 22. Buttock-Popliteal Length (Inches).

Group	Date of Survey	Refer-ence*	Number of Subjects	Age (Mean or Range)	Percentile					S.D.
					1	5	50	95	99	
Males										
Civilian population	1960-62	27	3,091	18-79	16.5	17.3	19.5	21.6	22.7	-
			411	18-24	16.5	17.4	19.5	21.6	22.9	-
			675	25-34	16.6	17.6	19.6	21.9	23.1	-
			703	35-44	16.5	17.4	19.5	21.8	22.7	-
			547	45-54	17.0	17.4	19.5	21.5	22.0	-
			418	55-64	16.4	17.2	19.4	21.5	22.2	-
			265	65-74	16.3	17.3	19.3	20.9	21.9	-
			72	75-79		17.0	18.9	21.2		-
Railroad travelers[a]	1944	20	1,959	38	16.6	17.4	18.9	20.8	21.5	-
Spanish-American War veterans	1959	8	131	81	16.5	16.9	18.5	20.3	21.1	1.00
Army drivers, white	1960	7	431	24	16.4	17.2	19.2	20.9	21.9	1.12
Army drivers, Negro	1960	7	79	27	17.3	17.7	20.0	21.9	22.2	1.08
Females										
Civilian population	1960-62	27	3,581	18-79	16.1	17.0	18.9	21.0	22.0	-
			534	18-24	16.1	16.9	18.8	21.1	21.9	-
			746	25-34	16.1	17.0	18.9	21.0	21.9	-
			784	35-44	16.2	17.1	18.9	21.1	22.4	-
			705	45-54	15.8	17.0	18.9	20.9	22.0	-
			443	55-64	16.1	17.1	18.9	21.0	22.0	-
			299	65-74	16.1	16.9	18.8	20.9	21.9	-
			70	75-79		17.0	18.7	20.0		-
Railroad travelers[a]	1944	20	1,908	35	16.0	16.8	18.2	20.0	20.6	-

[a]Subjects clothed.
*See references 1-43 on pp. 325-327.

Design Recommendations for Seat Length

For civilian men 16.3 inches will accommodate all but the smallest 1 per cent and 17.1 inches will accommodate all but the smallest 5 per cent. For civilian women the figures are 15.9 and 16.8 inches. For Army soldiers, 16.2 inches will fit the 1st percentile and 17.0 inches the 5th percentile. No data are available for Air Force flyers. These figures represent the nude percentiles (but see below) minus an arbitrary 0.2 inch.

Clothing: This makes no difference in recommended seat length, unless the clothing behind the buttocks greatly exceeds that behind the knee. In the usual case where seat length is fixed, trousers and coat push the popliteal region forward, about 0.2 inch for light clothing, 0.5 inch for heavy clothing, and 1.5 inches for the partial pressure suit.

Other factors: Seat angulation, slope of seat front. Sliding the buttocks forward or placing a cushion, clothing (see above), or a parachute behind the back will enable short-legged persons to adapt to longer seats.

Shoulder-Elbow Length

Definition: Distance from the top of the acromion process (at the uppermost point on the lateral edge of the shoulder) to the bottom of the elbow; subject sits erect, upper arm vertical at side and making a right angle with the forearm (Fig. 39 and Table 23).

Relevant Equipment or Workspace Dimension: Shoulder-Elbow—Distance between objects located at or above the shoulder and at or below the elbow—a "minimum" clearance.

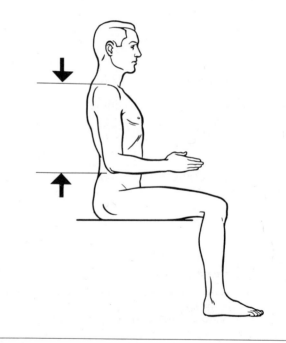

39. Shoulder-elbow length.

Design Recommendations for
Shoulder-Elbow Space

For civilian men, 16.5 inches will accommodate all but the largest 1 per cent and 16.1 inches will accommodate all but the largest 5 per cent. For women, subtract about 1.0 inch from these values. For Air Force flyers, 16.1 inches will fit all but the largest 1 per cent and 15.6 inches will fit all but the largest 5 per cent. Corresponding figures for Army soldiers are 16.5 and 15.9 inches. Note that these are nude dimensions plus a minimal clearance of 0.2 inch. The dimension for all groups except Air Force flyers in 1950 was measured to

Table 23. Shoulder-Elbow Length (Inches).

Group	Date of Survey	Reference*	Number of Subjects	Age (Mean or Range)	Percentile 1	5	50	95	99	S.D.
Males										
Truck and bus drivers	1951	26	311	36	13.3	13.8	14.8	15.9	16.3	0.81
Harvard students	pre-1930	2	477	18	12.8	13.3	14.5	15.7	16.1	0.66
Spanish-American War veterans	1959	8	131	81	13.4	13.5	14.5	15.6	16.4	0.66
Air Force flying personnel	1950	19	4,059	27	12.8	13.2	14.3	15.4	15.9	0.69
Air Force cadets	1942	32	2,955	23	13.2	13.6	14.7	15.8	16.3	-
Air Force gunners	1942	32	583	23	12.9	13.3	14.5	15.6	16.1	-
Army separatees, white	1946	28,37	24,556	23	12.3	12.9	14.3	15.6	16.3	0.81
Army separatees, Negro	1946	37	4,825	23	12.4	13.0	14.3	15.6	16.1	0.80
Army drivers, white	1960	7	431	24	13.0	13.5	14.6	15.8	16.3	0.71
Army drivers, Negro	1960	7	79	27	12.6	13.9	14.9	16.3	16.5	0.77
Army aviators	1959	43	500	30	13.4	13.9	15.0	16.1	16.5	0.70
Females										
Air Force pilots (WASPS)	1944	32	447	18-35	12.3	12.7	13.7	14.7	15.2	-
Air Force flying nurses	1943	32	151	-	12.3	12.7	13.6	14.8	15.3	-
WACs and Army nurses	1946	33,37	8,488	26	11.3	11.9	13.1	14.3	14.9	0.74

*See references 1-43 on pp. 325-327.

the top of the acromion process, not to the acromiale point, which accounts for the lower figures for the flyers. An additional 1.5 inches should be added for vertical distance from acromiale, or the top of the process, to the highest point between shoulder and neck, the more functional dimension.

Clothing: Add 0.2 inch for light clothing, 1.0 inch or more for heavy clothing.

Other factors: Hunching the shoulder increases the shoulder-elbow space needed. Angulation, whether flexion or extension, of the upper arm at the shoulder joint will decrease the amount of vertical clearance required, so that the present recommended values will suit even larger proportions of operating groups.

Forearm-Hand Length

Definition: Distance from tip of elbow to tip of longest finger; subject sits erect, upper arm vertical at side, forearm, hand, and fingers extended horizontally (Fig. 40 and Table 24).

Relevant Equipment or Workspace Dimension: Arm Reach from Elbow—Maximum fingertip reach from a fixed elbow point.

Design Recommendations for Arm Reach from Elbow

For civilian men 16.7 inches can be reached by all but the smallest 1 per cent, and 17.3 inches by all but the smallest 5 per cent. For women, subtract 2.0 inches (an estimate; no data are available). For Air Force flyers 17.0 inches can be reached by all but the smallest 1 per cent and 17.6 inches by all but the smallest 5 per cent. Corresponding figures for Army soldiers are 16.7 and

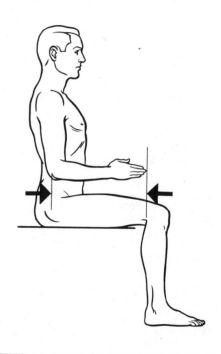

40. Forearm-hand length.

17.3 inches. These figures refer to distance between nude elbow and maximally extended fingertip, with the elbow fixed at a right angle and backed against a vertical surface.

Clothing: Add 0.2 inch for light clothing without gloves, 0.2 inch for light gloves, 0.5 inch for medium clothing and gloves, 0.8 to 1.0 inch for heavy clothing and gloves.

Other factors: For fingertip manipulation, subtract 0.5 inch for flip, 1.0 inch for push; for manipulation by thumb and forefinger, subtract 3.0 inches; for grasp by whole hand, subtract 5.0 inches. Angulation of the forearm shortens the forward distance required to reach a control, although the total elbow-fingertip distance remains constant (as the radius of a circle). If elbows are not fixed in the fore-and-aft plane, this dimension presents no problem. Note that Negroes exceed whites in this dimension.

Table 24. Forearm-Hand Length (Inches).

Group	Date of Survey	Refer-ence*	Number of Subjects	Age (Mean or Range)	Percentile					S.D.
					1	5	50	95	99	
Males										
Truck and bus drivers	1951	26	311	36	16.7	17.3	18.8	20.2	20.8	-
Harvard students	pre-1930	2	474	18	17.0	17.6	18.9	20.2	20.7	0.75
Spanish-American War veterans	1959	8	130	81	16.9	17.2	18.3	19.5	20.4	0.71
Air Force flying personnel	1950	19	4,059	27	17.0	17.6	18.9	20.2	20.7	0.81
Army separatees, white	1946	28,37	24,354	23	16.6	17.3	18.7	20.1	20.8	0.88
Army separatees, Negro	1946	37	4,781	23	17.3	18.0	19.6	21.4	22.1	0.94
Army aviators	1959	43	500	30	16.1	17.6	19.1	20.4	21.5	0.86
Army drivers, white	1960	7	431	24	16.8	17.3	18.7	20.2	20.9	0.85
Army drivers, Negro	1960	7	79	27	16.1	17.8	19.7	21.4	21.7	1.10

*See references 1-43 on pp. 325-327.

Arm Reach From Wall

Definition: Distance from wall to tip of longest finger; subject stands erect, with heels, buttocks, and shoulders (or interscapular region) pressed against a wall, right arm and hand extended forward horizontally and maximally. (Alternatively, both arms are extended equally, but only the right is measured. See Fig. 41 and Table 25 for various techniques used.)

Relevant Equipment or Workspace Dimension: Forward Arm Reach—Maximum forward distance reachable by the fingertips.

Design Recommendations for Forward Arm Reach

For civilian men, maximum reach, with the interscapular region against a wall, is 31.9 inches for the smallest 1 per cent, 33.0 inches for the smallest 5 per cent. For women subtract about 3.5 inches. For Air Force flyers, the smallest 1 per cent can reach 30.9 inches comfortably, 34.1 inches maximally (by shifting contact with the wall from both shoulders to only the left shoulder). The smallest 5 per cent can reach 31.9 inches comfortably, 35.4 inches maximally. Corresponding figures for Army soldiers are 30.4 inches comfortably for the 1st percentile, and 31.4 inches comfortably for the 5th percentile. These figures refer to nude dimensions with the arm reaching directly forward.

Clothing: Add 0.3 inch for light clothing, 0.2 inch for light gloves, 0.5 inch for medium and heavy clothing

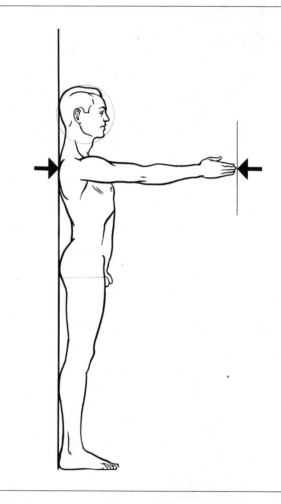

41. Arm reach from wall.

Table 25. Arm Reach from Wall[a] (Inches).

Group	Date of Survey	Refer-ence*	Number of Subjects	Age (Mean or Range)	Percentile					S.D.
					1	5	50	95	99	
Males										
Truck and bus drivers[b]	1951	26	312	36	31.9	33.0	35.8	38.4	39.5	-
Spanish-American War veterans[b]	1959	8	118	81	31.2	31.7	34.2	37.0	38.4	1.52
Air Force flying personnel	1950	19	4,062	27	30.9	31.9	34.6	37.3	38.6	1.7
"Maximum"[c]			4,055	27	34.1	35.4	38.6	41.7	43.2	1.9
"Functional"[d]			4,053	27	28.8	29.7	32.3	35.0	36.4	1.6
Air Force cadets	1942	32	2,959	23	31.6	32.7	35.2	37.8	38.8	-
Air Force gunners	1942	32	580	23	30.9	31.9	34.8	37.4	38.6	-
Army aviators[e]	1959	43	500	30	32.3	33.5	36.0	38.5	39.6	1.47
Army drivers, white	1960	7	431	24	30.4	31.4	34.1	37.0	38.5	1.68
Army drivers, Negro	1960	7	79	27	30.3	32.2	35.4	38.4	39.5	1.89
Navy enlisted men	1955	41	124	-	31.6	32.7	35.7	38.2	39.5	1.70
Navy, mixed sample	1946	24	141	23	30.0	31.1	33.7	36.3	37.4	1.57
Navy aviation cadets	1955	41	472	-	31.7	32.8	35.4	38.1	39.2	-
Females										
Air Force pilots (WASPS)	1944	32	447	18-35	29.2	29.7	31.8	34.1	34.9	-
Air Force flying nurses	1943	32	152	-	27.9	28.7	31.0	33.5	34.4	-

[a]Measured with both shoulders against wall.

[b]Interscapular region (not shoulders) against wall.

[c]Right shoulder thrust as far forward as possible.

[d]Shoulders against wall, tips of thumb and forefinger pressed together.

[e]Subject sitting, both shoulders against wall.

*See references 1-43 on pp. 325-327.

and gloves. Heavy clothing pushes the man forward but may hamper maximum reach.

Other factors: For fingertip manipulation, subtract 0.5 inch for flip, 1.0 inch for push; for manipulation by thumb and forefinger, subtract 3.0 inches; for grasp by whole hand, subtract 5.0 inches. Any deviation of the arm from the strict horizontal forward position shortens forward reach. (See section 6 of this chapter.) Since shoulder mobility varies with position, maximum arm reach varies as well, and should be measured for each combination of seat position, restraining harness, clothing, and direction of reach. (See King, Morrow, and Vollmer, 1947; Kennedy, 1964.)

Old men, though shorter in height and arm length than younger men, are pushed forward by kyphosis (humpback), which increases their forward arm reach.

Elbow-to-Elbow Breadth
(*Bi-epicondylar Breadth, Elbows*)

Definition: Maximum horizontal distance across the lateral surface of the elbows; subject sits erect, upper arms vertical and touching the sides, forearms extended horizontally. In some surveys—Air Force cadets, gunners, female pilots (WASPS), flight nurses, and civilian bus and truck drivers—the hands rested on the thighs, and elbows were pressed in maximally (Fig. 42 and Table 26).

Relevant Equipment or Workspace Dimension: Breadth across Elbows—Distance between objects located beside both elbows. A "minimum" clearance dimension.

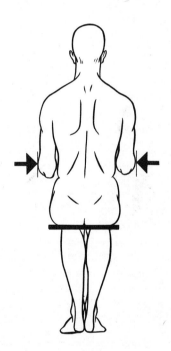

42. Elbow-to-elbow breadth.

Table 26. Elbow-to-Elbow Breadth (Inches).[a]

Group	Date of Survey	Reference*	Number of Subjects	Age (Mean or Range)	1	5	50	95	99	S.D.
Males										
Civilian population	1960-62	27	3,091	18-79	13.0	13.7	16.5	19.9	21.4	-
			411	18-24	12.3	13.1	15.4	19.4	20.8	-
			675	25-34	13.1	13.7	16.3	19.7	21.4	-
			703	35-44	13.1	14.1	16.7	20.0	21.5	-
			547	45-54	13.2	14.1	16.8	20.0	21.8	-
			418	55-64	13.2	14.1	16.7	20.0	22.0	-
			265	65-74	13.2	14.0	16.8	19.9	21.0	-
			72	75-79		14.0	16.4	19.5		-
Truck and bus drivers[b]	1951	26	282	36	13.8	14.9	17.5	20.7	22.2	-
Spanish-American War veterans[b]	1959	8	132	81	15.0	15.5	17.8	20.1	21.0	1.32
Air Force flying personnel	1950	19	4,060	27	14.5	15.2	17.2	19.8	20.9	1.42
Air Force cadets[b]	1942	32	2,955	23	14.4	15.1	16.7	18.4	19.1	-
Air Force gunners[b]	1942	32	584	23	13.9	14.6	16.4	18.2	18.9	-
Army separatees, white	1946	28,37	24,590	23	14.4	15.3	17.4	20.3	21.8	1.54
Army separatees, Negro	1946	37	4,822	23	14.4	15.1	16.9	19.3	20.4	1.28
Army drivers, white	1960	7	431	24	14.6	15.7	17.6	20.6	21.7	1.52
Army drivers, Negro	1960	7	79	27	15.0	15.5	17.3	21.3	21.8	1.79
Females										
Civilian population	1960-62	27	3,581	18-79	11.4	12.3	15.1	19.3	21.2	-
			534	18-24	11.0	11.7	13.8	16.9	20.0	-
			746	25-34	11.4	12.2	14.2	18.3	20.6	-
			784	35-44	11.7	12.5	14.9	19.3	21.5	-
			705	45-54	11.6	12.7	15.5	19.7	21.7	-
			443	55-64	12.3	13.4	16.3	20.2	21.8	-
			299	65-74	12.4	13.7	16.4	19.7	20.8	-
			70	75-79		13.1	15.7	19.1		-
Air Force pilots (WASPS)[b]	1944	32	445	18-35	12.8	13.3	15.1	17.1	18.5	-
Air Force flying nurses	1943	32	152	-	13.0	13.5	14.9	16.7	17.3	-

[a]Variations in the extent to which the elbows are pressed to the sides result in some intergroup differences.

[b]Hands on thighs, elbows pressed in maximally.

*See references 1-43 on pp. 325-327.

Design Recommendations for
Breadth across Elbows

For civilian men, 21.6 inches will accommodate all but the largest 1 per cent, and 20.1 inches will accommodate all but the largest 5 per cent. For civilian women the comparable figures are 21.4 and 19.5 inches. All but the largest 1 per cent of Air Force flyers will be accommodated by 21.1 inches in this dimension, and all but the largest 5 per cent by 20.0 inches. For Army soldiers allow 21.9 inches for the 99th percentile and 20.5 inches for the 95th percentile. These figures refer to nude dimensions, taken in the rigid, upright posture with elbows touching the trunk, plus 0.2 inch for minimum clearance.

Clothing: Add 0.5 inch for light clothing, 2.0 inches for medium, 4.5 inches for heavy clothing, and 11.0 inches for the partial pressure suit.

Other factors: Men neither work nor rest with elbows pressed tightly to the sides. The amount added to breadth across elbows by relaxation is about 2.7 inches, which is also the amount added to standing elbow breadth when the subject stands relaxed. Additional space should be provided for the trunk and arm movements required to operate controls, the amount to be determined for various types of workspace and control.

Hip Breadth, Sitting

Definition: Maximum horizontal distance across the hips; subject sits erect, knees and ankles supported at right angles, knees and heels together (Fig. 43 and Table 27).

Relevant Workspace or Equipment Dimension: Seat Breadth—Distance across seat surface or between the arms or sides of the seat.

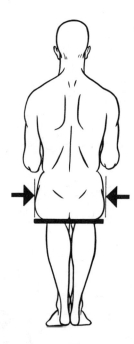

43. Hip breadth, sitting.

Table 27. Hip Breadth, Sitting (Inches).

Group	Date of Survey	Refer-ence*	Number of Subjects	Age (Mean or Range)	Percentile 1	5	50	95	99	S.D.
Males										
Civilian population[a]	1960-62	27	3,091	18-79	11.5	12.2	14.0	15.9	17.0	—
			411	18-24	11.3	12.0	13.5	15.8	17.3	—
			675	25-34	11.7	12.2	14.0	16.0	17.4	—
			703	35-44	12.0	12.4	14.1	15.9	17.1	—
			547	45-54	11.5	12.2	14.2	16.0	16.9	—
			418	55-64	11.6	12.2	14.0	15.9	16.9	—
			265	65-74	11.4	12.2	13.9	15.7	16.6	—
			72	75-79		12.1	13.6	15.5		—
Railroad travelers[b]	1944	20	1,959	38	12.9	13.7	15.3	17.4	18.1	—
Truck and bus drivers	1951	26	308	36	12.4	13.2	14.5	16.3	16.8	0.94
Spanish-American War veterans	1959	8	131	81	13.2	13.5	14.8	16.7	17.2	0.87
Air Force flying personnel	1950	19	4,058	27	12.2	12.7	13.9	15.4	16.2	—
Air Force cadets	1942	32	2,954	23	12.6	13.1	14.2	15.5	15.9	—
Air Force gunners	1942	32	583	23	12.1	12.7	13.8	15.1	15.5	—
Army separatees, white	1946	28,37	24,575	23	12.2	12.7	13.9	15.5	16.7	0.90
Army separatees, Negro	1946	37	4,842	23	11.6	12.1	13.4	15.0	15.8	0.84
Army aviators	1959	43	500	30	12.4	12.8	14.2	15.7	16.3	0.87
Army drivers, white	1960	7	431	24	12.2	12.6	13.9	15.7	16.9	1.01
Army drivers, Negro	1960	7	79	27	11.9	12.1	13.5	15.9	16.7	1.15
Navy enlisted men	1955	41	124	—	12.4	13.0	14.8	16.4	17.2	1.05
Navy aviation cadets	1955	41	472	23	13.4	14.0	15.4	16.8	17.3	—
Females										
Civilian population[a]	1960-62	27	3,581	18-79	11.7	12.3	14.3	17.1	18.8	—
			534	18-24	11.3	12.1	13.8	15.9	18.4	—
			746	25-34	11.5	12.2	14.0	16.8	19.0	—
			784	35-44	12.0	12.4	14.5	17.3	19.2	—
			705	45-54	12.0	12.4	14.6	17.6	19.0	—
			443	55-64	12.1	12.9	14.7	17.4	18.7	—
			299	65-74	12.1	12.4	14.6	17.3	18.2	—
			70	75-79		11.7	14.0	16.8		—
Railroad travelers[b]	1944	20	1,908	35	12.2	13.1	14.6	17.2	17.8	—
Air Force pilots (WASPS)	1944	32	447	18-35	13.0	13.5	15.0	16.9	18.1	—
Air Force flying nurses	1943	32	152	—	13.1	13.5	15.1	16.6	17.1	—

[a]Subjects lightly clothed, measured with firm pressure.

[b]Subjects fully clothed.

*See references 1-43 on pp. 325-327.

Design Recommendations for Seat Breadth

For civilian men 17.2 inches will accommodate all but the largest 1 per cent, and 16.1 inches will accommodate all but the largest 5 per cent. Comparable figures for civilian women are 19.0 and 17.3 inches. For Air Force flyers 16.4 inches will accommodate all but the largest 1 per cent and 15.6 inches all but the largest 5 per cent. For Army soldiers allow 17.0 and 15.8 inches, respectively. These figures refer to rigidly posed subjects, plus 0.2 inch for clearance. The military groups were measured nude, the civilians clothed. However, for the latter, firm pressure was exerted when the measurement was taken in order to compress the clothing.

Clothing: Add 0.5 inch for light clothing, 1.0 inch for medium, 2.0 inches for heavy clothing; and for the partial pressure suit, 3.0 inches uninflated and 5.5 inches inflated.

Other factors: Relaxation adds an unknown amount, perhaps 0.5 to 1.0 inch. Additional space should be provided to permit change of position—the amount, again, is unknown.

Knee-to-Knee Breadth, Sitting

Definition: Maximum horizontal distance across the lateral surfaces of the knees; subject sits erect, knees at right angles and pressed together (lightly in some surveys, maximally in others; see Fig. 44 and Table 28).

Relevant Equipment or Workspace Dimension: Lateral Knee Space—Space across one or both knees. A "minimum" clearance dimension.

Design Recommendations for Lateral Knee Space

For men, civilian or military, 9.7 inches will accommo-

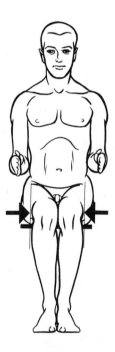

44. Knee-to-knee breadth, sitting.

date all but the largest 1 per cent, 9.0 inches, all but the largest 5 per cent of all series. For distance across one knee, halve the figure. These values refer to nude dimensions, with the knees touching. For women, the same figures hold.

Clothing: Add 0.5 inch for light clothing, 1.0 inch for medium, 2.0 inches for heavy clothing, and 9.5 inches for the partial pressure suit.

Other factors: This is an absolute minimum clearance dimension, as regards both knees, since pressing the knees together for any period of time is difficult or painful. The amount added by relaxation is unknown, but the knees can be comfortable considerably closer together than the normal resting position, if supported laterally. Some degree of additional lateral clearance must, however, be provided beyond those recommended here for both knees, to permit relaxation. No such problem arises for the single knee. Space is also desirable for lateral knee movement.

Table 28. Knee-to-Knee Breadth, Sitting (Inches).

Group	Date of Survey	Refer-ence*	Number of Subjects	Age (Mean or Range)	Percentile					S.D.
					1	5	50	95	99	
Males										
Truck and bus drivers	1951	26	311	36	6.8	7.3	8.1	9.2	9.5	-
Spanish-American War veterans	1959	8	129	81	7.3	7.5	8.0	8.7	10.1	0.52
Air Force flying personnel	1950	19	4,056	27	7.0	7.2	7.9	8.8	9.4	0.52
Air Force cadets	1942	32	2,955	23	6.8	7.1	7.7	8.4	8.7	-
Air Force gunners	1942	32	581	23	6.7	6.9	7.6	8.2	8.5	-
Army drivers, white	1960	7	431	24	6.9	7.2	7.8	8.8	9.5	0.52
Army drivers, Negro	1960	7	79	27	6.9	7.1	7.8	9.8	10.1	0.72
Females										
Air Force pilots (WASPS)	1944	32	444	18-35	6.5	6.7	7.6	8.6	9.6	-
Air Force flying nurses	1943	32	152	-	6.6	6.8	7.5	8.4	9.6	-

*See references 1-43 on pp. 325-327.

Shoulder Breadth
(*Bideltoid Diameter*)

Definition: Maximum horizontal distance across the deltoid muscles; subject sits erect, upper arms vertical and touching the sides, forearms extended horizontally. (Several surveys have taken this measurement with the subject standing, pressing arms into sides of body, and with palms forward; see Fig. 45 and Table 29.)

Relevant Equipment or Workspace Dimension: Breadth across Shoulders—Space available across shoulders.

Design Recommendations for Breadth across Shoulders

For civilian men, 20.7 inches will accommodate all but the largest 1 per cent and 20.1 inches will accommodate all but the largest 5 per cent. For women, subtract about 2.0 inches. For Air Force flyers, 20.3 inches will fit all but the largest 1 per cent and 19.6 inches will fit all but the largest 5 per cent. Comparable figures for Army soldiers are 21.1 and 19.9 inches. These figures comprise nude percentiles plus 0.2 inch of minimal clearance, with the arms held against the sides.

Clothing: Add 0.3 inch for light clothing, 1.5 inches for heavy clothing, 0.4 inch for the partial pressure suit uninflated, 6.0 inches inflated.

Other factors: Shoulder breadth increases, to a vari-able and unknown extent, as the arms move away from the body during rest or work. Trunk or shoulder movement may also increase the space needed across the shoulders.

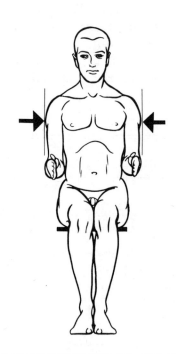

45. Shoulder breadth.

Table 29. Shoulder Breadth (Inches).

Group	Date of Survey	Refer- ence*	Number of Subjects	Age (Mean or Range)	Percentile					S.D.
					1	5	50	95	99	
Males										
Railroad travelers[a]	1944	20	1,959	38	15.7	16.4	17.6	19.2	19.8	–
Truck and bus drivers[b]	1951	26	310	36	16.2	16.9	18.3	19.9	20.5	–
Harvard students	pre-1930	2	476	18	15.1	15.7	17.2	18.7	19.3	0.86
Spanish-American War veterans	1959	8	129	81	15.3	15.6	17.0	18.5	19.1	0.90
Air Force flying personnel	1950	19	4,057	27	15.9	16.5	17.9	19.4	20.1	0.91
Air Force cadets[b]	1942	32	2,955	23	16.1	16.7	18.0	19.3	19.9	–
Air Force gunners[b]	1942	32	584	23	16.0	16.5	17.7	19.0	19.5	–
Army separatees, white	1946	28,37	24,461	23	15.8	16.4	17.9	19.6	20.6	0.99
Army separatees, Negro	1946	37	6,620	23	15.8	16.4	17.9	19.4	20.0	0.89
Army aviators	1959	43	500	30	16.4	16.8	18.2	20.0	20.5	0.88
Army drivers, white	1960	7	431	24	15.8	16.5	18.1	19.9	21.2	1.07
Army drivers, Negro	1960	7	79	27	16.3	16.9	18.3	20.9	21.7	1.23
Navy, mixed sample	1946	24	141	23	15.1	15.8	17.6	19.4	20.2	1.09
Females										
Railroad travelers[a]	1944	20	1,908	35	13.7	14.4	15.7	17.6	18.2	–
Air Force pilots (WASPS)[b]	1944	32	447	18-35	14.3	14.9	16.1	17.6	18.0	–
Air Force flying nurses	1943	32	152	–	14.1	14.5	15.7	16.8	17.2	–

[a]Subject clothed.
[b]Subject stands with arms at sides, palms forward.
*See references 1-43 on pp. 325-327.

Chest Breadth

Definition: Horizontal distance across the chest at nipple level; subject stands erect, breathing normally, arms hanging naturally at sides (Fig. 46 and Table 30).

Relevant Workspace or Equipment Dimension: Breadth across Chest—Lateral space available at chest level. This dimension is of minor importance in human engineering, since only rarely will breadth across the chest present a problem independently of the much larger shoulder breadth. Body armor or chest respirators are such possibilities.

Design Recommendations for
Breadth across Chest

For civilian men 14.1 inches will accommodate all but the largest 1 per cent, and 13.7 inches will accommodate all but the largest 5 per cent. For women subtract about 1.5 inches. For Air Force flyers, 14.3 inches will fit 99 per cent and 13.6 inches will fit 95 per cent. Comparable figures for Army soldiers are 14.0 and 13.3 inches. These figures refer to nude chest breadth, plus 0.2 inch for clearance.

Clothing: Add 0.3 inch for light clothing, 0.6 inch for heavy clothing, and 2.5 inches for the partial pressure suit, inflated.

Other factors: Chest breadth expands with inspiration and decreases with expiration, by unknown amounts. An average value for chest *circumference* expansion, between maximum inspiration and expiration, is 3.0 inches in young men. The chest is less expansile after age 60.

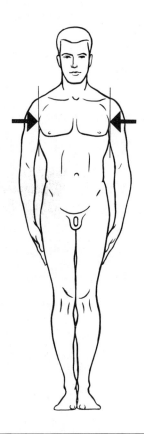

46. Chest breadth.

Table 30. Chest Breadth (Inches).

Group	Date of Survey	Reference*	Number of Subjects	Age (Mean or Range)	Percentile 1	5	50	95	99	S.D.
Males										
Truck and bus drivers	1951	26	311	36	9.6	10.2	11.8	13.5	13.9	-
Harvard students	1938-42	18	258	19	9.9	10.4	11.5	12.7	13.1	0.67
Univ. Chicago students	1941	4	119	18	9.3	9.9	11.1	12.4	12.9	0.79
Spanish-American War veterans	1959	8	133	81	9.9	10.2	11.7	13.0	13.4	0.81
Air Force flying personnel	1950	19	4,058	27	10.4	10.8	12.0	13.4	14.1	0.80
Air Force basic trainees	1952	9	3,326	18	9.7	10.2	11.4	13.0	14.3	0.91
Air Force cadets	1942	32	2,957	23	9.8	10.3	11.3	12.4	12.8	-
Air Force gunners	1942	32	581	23	9.7	10.1	11.1	12.1	12.5	-
Army separatees	1946	28,37	24,574	23	9.3	10.0	11.1	12.4	13.2	0.77
Army drivers, white	1960	7	431	24	9.1	10.4	11.6	13.1	13.8	0.86
Army drivers, Negro	1960	7	79	27	8.4	10.2	11.5	13.1	13.8	0.93
Females										
College students, Midwest	1939	13	1,013	17-21	8.6	9.0	10.1	11.1	11.5	0.64
Pembroke College students	1927	3	109	19	8.3	8.7	9.7	10.7	11.1	0.59
WAF basic trainees	1952	10	834	19	8.9	9.1	9.9	10.9	11.3	0.55

*See references 1-43 on pp. 325-327.

Hip Breadth, Standing

Definition: Maximum horizontal distance across hips; subject stands erect, heels together (Fig. 47 and Table 31).

Relevant Workspace or Equipment Dimension: Breadth across Hips, Standing—Lateral space available at level of hips, with operator standing. A minor human engineering measurement, since shoulder breadth greatly exceeds hip breadth, and workspace layouts are quite rare in which standing hip breadth poses a problem apart from shoulder breadth. Seated hip breadth, or "seat breadth," is much more important than standing hip breadth.

Design Recommendations for
Breadth across Hips, Standing

For Air Force flyers, 15.4 inches will accommodate all but the largest 1 per cent and 14.6 inches all but the largest 5 per cent. For Air Force ground personnel, corresponding figures are 16.5 and 15.2 inches. No other recent groups have been measured, but civilians will be at least as broad-hipped as the Air Force ground troops; for older men (mid-30's), add 1 inch to the Air Force ground troop figures. For women, add 0.5 inch. These figures refer to nude, standing hip breadth, plus 0.2 inch for minimal clearance.

Clothing: Add 0.5 inch for light clothing, 1.5 inches or more for heavy clothing.

Other factors: Space needed across the hips increases as the legs move apart, as the pelvis rotates, and as objects are carried in the pockets.

Chest Depth

Definition: Horizontal distance from front to back of chest at nipple level. (On females, above the breasts, at the level where the 4th rib meets the sternum or breastbone.) Subject stands erect, breathing normally (Fig. 48 and Table 32).

Relevant Equipment or Workspace Dimension: Chest Depth—Fore-and-aft space available at chest level.

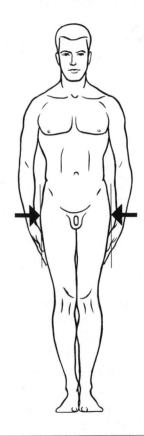

47. Hip breadth.

Table 31. Hip Breadth, Standing (Inches)

Group	Date of Survey	Refer-ence*	Number of Subjects	Age (Mean or Range)	Percentile					S.D.
					1	5	50	95	99	
Males										
Harvard students	pre-1930	2	479	18	11.4	11.8	13.0	14.2	14.7	0.67
Air Force flying personnel	1950	19	4,062	27	11.3	12.1	13.2	14.4	15.2	0.73
Air Force basic trainees	1952	9	3,328	18	11.5	12.1	13.3	15.0	16.3	0.94
Females										
WAF basic trainees	1952	10	850	19	12.2	12.5	13.5	15.4	16.9	0.95

*See references 1-43 on pp. 325-327.

Design Recommendations for Chest Depth

For men, both civilian and military, 11.3 inches will accommodate all but the largest 1 per cent of most groups and 10.7 inches will accommodate all but the largest 5 per cent. These figures refer to nude dimensions during quiet breathing in the erect posture, plus 0.2 inch for minimal clearance. For women, the proper "human engineering" dimension, maximum body depth at chest level, including the breasts, has not been measured.

Clothing: Add 0.5 inch for light clothing, 2 inches for heavy clothing, and 4.5 inches for the inflated partial pressure suit.

Other factors: Inspiration increases chest depth; expiration decreases it. Chest depth in normal sitting exceeds the standing value, because of rounding of the shoulders away from the seat back, but by an unknown amount. (Abdominal depth, seated, increases by 1.0 inch over the standing dimension on sitting bolt upright and by more when sitting normally.) In young, slender, or athletic men, chest depth equals or may exceed waist or abdominal depth, but with age, abdominal depth exceeds chest depth even in muscular groups like bus and truck drivers. The excess of waist depth over chest depth is 2.0 or more inches for most older, sedentary people. Moreover, sitting usually pushes the whole abdomen forward, so that abdominal rather than chest depth becomes the critical fore-and-aft trunk dimension for locating control sticks, steering wheels, or work surfaces. Kyphosis in the aged increases chest depth.

Waist Depth

Definition: Horizontal distance between the back and abdomen at the level of the greatest lateral indentation at the waist (if this is not apparent, at the level at which the belt is worn); subject stands erect, abdomen relaxed (Fig. 49). Alternative techniques are noted in Table 33. Still another site is at umbilical level, as in the NATO survey (Hertzberg *et al.,* 1963). More useful for the engineer than any of these is the functional dimension, Maximum Body Depth, Standing, of Table 46. Most useful of all is the seated equivalent, as yet measured only for Army drivers.

   Relevant Equipment or Workspace Dimension: Abdominal Depth—Fore-and-aft space at abdominal level. Clearances between the operator's back and control stick, steering wheel, and work surface are examples.

Design Recommendations for Abdominal Depth

For civilian men, standing erect, 13.3 inches will accommodate all but the largest 1 per cent of the population and 12.3 inches will accommodate all but the largest

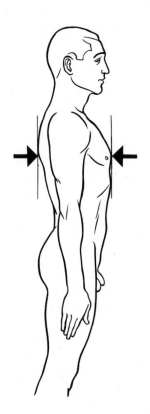

48. Chest depth.

Table 32. Chest Depth (Inches).

Group	Date of Survey	Reference*	Number of Subjects	Age (Mean or Range)	Percentile					S.D.
					1	5	50	95	99	
Males										
Truck and bus drivers	1951	26	312	36	7.1	7.6	8.9	10.5	11.1	-
Harvard students	1938-42	18	258	19	6.5	6.9	7.9	8.9	9.3	0.55
Univ. Chicago students	1941	4	119	18	6.4	6.9	8.0	9.2	9.7	0.71
Spanish-American War veterans	1959	8	133	81	7.9	8.2	9.6	10.8	11.2	0.78
Air Force flying personnel	1950	19	4,063	27	7.6	8.0	9.0	10.4	11.1	0.75
Air Force cadets	1942	32	2,959	23	6.8	7.2	8.2	9.3	9.7	-
Air Force gunners	1942	32	583	23	6.7	7.1	8.2	9.2	9.6	-
Army separatees	1946	28,37	24,558	23	6.8	7.2	8.3	9.6	10.5	0.75
Army drivers, white	1960	7	431	24	6.4	7.0	8.4	10.8	11.7	0.90
Army drivers, Negro	1960	7	79	27	6.9	7.2	8.4	11.4	12.2	1.00
Army aviators	1959	43	500	30	7.4	7.9	8.9	10.4	11.0	0.79
Females										
College students, Midwest	1939	13	1,013	17-21	6.0	6.4	7.3	8.2	8.6	0.56
Pembroke College students	1927	3	109	19	5.8	6.3	7.4	8.6	9.0	0.68

*See references 1-43 on pp. 325-327.

5 per cent. These same values should be adequate for even larger percentages of Air Force and Army personnel. In the seated position, 12.8 inches will accommodate all but the largest 1 per cent of Army drivers, and 11.9 inches all but the largest 5 per cent of the same group. No data are available for women on any of these measurements.

Clothing: Add 1.0 inch for light clothing, 2.0 inches for medium, 2.5 inches for heavy clothing, and 5.0 inches for the inflated partial pressure suit.

Other factors: Sitting bolt upright adds 1.0 inch to the nude, standing waist or abdominal depth; normal sitting adds an additional, unknown amount. Breathing in many sedentary or elderly persons is abdominal rather than thoracic, so that inspiration increases abdominal depth, again by an unknown amount.

In young, slender, or athletic men, abdominal depth may be less than chest depth, but in older men, abdominal depth exceeds chest depth, by 2.0 or more inches for most persons.

Buttock Depth

Definition: Horizontal distance between the buttocks and the abdomen at the level of the maximum protrusion of the buttocks; subject stands erect (Fig. 50 and Table 34).

Related Equipment or Workspace Dimension: Buttock Depth—Fore-and-aft space at buttock level, with operator standing.

Design Recommendations for Buttock Depth

For men, 11.1 inches will fit 99 per cent and 10.4 inches will fit 95 per cent of Air Force flyers. No other groups have been measured. These figures refer to the nude, standing percentiles plus 0.2 inch for minimal clearance. For women, no data are available.

Clothing: Add 1.0 inch for light clothing, 2.0 inches for medium, and 2.5 inches for heavy clothing. (These are tentative values, since no direct data are available. They are assumed to be the same increments as for waist or abdominal depth.)

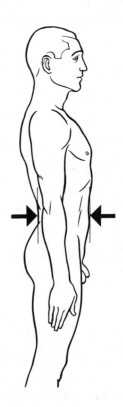

49. Waist depth.

Table 33. Waist Depth (Inches).[a]

Group	Date of Survey	Refer-ence*	Number of Subjects	Age (Mean or Range)	Percentile					S.D.
					1	5	50	95	99	
Males										
Truck and bus drivers[b]	1951	26	311	36	7.3	7.8	9.5	12.1	13.1	–
Spanish-American War veterans[b]	1959	8	126	81	8.4	8.6	10.8	13.2	14.0	1.32
Air Force flying personnel[c]	1950	19	4,062	27	6.3	6.7	7.9	9.5	10.3	0.88
Air Force cadets[b]	1942	32	2,958	23	6.7	7.2	8.2	9.3	9.8	–
Air Force gunners[b]	1942	32	584	23	6.7	7.2	8.2	9.3	9.8	–
Army separatees[d]	1946	28,37	24,588	23	7.5	7.9	9.0	10.5	11.5	0.81
Army drivers, white[e]	1960	7	431	24	7.8	8.3	9.5	11.7	12.6	1.19
Army drivers, Negro[e]	1960	7	79	27	7.8	8.2	9.4	12.9	13.4	1.29

[a]Waist depth, sitting, is about one inch greater.

[b]Maximum abdominal depth, wherever found.

[c]At belt line.

[d]Maximum abdominal or thoracic depth, whichever is greater.

[e]Subject seated, back against wall, maximum depth whenever found, measured from wall.
*See references 1-43 on pp. 325-327.

Table 34. Buttock Depth (Inches).

Group	Date of Survey	Refer-ence*	Number of Subjects	Age (Mean or Range	Percentile					S.D.
					1	5	50	95	99	
Males										
Air Force flying personnel	1950	19	4,058	27	7.2	7.6	8.8	10.2	10.9	0.82

*See references 1-43 on pp. 325-327.

Other factors: Space required for buttock depth is increased by age, sedentary habits, lumbosacral angulation ("lordosis")—encountered in many Negroes—and by objects in back pockets.

Foot Length

Definition: Distance, parallel to long axis of foot, from the back of the heel to the tip of the longest toe; subject stands with weight equally distributed on both feet (Fig. 51 and Table 35).
Relevant Equipment or Workspace Dimension: Heel to Toe Distance—Fore-and-aft space available for foot.

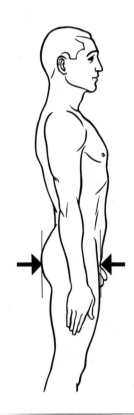

50. Buttock depth.

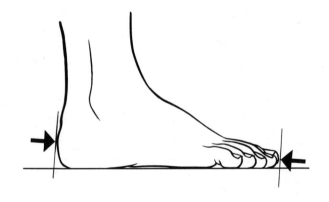

51. Foot length.

Table 35. Foot Length (Inches).

Group	Date of Survey	Refer-ence*	Number of Subjects	Age (Mean or Range)	Percentile					S.D.
					1	5	50	95	99	
Males										
Truck and bus drivers	1951	26	293	36	9.2	9.6	10.4	11.3	11.6	-
Harvard students	pre-1930	2	477	18	9.2	9.4	10.3	11.1	11.4	0.48
Spanish-American War veterans	1959	8	132	81	9.2	9.7	10.2	10.9	11.3	0.39
Air Force flying personnel	1950	19	4,063	27	9.5	9.8	10.5	11.3	11.6	0.45
Air Force cadets	1942	32	2,959	23	9.5	9.8	10.5	11.3	11.6	-
Air Force gunners	1942	32	583	-	9.3	9.6	10.4	11.1	11.4	-
Air Force basic trainees	1952	9	3,331	18	9.2	9.5	10.3	11.2	11.5	0.50
Army men	1945	34	5,575	-	9.3	9.6	10.4	11.1	11.5	0.47
Army separatees, white	1946	28,37	24,372	23	9.3	9.7	10.4	11.2	11.5	0.48
Army separatees, Negro	1946	37	6,636	23	9.6	9.9	10.8	11.6	12.0	0.50
Army aviators	1959	43	500	30	9.5	9.9	10.6	11.5	11.9	0.49
Army drivers, white	1960	7	431	24	9.2	9.7	10.5	11.3	11.7	0.50
Army drivers, Negro	1960	7	79	27	9.6	10.1	10.6	11.8	12.0	0.53
Females										
Pembroke College students	1927	3	109	19	8.3	8.7	9.5	10.3	10.7	0.45
WAF basic trainees	1952	10	850	19	8.4	8.7	9.4	10.2	10.5	0.46
Air Force pilots (WASPS)	1944	32	445	18-35	8.6	8.9	9.6	10.2	10.5	-
Air Force flying nurses	1943	32	152	-	8.7	8.9	9.6	10.3	10.5	-
WACs and Army nurses	1946	33,37	1,949	26	8.4	8.7	9.4	10.2	10.5	0.44

*See references 1-43 on pp. 325-327.

Design Recommendations for Heel to Toe Distance

For men, civilian and military, 11.8 inches will accommodate all but the largest 1 per cent of almost all white groups and 11.4 inches will accommodate all but the largest 5 per cent. For Negroes the corresponding values are 12.1 and 11.9 inches. For women 10.8 inches will accommodate all but the largest 1 per cent, and 10.4 inches all but the largest 5 per cent. These figures refer to nude percentiles plus a minimal 0.1 inch for clearance.

Clothing: Add 1.2 inches for men's street shoes, 1.6 inches for military boots, 2.7 inches for heavy flying boots.

Other factors: Race—Negroes, as noted, have longer feet than white persons; Orientals have shorter feet. *Age*—As stature has increased over the past 50–100 years, so has foot length. Younger groups will therefore have longer feet than older ones. *Posture*—Standing increases foot length over the seated value, by an unknown amount.

Foot Breadth

Definition: Maximum horizontal distance across the foot wherever found, at right angles to the long axis; subject stands with weight equally distributed on both feet (Fig. 52 and Table 36).

Relevant Equipment or Workspace Dimension:

Breadth across Foot—Lateral space available for the foot.

Design Recommendations for Breadth across Foot

For men, civilian or military, 4.8 inches will fit 99 per cent or more and 4.5 inches will fit 95 per cent or more, of all groups. For women, comparable figures are 4.2

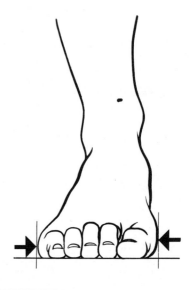

52. Foot breadth.

and 4.1 inches. This dimension relates to nude foot breadth plus 0.1 inch for clearance.

Clothing: Add 0.3 inch for street shoes and military boots, 1.2 inches or more for heavy flying boots.

Other factors: Age, race, posture—as for Foot Length.

Table 36. Foot Breadth (Inches).

Group	Date of Survey	Refer-ence*	Number of Subjects	Age (Mean or Range)	Percentile					S.D.
					1	5	50	95	99	
Males										
Truck and bus drivers	1951	26	312	36	3.6	3.7	4.0	4.3	4.4	-
Spanish-American War veterans	1959	8	119	81	3.5	3.6	3.9	4.3	4.3	0.19
Air Force flying personnel	1950	19	4,060	27	3.4	3.5	3.8	4.1	4.4	0.19
Air Force basic trainees	1952	9	3,327	18	3.5	3.6	4.0	4.4	4.7	0.25
Air Force cadets	1942	32	2,959	23	3.5	3.6	3.9	4.2	4.3	-
Air Force gunners	1942	32	584	23	3.3	3.5	3.8	4.2	4.3	-
Army men	1945	34	5,575	-	3.4	3.5	3.9	4.2	4.3	0.20
Army separatees, white	1946	28,37	24,466	23	3.3	3.5	3.9	4.3	4.4	0.25
Army separatees, Negro	1946	37	6,641	23	3.4	3.6	4.0	4.4	4.6	0.25
Army aviators	1959	43	500	30	3.5	3.6	4.0	4.4	4.5	0.21
Army drivers, white	1960	7	431	24	3.4	3.6	3.9	4.3	4.4	0.21
Army drivers, Negro	1960	7	79	27	3.6	3.6	4.0	4.5	4.7	0.24
Females										
WAF basic trainees	1952	10	849	19	3.1	3.2	3.6	3.9	4.0	0.20
Air Force pilots (WASPS)	1944	32	447	18-35	3.2	3.3	3.6	3.9	4.1	-
Air Force flying nurses	1943	32	152	-	3.2	3.3	3.6	3.9	4.1	-
WACs and Army nurses	1946	33,37	1,945	26	3.1	3.2	3.6	4.0	4.1	0.22

*See references 1-43 on pp. 325-327.

Hand Length

Definition: Distance from the proximal edge of the navicular bone at the wrist (base of thumb) to middle fingertip; right hand, held straight and stiff (Fig. 53 and Table 37).

Relevant Equipment or Workspace Dimension: Fingertip Reach, from Wrist—maximum fingertip reach from the wrist.

*Design Recommendations for
Fingertip Reach, from Wrist*

For men, civilian or military, 6.5 inches can be reached by all but the smallest 1 per cent or fewer and 6.9 inches by all but the smallest 5 per cent of all groups. For women, the comparable figures are 6.0 and 6.2 inches. These dimensions refer to maximum stretch by the ungloved hand.

Clothing: No increment, since gloves do not effectively increase functional hand reach.

Other factors: Race—Negroes have longer hands, by 0.2 to 0.5 inch, and Orientals shorter hands than white persons. *Posture*—The normal resting position of the hand is partial flexion, which reduces wrist-fingertip distance by about 2.0 inches, as for comfortably flicking switches or pushing buttons. For manipulating objects with thumb and forefinger, subtract 3.0 inches from hand length; for grasping objects with the whole hand, subtract 5.0 inches. These figures are tentative.

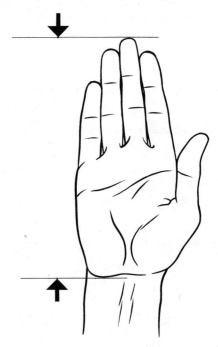

53. Hand length.

Table 37. Hand Length (Inches).

Group	Date of Survey	Refer-ence*	Number of Subjects	Age (Mean or Range)	Percentile					S.D.
					1	5	50	95	99	
Males										
Truck and bus drivers	1951	26	314	36	6.9	7.1	7.6	8.1	8.3	-
Spanish-American War veterans	1959	8	130	81	6.7	7.0	7.4	8.0	8.2	0.31
Air Force flying personnel	1950	19	4,063	27	6.7	6.9	7.5	8.0	8.3	0.34
Air Force basic trainees	1952	9	3,328	18	6.7	6.9	7.5	8.2	8.5	0.38
Air Force cadets	1942	32	2,952	23	6.8	7.1	7.6	8.2	8.4	-
Air Force gunners	1942	32	582	23	6.6	6.9	7.5	8.1	8.4	-
Army separatees, white	1946	28,37	24,487	23	6.7	7.0	7.6	8.2	8.5	0.36
Army separatees, Negro	1946	37	6,669	23	7.0	7.3	8.0	8.7	9.0	0.42
Army aviators	1959	43	500	30	6.7	6.9	7.5	8.1	8.3	0.34
Army drivers, white	1960	7	431	24	6.5	6.9	7.4	8.0	8.2	0.35
Army drivers, Negro	1960	7	79	27	6.7	7.1	7.7	8.3	8.4	0.36
Females										
Pembroke College students	1927	3	109	19	6.0	6.2	6.7	7.2	7.4	0.31
WAF basic trainees	1952	10	851	19	6.0	6.2	6.8	7.3	7.6	0.34
Air Force pilots (WASPS)	1944	32	437	18-35	6.2	6.4	6.9	7.5	7.7	-
Air Force flying nurses	1944	32	142	-	6.3	6.5	6.9	7.4	7.6	-
WACs and Army nurses	1946	33,37	8,488	26	6.1	6.4	6.9	7.4	7.7	0.33

*See references 1-43 on pp. 325-327.

Hand Breadth at Metacarpale

Definition: Maximum breadth across the distal ends of the metacarpal bones (where fingers join palm) of index and little fingers; right hand held straight and stiff, fingers together. Measure with firm pressure (Fig. 54 and Table 38).

Relevant Equipment or Workspace Dimension: Palm Breadth—Breadth available for palm, with fingers extended. Palm breadth is less important for equipment design than functional hand breadth, with fingers normally relaxed or clenched in a fist.

Design Recommendations for Palm Breadth

For men, 4.1 inches will accommodate all but the largest 1 per cent or fewer and 3.9 inches will fit all but the largest 5 per cent or fewer, of all groups, military and civilian. For women, comparable figures are 3.7 and 3.5 inches. These figures refer to the bony breadth across the ungloved palm (excluding thumb) with fingers extended and in contact, plus 0.1 inch for clearance.

Clothing: Add 0.3 inch for woolen or leather gloves, and about 1.0 inch for arctic handwear.

Other factors: Functional palm breadth exceeds anthropometric palm breadth, since the functional dimension is a skin contact rather than a bony dimension, and since the normal resting or working hand is not held rigid but is flexed or made into a fist. Muscular persons have broader hands, for a given length, than slender ones. Gloves are compressible.

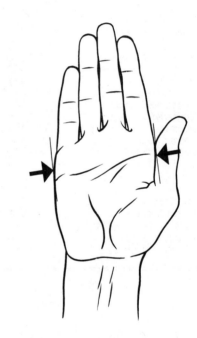

54. Hand breadth at metacarpale.

Table 38. Hand Breadth at Metacarpale (Inches).

Group	Date of Survey	Refer-ence*	Number of Subjects	Age (Mean or Range)	Percentile					S.D.
					1	5	50	95	99	
Males										
Truck and bus drivers	1951	26	311	36	3.1	3.2	3.5	3.8	4.0	-
Spanish-American War veterans	1959	8	129	81	3.0	3.1	3.3	3.6	3.7	0.15
Air Force flying personnel	1950	19	4,058	27	3.1	3.2	3.5	3.7	3.9	0.16
Air Force cadets	1942	32	2,955	23	3.0	3.1	3.4	3.7	3.8	-
Air Force gunners	1942	32	582	23	3.0	3.1	3.4	3.6	3.7	-
Air Force basic trainees	1952	9	3,317	18	3.0	3.2	3.5	3.7	3.9	0.18
Army separatees, white	1946	28,37	24,488	23	2.9	3.1	3.4	3.8	3.9	0.19
Army separatees, Negro	1946	37	6,642	23	3.0	3.2	3.5	3.8	4.0	0.20
Army aviators	1959	43	500	38	3.1	3.2	3.5	3.8	3.9	0.16
Army drivers, white	1960	7	431	24	2.8	2.9	3.2	3.5	3.6	0.18
Army drivers, Negro	1960	7	79	27	2.9	2.9	3.2	3.6	3.8	0.20
Females										
WAF basic trainees	1952	10	851	19	2.6	2.7	3.0	3.4	3.6	0.19
Air Force pilots (WASPS)	1944	32	440	18-35	2.8	2.8	3.0	3.3	3.4	-
Air Force flying nurses	1943	32	142	-	2.7	2.8	3.0	3.2	3.3	-
WACS and Army nurses	1946	33,37	8,505	26	2.6	2.7	3.0	3.4	3.6	0.20

*See references 1-43 on pp. 325-327.

Hand Breadth at Thumb

Definition: Maximum breadth across the palm at right angles to the long axis of the hand, at the proximal knuckle of the thumb (joint between metacarpal bone and first phalanx); right hand, fingers extended, thumb lying alongside and in plane of hand (Fig. 55 and Table 39).

Relevant Equipment or Workspace Dimension: Hand Breadth at Thumb—Space available across entire palm, at level of base of thumb. This hand measurement is less important for equipment design than functional hand breadth at thumb, with the thumb in the normally abducted position away from the palm. Functional hand breadth has not yet been measured.

Design Recommendations for
Hand Breadth at Thumb

For men, 4.7 inches will accommodate all but the largest 1 per cent and 4.5 inches, all but the largest 5 per cent of Air Force flyers and basic trainees. For women subtract 0.5 inch from these figures. No other groups have been measured. The figures refer to ungloved percentiles, with the hand in the artificial position specified, plus 0.1 inch for clearance.

Clothing: Add 0.3 inch for wool or leather gloves, about 1.0 inch for arctic handwear.

Other factors: As noted, this dimension falls between hand breadth with the thumb abducted as for grasping or resting, and the much smaller breadth obtainable by tucking the thumb inside the palm (adducting or "opposing" the thumb to the other fingertips). Abduction increases hand breadth; opposing the thumb decreases hand breadth but increases thickness.

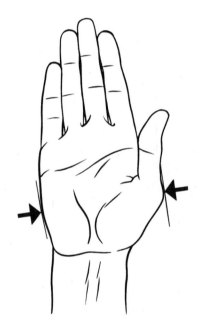

55. Hand breadth at thumb.

Table 39. Hand Breadth at Thumb (Inches).

Group	Date of Survey	Refer-ence*	Number of Subjects	Age (Mean or Range)	Percentile					S.D.
					1	5	50	95	99	
Males										
Air Force flying personnel	1950	19	4,062	27	3.6	3.7	4.1	4.4	4.6	0.21
Air Force basic trainees	1952	9	3,316	18	3.5	3.7	4.1	4.5	4.7	0.25
Females										
WAF basic trainees	1952	10	844	19	3.1	3.2	3.6	4.0	4.1	0.23

*See references 1-43 on pp. 325-327.

Hand Thickness at Metacarpale III

Definition: Maximum distance between the dorsal and palmar surfaces of the knuckle of the middle finger where it joins the palm; right hand, fingers extended (Fig. 56 and Table 40).

Related Equipment or Workspace Dimension: Palm Thickness—Space available for the flat palm. More space is required for the whole hand than for the palm alone, since the muscle masses at the bases of the thumb and little finger—the thenar and hypothenar eminences—increase hand thickness.

Design Recommendations for Palm Thickness

For men, 1.5 inches will accommodate 99 per cent and 1.4 inches about 95 per cent of Air Force flying and ground personnel. For women, subtract 0.2 inch from these figures. No other groups have been measured. The figures refer to percentiles of the ungloved hand held rigidly flat, plus 0.1 inch for clearance.

Clothing: Add 0.2 inch for wool or leather gloves, about 1.5 inches for arctic handwear.

Other factors: This is a minimum dimension; the normally bent resting or working hand is thicker in the palm (by an unknown amount) than the hand held flat. Also, as mentioned, the hand is thicker at the base of the thumb or fifth finger than at the third metacarpal. Fat or muscular persons have thicker hands, for a given length, than slender ones.

Table 40. Hand Thickness at Metacarpale III (Inches).

Group	Date of Survey	Refer- ence*	Number of Subjects	Age (Mean or Range)	Percentile					S.D.
					1	5	50	95	99	
Males										
Air Force flying personnel	1950	19	4,061	27	1.0	1.1	1.2	1.3	1.4	0.07
Air Force basic trainees	1952	9	2,019	18	1.0	1.1	1.2	1.4	1.4	0.09
Females										
WAF basic trainees	1952	10	850	19	0.8	0.8	1.0	1.1	1.2	0.09

*See references 1-43 on pp. 325-327.

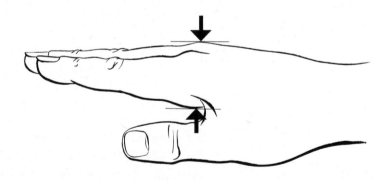

56. Hand thickness at metacarpale III.

Head Length

Definition: Distance between glabella (the most anterior point on the forehead between the brow ridges) and the most posterior point on the occiput (back of head), in the midline (Fig. 57 and Table 41).

Relevant Equipment or Workspace Dimension: Head Length—Space available between front and back of head. More space is needed between the occiput and the tip of the nose, probably the more important engineering dimension. See Head Length, Maximum, below.

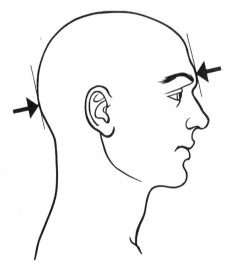

57. Head length.

Design Recommendations for Head Length

For men, civilian or military, 8.6 inches will accommodate 99 per cent or more and 8.4 inches, 95 per cent or more of most groups. Corresponding figures for women are 8.0 and 7.9 inches. These figures represent the indicated percentile plus 0.1 inch for minimum clearance.

Clothing: Adds varying amounts, depending on headgear. Military and flying helmets add about 3.5 inches.

Other factors: As mentioned, more space than for head length alone is needed from back of head to tip of nose, goggles, or oxygen mask. Raising or lowering the head increases the fore-and-aft space required; looking directly upward, the average subject needs 3.0 inches more space behind his head. The eye is located, on the average (among Air Force flyers), 7.0 inches forward of the occiput and 0.7 inch behind the leading point on the forehead. The normal standing or sitting position, or equipment worn on the back, advances the head anterior to the plane of the back support, increasing the required head space.

Head Length, Maximum

Definition: Distance between the most anterior point on the tip of the nose and the most posterior point on the occiput (back of head), in the midline (Fig. 58 and Table 42).

Relevant Equipment or Workspace Dimension: Maximum Head Length—Space available between the tip of the nose and the back of the head.

Design Recommendations for Maximum Head Length

For men, 9.7 inches will accommodate 99 per cent or

Table 41. Head Length (Inches).

Group	Date of Survey	Refer- ence*	Number of Subjects	Age (Mean or Range)	Percentile 1	5	50	95	99	S.D.
Males										
Harvard students	pre-1930	2	475	18	6.9	6.9	7.6	8.2	8.4	0.32
	1940	4	174	18	7.1	7.3	7.7	8.2	8.4	0.27
Univ. Chicago students	1941	4	119	18	7.1	7.3	7.7	8.2	8.4	0.27
Spanish-American War veterans	1959	8	133	81	7.1	7.3	7.7	8.1	8.3	0.25
Air Force flying personnel	1950	19	4,063	27	7.2	7.3	7.7	8.2	8.3	0.25
Air Force basic trainees	1952	9	3,317	18	7.0	7.2	7.6	8.1	8.3	0.28
Air Force cadets	1942	32	1,231	23	7.2	7.4	7.8	8.2	8.4	0.26
Army separatees	1946	28,37	24,471	23	7.0	7.2	7.7	8.1	8.3	0.28
Army aviators	1959	43	500	30	7.2	7.3	7.8	8.2	8.5	0.27
Army drivers, white	1960	7	431	24	7.1	7.2	7.6	8.1	8.4	0.28
Army drivers, Negro	1960	7	79	27	7.0	7.2	7.8	8.4	8.4	0.31
Navy pilots	1958	21	1,190	-	7.0	7.2	7.8	8.6	8.8	0.39
Females										
Normal working women	1932	1	100	36	6.8	7.0	7.4	7.7	7.9	0.23
Pembroke College students	1927	3	109	19	6.8	7.0	7.4	7.7	7.9	0.24
Univ. Tennessee students	1930	3	161	19	6.8	7.0	7.4	7.8	7.9	0.23
WAF basic trainees	1952	10	847	19	6.1	6.4	6.9	7.3	7.5	0.30
WACs and Army nurses	1946	33,37	8,451	26	6.7	6.8	7.2	7.7	7.8	0.26

*See references 1-43 on pp. 325-327.

Table 42. Head Length, Maximum (Inches)

Group	Date of Survey	Reference*	Number of Subjects	Age (Mean)	Percentile					S.D.
					1	5	50	95	99	
Males										
Army drivers, white	1960	7	431	24	7.9	8.1	8.7	9.2	9.5	0.32
Army drivers, Negro	1960	7	79	27	8.3	8.4	8.8	9.5	9.6	0.31

*See references 1-43 on pp. 325-327.

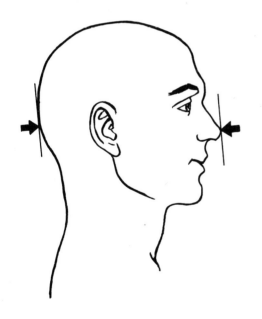

58. Head length, maximum.

more and 9.6 inches, 95 per cent or more of Army drivers. No data are available for any other groups. These figures include 0.1 inch for minimum clearance.

Clothing: Varying amounts are added by headgear. Flying helmets can add 3.5 inches or more.

Other factors: Raising or lowering the head increases the fore-and-aft space required; looking directly upward the average subject needs 3.0 inches more space behind his head. The eye is located, on the average, 7.0 inches forward of the occiput. The normal standing or sitting position, or equipment worn on the back, advances the head anterior to the plane of the back support.

Head Breadth

Definition: Maximum head breadth above the ears, at right angles to the midsagittal plane (Fig. 59 and Table 43).

Relevant Equipment or Workspace Dimension: Breadth across Head—Lateral space available for the head (above the ears). Breadth across the ears at greatest protrusion is the more useful engineering dimension—head breadth is useful chiefly to the headgear designer.

Design Recommendations for Breadth across Head

For men, 6.9 inches will accommodate at least 99 per cent and 6.6 inches at least 95 per cent of most groups. For women, the corresponding figures are 6.4 and 6.3 inches. These values represent the indicated percentile plus 0.1 inch minimal clearance.

Clothing: Increments vary, depending on headgear. Add 3.5 inches for steel helmets, 4.3 inches or more for flying helmets.

Other factors: For space needed across the ears, add 1.3 inches to head breadth. Any turning of the head away from the midline increases lateral head space needed, and such movements are the rule. The greatest increment, 1.6 inches—difference between head length and breadth—would occur with the head turned fully to either side.

Interpupillary Distance

Definition: Distance between the centers of the pupils; subject looks straight ahead (Fig. 60 and Table 44).

Relevant Equipment or Workspace Dimension: Binocular Spacing—Separation distance between binocular eyepieces, as in sighting stations, stereoscopic photogrammetry, microscopy, and the like.

Design Recommendations for Binocular Spacing

For Air Force flyers, 2.5 inches will fit the average, or

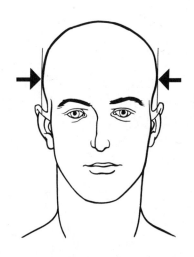

59. Head breadth.

50th percentile; a range from 2.3 to 2.7 inches will accommodate the 90 per cent midrange; and from 2.2 to 2.8 inches, the 98 per cent midrange. For Army soldiers, 2.3 inches will fit the average; a range from 2.2 to 2.5 inches will accommodate the middle 90 per cent;

and a range from 2.1 to 2.6 inches will accommodate the middle 98 per cent.

Clothing: Not applicable.

Other factors: None.

Table 43. Head Breadth (Inches).

Group	Date of Survey	Refer- ence*	Number of Subjects	Age (Mean)	Percentile					S.D.
					1	5	50	95	99	
Males										
Harvard students	pre-1930	2	475	18	5.4	5.6	5.9	6.3	6.4	0.22
	1940	4	174	18	5.5	5.7	6.0	6.4	6.5	0.22
Univ. Chicago students	1941	4	119	18	5.6	5.8	6.1	6.5	6.6	0.20
Spanish-American War veterans	1959	8	133	81	5.6	5.8	6.1	6.4	6.5	0.20
Air Force flying personnel	1950	19	4,059	27	5.6	5.7	6.1	6.4	6.6	0.20
Air Force basic trainees	1952	9	2,019	18	5.4	5.6	5.9	6.3	6.5	0.23
Air Force cadets	1942	32	1,231	23	5.6	5.7	6.1	6.4	6.6	0.21
Army separatees	1946	28,37	24,447	23	5.4	5.6	6.0	6.4	6.6	0.23
Army aviators	1959	43	500	30	5.6	5.7	6.1	6.5	6.8	0.21
Army drivers, white	1960	7	431	24	5.5	5.7	6.0	6.3	6.5	0.22
Army drivers, Negro	1960	7	79	27	5.6	5.7	6.0	6.5	6.6	0.24
Navy pilots	1958	21	1,190	-	5.3	5.5	6.1	6.6	6.7	0.39
Females										
Normal working women	1932	1	100	36	5.5	5.6	5.9	6.1	6.3	0.17
Pembroke College students	1927	3	109	19	5.4	5.5	5.8	6.2	6.3	0.20
Univ. Tennessee students	1930	3	161	19	5.4	5.5	5.8	6.1	6.2	0.18
WAF basic trainees	1952	10	847	19	5.3	5.4	5.7	6.1	6.2	0.20
WACs and Army nurses	1946	33,37	8,500	26	5.2	5.4	5.7	6.1	6.2	0.22

*See references 1-43 on pp. 325-327.

Table 44. Interpupillary Distance (Inches).

Group	Date of Survey	Reference*	Number of Subjects	Age (Mean)	Percentile					S.D.
					1	5	50	95	99	
Males										
Air Force flying personnel	1950	19	4,057	27	2.19	2.27	2.49	2.74	2.84	0.14
Army drivers, white	1960	7	431	24	2.05	2.13	2.32	2.52	2.60	0.12
Army drivers, Negro	1960	7	79	27	2.13	2.28	2.44	2.80	2.83	0.15

*See references 1-43 on pp. 325-327.

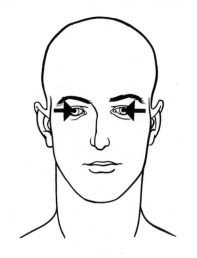

60. Interpupillary distance.

Body Circumferences

Circumferences (Fig. 61) provide important information on body size and shape, but their relevance to equipment design is limited to personal equipment. They do not enter into workspace design and are therefore presented here in abbreviated form and without discussion, for reference purposes. Further information can be found in the studies cited in Table 45.

Definitions

Head: maximum circumference above the brow ridges.
Neck: circumference below the laryngeal prominence (Adam's apple), perpendicular to the axis of the neck.
Shoulder: maximum horizontal circumference over the

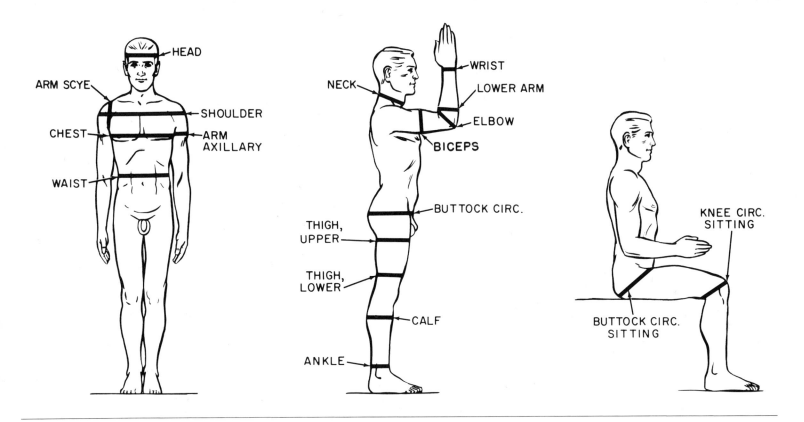

61. Body circumferences.

Table 45. Body Circumferences for Selected Populations (Inches).

For sources of data (indicated by numbers in parentheses), see references on pages 325-327.

Males	Air Force Flying Personnel (19) Mean Age = 27 N = 4,000+		Air Force Basic Trainees (9) Mean Age = 18 N = 1,400 - 3,300+		Army Separatees (28) Mean Age = 24 N = 24,000+	
Circumference	Mean	S.D.	Mean	S.D.	Mean	S.D.
Head	22.5	0.6	22.2	0.6	22.3	0.6
Neck	15.0	0.7	14.3	0.8	14.5	0.7
Shoulder	45.3	2.4	42.6	2.2	-	-
Chest	38.8	2.5	35.6	2.4	36.4	2.3
Waist	32.0	3.0	30.3	2.7	30.6[a]	2.8
Buttock	37.8	2.3	36.3	2.4	36.7[c]	2.2
Buttock, sitting	41.7	2.8	39.6	2.9	-	-
Thigh - upper	22.4	1.7	21.4	2.0	21.4	1.7
Thigh - lower	17.3	1.4	14.9	1.2	14.7	1.1
Knee - sitting	15.4	0.9	15.1	1.1	-	-
Calf	14.4	1.0	13.9	1.1	14.1[e]	0.9
Ankle	8.9	0.6	8.8	0.6	10.4[f]	0.6
Arm scye	18.1	1.4	16.8	1.5	17.1	1.2
Arm - axillary	12.5	1.1	12.4	1.3	12.1	1.1
Biceps	12.8	1.1	12.2	1.1	11.0[g]	0.9
Elbow	12.3	0.8	11.8	0.9	-	-
Lower arm	11.5	0.7	10.7	0.7	9.5[h]	0.9
Wrist	6.9	0.4	6.9	0.4	6.7	0.4

Females	WAF Basic Trainees (10) Mean Age = 19 N = 800+		WACs and Army Nurses (33,37) Mean Age = 25 N = 8,000+		Civilian Women (29) Mean Age = 34 N = 10,000+	
Circumference	Mean	S.D.	Mean	S.D.	Mean	S.D.
Head	21.5	0.7	21.7	0.6	-	-
Neck	13.0	0.7	12.5	0.7	-	-
Shoulder	37.8	1.7	-	-	-	-
Chest	33.7	2.0	35.1	3.0	35.6	3.9
Waist	25.9	1.6	26.5[a]	2.6	29.2[b]	4.5
Buttock	36.9	2.1	37.5[c]	2.7	38.8[c]	3.3
Buttock, sitting	39.2	2.2	-	-	-	-
Thigh - upper	21.9	1.7	22.1	2.1	22.2	2.3
Thigh - lower	14.8	1.1	19.4[d]	2.0	19.6[e]	2.0
Knee - sitting	14.7	1.1	-	-	14.3	1.4
Calf	13.4	1.0	-	-	13.5	1.2
Ankle	8.5	0.5	-	-	8.3	0.7
Arm scye	14.4	1.2	15.0	1.3	16.1	1.6
Arm - axillary	11.0	1.0	-	-	11.4	1.5
Biceps	10.5	0.9	10.1[g]	1.1	-	-
Elbow	11.2	1.1	-	-	10.4	0.9
Lower arm	9.5	0.8	8.9[i]	0.7	9.8[j]	0.8
Wrist	6.1	0.4	5.8	0.3	6.0	0.4

[a]—[j] For explanation of superscripts, see corresponding letters in list on pages 132-133.

deltoid muscles.

Chest: horizontal circumference at nipple level during normal breathing.

Waist: horizontal circumference at level of the greatest lateral indentations of trunk.

Exceptions:

(a) for "Army Separatees" and "WACs and Army Nurses"—midway between the lower edge of the lowest rib and the upper margin of the lateral iliac crests;

(b) for "Civilian Women"—at the level of the lower edge of the lowest palpable rib.

Buttock: horizontal circumference at the level of the rearmost protrusion of the buttocks.

Exception:

(c) for "Army Separatees," "WACs and Army Nurses" and "Civilian Women"—at the level of the most lateral protrusion of the greater trochanters.

Buttock, Sitting: Circumference encompassing the rearmost contact of buttocks with the sitting surface, and the furrow between torso and legs, measuring tape held at about a 45-degree angle.

Thigh, Upper: horizontal circumference just below the gluteal furrow.

Thigh, Lower: horizontal circumference just above the knee.

Exceptions:

(d) for "WACs and Army Nurses"—midway between crotch and knee;

(e) for "Civilian Women"—midway between the most lateral protrusion of the greater trochanter and the highest point on the tibia.

Knee, Sitting: circumference encompassing the popliteal area and the front of the knee, knee at right angles—measuring tape held at about a 45-degree angle.

Calf: maximum horizontal circumference, wherever found.

Ankle: minimum horizontal circumference above the projections of the ankle bones (external and internal malleoli).

Exception:

(f) for "Army Separatees"—at and over the projections of the ankle bones; not a minimum dimension.

Arm Scye: circumference encompassing the highest skeletal point on the lateral edge of the acromion process and the highest point in the axilla (armpit).

Arm, Axillary: horizontal circumference at the armpit, arm hanging loosely at the side.

Biceps: maximum circumference with elbow bent at 90 degrees and biceps maximally flexed.

Exceptions:

(g) for "Army Separatees," and "WACs and Army Nurses"—midway between the armpit and the elbow, with arm hanging loosely at the side.

Elbow: circumference encompassing the elbow tip and the elbow crotch, with the elbow at a right angle and with biceps and forearm muscles maximally tensed.

Lower Arm: the maximum circumference, wherever found, with the upper arm horizontal, forearm vertical, and the elbow at 90 degrees, muscles maximally tensed.

Exceptions:

(h) for "Army Separatees"—horizontal circumference midway between elbow and wrist, with arm hanging loosely at the side.

(i) for "WACs and Army Nurses"—maximum horizontal circumference, arm hanging loosely at side.

(j) for "Civilian Women"—maximum horizontal circumference, with upper arm vertical and elbow at a right angle.

Wrist: minimum circumference above the wrist bones (styloid processes of radius and ulna), with the measuring tape at right angles to the long axis of the forearm.

6. Dynamic Human Body Dimensions

Dynamic anthropometry deals with the dimensions of the "workspace or work envelope needed by men as they perform their work" (Hertzberg, 1960). Unlike static body dimensions, which are measured with the subject in a rigid, standardized position, dynamic measurements are made in working positions and vary accordingly. The static dimensions corresponding most closely to functional reaches, the most important dynamic dimensions, would be anatomical arm and leg lengths. Dynamic dimensions in equipment design relate more to performance than to problems of static "fit." For further discussion of dynamic anthropometry, see Dempster (1955a,b) and Dempster, Gabel, and Felts (1959).

THE ANTHROPOMETRY OF WORKING POSITIONS

The nine body measurements in Table 46 are related to the design of spatially restricted areas for such workers as mechanics, repairmen for heavy equipment, plumbers, or pipe fitters. All measurements were made on a small group of college and young Air Force personnel, the only ones measured to date in working positions. Older workers will be larger in body breadths and depths, to an unknown extent. Definitions of the measurements are as follows (Fig. 62 a-f and Table 46):

(a) Maximum body depth. Subject stands erect, arms at sides. Measurements are made, from a lateral photograph, of the maximum horizontal distance between vertical planes passing through the most anterior and posterior points on the trunk. The anterior points are on the chest or abdomen; the posterior points, in the shoulder or buttock region.

(b) Maximum body breadth. Subject stands erect with arms hanging relaxed at sides. The maximum breadth across the body including arms is measured.

(c) Overhead reach. Subject stands erect, grasps a horizontal bar above his right side with his right hand (at the bend formed by palm and fingers), and raises the bar to the highest position attainable without strain.

(d) Kneeling length and height. Subject kneels with knees and feet together, fists clenched and on floor in front of knees, arms roughly vertical, head in line with long axis of body. Length (A) is measured from the rearmost point on the foot to the foremost point on the head. Height (B) is measured vertically from the floor to the highest point on the head.

(e) Crawling length and height. Subject rests on knees and flattened palms, arms and thighs perpendicular to the floor, feet comfortably extended and spaced, body straight, head in line with the long axis of the body. Length (A) is measured from the rearmost point on the foot to the foremost point on the head. Height (B) is measured from the floor to the highest point on the head.

(f) Prone length and height. Subject lies prone, feet together and comfortably extended, arms extended forward in a maximum unstrained position, fists clenched, backs of hands facing to sides. Height (B) is measured vertically from the floor to the highest point on the head with the head raised as high as possible and chest on floor. Length (A) is measured horizontally from the rearmost point on the foot to the foremost point on the fists.

Table 46. Anthropometry of Working Positions (inches).

(Hertzberg, Emanuel and Alexander, 1956)

Measurement	Number of Subjects	5th Percentile	Mean	95th Percentile	S.D.
Maximum body depth	118	10.1	11.5	13.0	0.88
Maximum body breadth	40	18.8	20.9	22.8	1.19
Overhead reach	40	76.8	82.5	88.5	3.33
Kneeling height	40	29.7	32.0	34.5	1.57
Kneeling length	40	37.6	43.0	48.1	3.26
Crawling height	40	26.2	28.4	30.5	1.30
Crawling length	40	49.3	53.2	58.2	2.61
Prone height	40	12.3	14.5	16.4	1.28
Prone length	40	84.7	90.1	95.8	3.41

FUNCTIONAL REACH MEASUREMENTS

Functional arm reach data are used to determine the outer limits of the workspace or "space envelope" for the placement of controls, tools, or materials to be handled. Reach measurements to any peripheral point of the workspace or space envelope vary with body size and position, biomechanical abilities, clothing and personal equipment, as well as with the nature of the task or control. A change in any one of these factors alters functional reach. It is therefore difficult to derive generally applicable design recommendations from functional reach data, but it is possible to present functional reach measurements for a given task at specified positions in elevation and azimuth. Such data can be used as a guide in the design and placement of items within the workspace.

For purposes of standardization, body reach dimensions are measured with reference to a fixed locus within the workspace. Most commonly used is the Seat Reference Point (SRP), defined as the intersection of the plane of the seat surface with the plane of the backrest surface in the mid-line (from side to side) of the subject's body. Functional reach is measured horizontally from a vertical line projected through this point (the SRV, or Seat Reference Vertical). Although such a dimension does not measure anatomical reach, it is adequate for design purposes, being functional and easily measured.

62. Anthropometry of working positions:

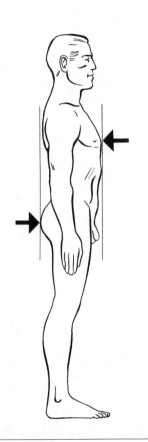

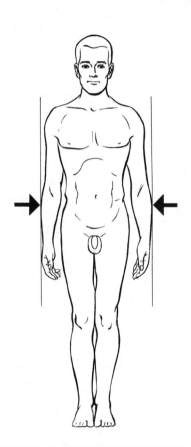

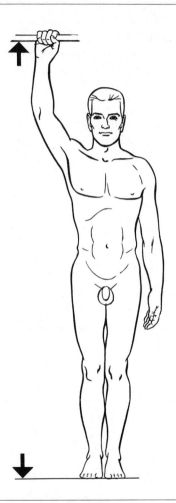

a) maximum body depth;   b) maximum body breadth;   c) overhead reach;

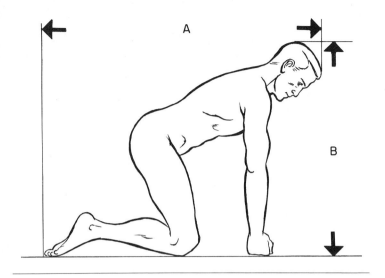

d) kneeling length and height;

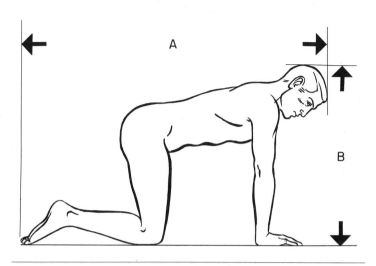

e) crawling length and height;

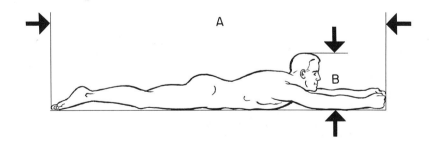

f) prone length and height.

The Data

Functional arm reach data are available for two military populations, the Navy (King, Morrow, and Vollmer, 1947), and the Air Force (Dempsey and Emanuel, n.d.; Kennedy, 1964). There are no comparable studies on functional leg reach, which has been investigated from the standpoint of forces exertable with various leg lengths and at various knee and ankle angles—but not from the standpoint of the spatial limits of operation. The arm reach data in Tables 47 to 56 and Figs. 63 to 72 are from the Kennedy study, based on twenty subjects specially selected to represent the range in body size of U.S. Air Force personnel. Each subject, lightly clothed and in shirt sleeves, sat in a hard seat with a backrest angle of 13 degrees and a seat angle of 6 degrees. Encircling the seat was a large wooden arch from which a series of measuring staves, held in place by friction, projected at 15-degree intervals. Each of the staves pointed toward the approximate joint center of the right shoulder of the seat occupant. The subject then grasped between the right thumb and forefinger (second segment) small knobs mounted on the ends of each of these staves, and pushed them away until the arm was fully extended without pulling the right shoulder away from the seat back. The seat was then rotated 15 degrees under the arch, and a new set of reaches was made from this position. In each case the measurement of reach was made horizontally from the knob to the SRV, a vertical line through the SRP at the intersection of seat surface and backrest. The measurements were made laterally at 15-degree intervals throughout 360 degrees at each of twelve horizontal levels. In Figs. 63–72, SRL (Seat Reference Line) is the horizontal line through the SRP.

In Tables 47 to 56 the "minimum" and 5th percentile values will be most useful, since reach dimensions should be based primarily on the arm reaches of the smallest rather than the largest persons. The "minimum" approximates the reach of the smallest man and will accommodate 99+ per cent of the U.S. Air Force population, and the 5th percentile reaches will assure accommodation for 95 per cent of this population. While the 50th and 95th percentiles will be less useful, they could, for example, help in locating items outside the reach area of large persons, to prevent such items from impeding normal arm movements.

Note: When applying these data, three points should be remembered: (1) They are intended to be used only under workspace conditions similar to those in which the data were gathered. Changes in any of these conditions, *e.g.,* angulation of seat or backrest, amount of body movement permitted, kind of clothing worn, or type of grasp used, will alter the arm reach values given here. (2) The right arm was used; on the average, the right arm is slightly longer than the left. (3) The data are applicable only to populations closely approximating the U.S. Air Force in body size (see Tables 7 to 45). Smaller groups will have shorter reaches.

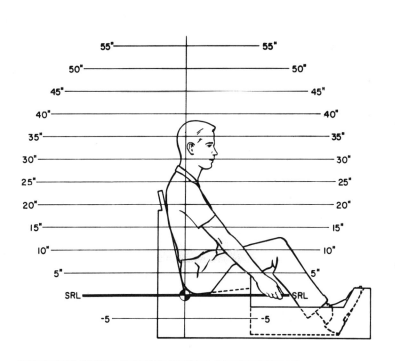

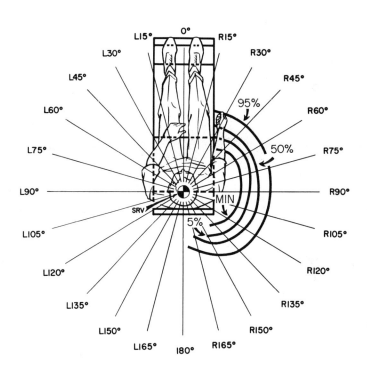

63. Grasping reach, in inches, to a horizontal plane at the seat reference point. Right arm.

Table 47. Grasping Reach, in Inches, to a Horizontal Plane at the Seat Reference Point (Kennedy, 1964).
(right arm)

Angle to Left or Right	Minimum	Percentiles		
		5	50	95
L 165				
L 150				
L 135				
L 120				
L 105				
L 90				
L 75				
L 60				
L 45				
L 30				
L 15				
0				
R 15				
R 30		17.50	20.75	25.00
R 45	16.25	19.50	21.75	26.00
R 60	17.50	20.50	22.25	26.25
R 75	17.25	20.00	22.25	26.00
R 90	17.00	19.50	22.25	25.50
R 105	16.25	18.75	22.00	25.25
R 120	15.00	18.25	20.75	24.50
R 135	13.00	16.50	19.00	23.50
R 150		14.00	16.50	20.25
R 165			13.00	17.00
180				

Table 48. Grasping Reach, in Inches, to a Horizontal Plane 5 Inches Above the Seat Reference Point (Kennedy, 1964).
(right arm)

Angle to Left or Right	Minimum	Percentiles		
		5	50	95
L 165				
L 150				
L 135				
L 120				
L 105				
L 90				
L 75				
L 60				
L 45				
L 30				
L 15				
0				
R 15				
R 30	22.00	23.75	26.00	29.50
R 45	23.50	25.25	27.25	30.00
R 60	23.75	25.75	27.75	30.00
R 75	24.00	25.75	27.50	30.25
R 90	24.00	25.75	27.50	30.75
R 105	23.75	25.25	27.00	30.00
R 120	23.00	24.50	26.50	29.00
R 135	21.50	22.75	25.00	28.00
R 150			22.25	25.75
R 165			19.25	21.25
180				

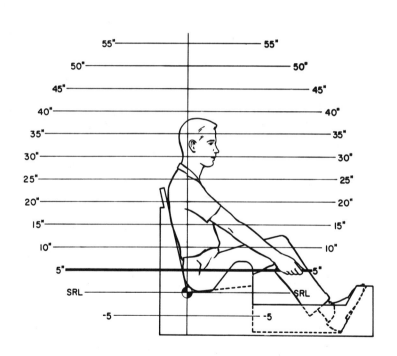
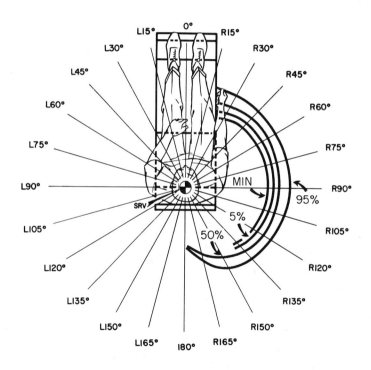

64. Grasping reach, in inches, to a horizontal plane 5 inches above the seat reference point. Right arm.

Table 49. Grasping Reach, in Inches, to a Horizontal Plane 10 Inches Above the Seat Reference Point (Kennedy, 1964). (right arm)

Angle to Left or Right	Minimum	Percentiles		
		5	50	95
L 165				
L 150				
L 135				
L 120				
L 105				
L 90				13.50
L 75				17.25
L 60			16.50	21.00
L 45			19.50	23.25
L 30			21.00	24.75
L 15			22.00	26.25
0				
R 15				
R 30	26.25	27.00	29.25	33.00
R 45	27.25	28.25	30.50	33.75
R 60	28.00	29.00	30.75	33.50
R 75	28.25	29.25	30.75	33.50
R 90	28.25	29.25	31.00	33.50
R 105	27.75	28.75	30.50	32.75
R 120	26.75	27.75	29.75	31.50
R 135		26.25	28.25	30.75
R 150			25.25	28.75
R 165				
180				

Table 50. Grasping Reach, in Inches, to a Horizontal Plane 15 Inches Above the Seat Reference Point (Kennedy, 1964). (right arm)

Angle to Left or Right	Minimum	Percentiles		
		5	50	95
L 165				
L 150				
L 135				
L 120				
L 105				
L 90				17.50
L 75				20.00
L 60			19.25	23.00
L 45		19.00	21.50	25.75
L 30	21.00	21.75	24.00	27.25
L 15	22.50	23.25	26.00	28.75
0	24.25	24.75	28.75	31.00
R 15	26.00	26.50	30.50	34.00
R 30	28.25	28.50	31.50	35.00
R 45	29.50	30.00	32.75	35.50
R 60	30.00	31.00	32.50	34.75
R 75	30.00	31.50	32.50	34.75
R 90	30.25	31.00	32.50	34.75
R 105	30.00	30.75	32.25	34.50
R 120	29.00	29.50	32.00	33.75
R 135			30.00	32.50
R 150				29.50
R 165				
180				

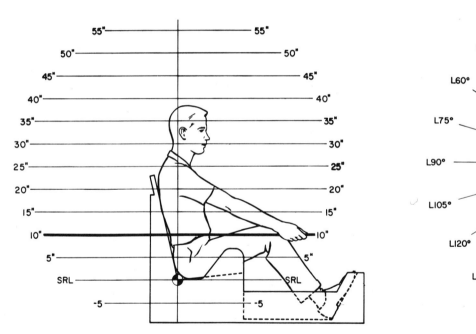

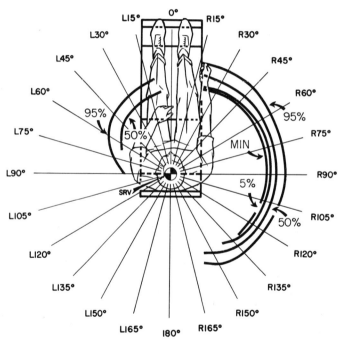

65. Grasping reach, in inches, to a horizontal plane 10 inches above the seat reference point. Right arm.

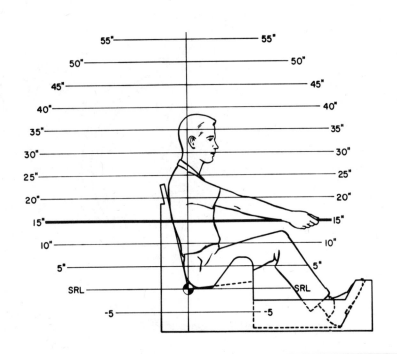
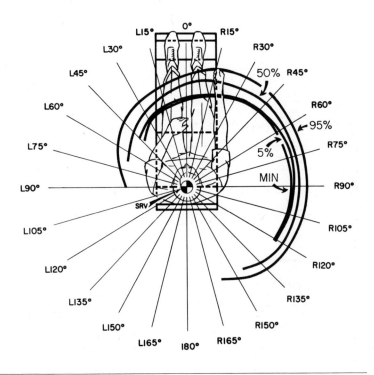

66. Grasping reach, in inches, to a horizontal plane 15 inches above the seat reference point. Right arm.

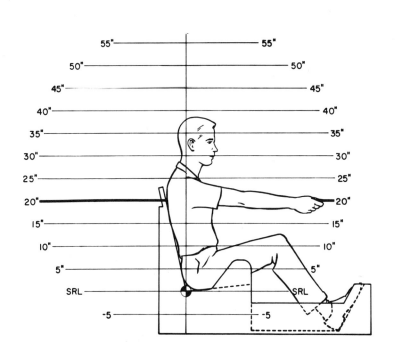

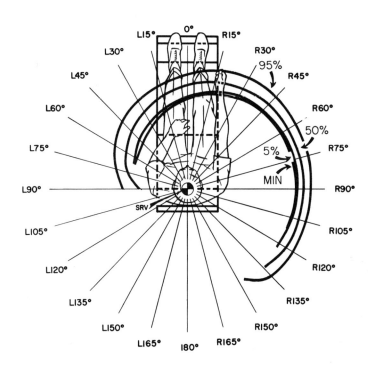

67. Grasping reach, in inches, to a horizontal plane 20 inches above the seat reference point. Right arm.

Table 51. Grasping Reach, in Inches, to a Horizontal Plane 20 Inches Above the Seat Reference Point (Kennedy, 1964).
(right arm)

Angle to Left to Right	Minimum	Percentiles		
		5	50	95
L 165				
L 150				
L 135				
L 120				
L 105				
L 90			14.00	18.75
L 75			18.00	21.50
L 60	17.00	17.50	20.50	24.50
L 45	18.25	19.50	22.75	26.75
L 30	20.25	21.50	24.75	28.25
L 15	22.50	23.50	26.75	29.75
0	25.00	25.50	28.75	31.75
R 15	27.25	28.00	30.50	34.00
R 30	29.00	30.00	32.00	35.75
R 45	30.50	31.00	33.50	36.25
R 60	31.50	32.00	33.75	36.25
R 75	31.50	32.25	34.00	36.50
R 90	31.75	32.25	34.00	36.00
R 105	31.50	31.75	33.50	35.75
R 120		30.50	33.00	35.50
R 135				34.50
R 150				
R 165				
180				

Table 52. Grasping Reach, in Inches, to a Horizontal Plane 25 Inches Above the Seat Reference Point (Kennedy, 1964).
(right arm)

Angle to Left to Right	Minimum	Percentiles		
		5	50	95
L 165				
L 150				
L 135				
L 120				
L 105				17.75
L 90			15.75	20.25
L 75			19.25	22.25
L 60	17.75	18.25	21.50	24.75
L 45	19.25	20.00	23.25	27.25
L 30	21.50	22.50	25.00	28.50
L 15	23.25	24.00	27.00	29.75
0	25.00	26.25	28.50	31.50
R 15	27.25	28.25	30.25	33.50
R 30	29.25	30.25	32.50	35.25
R 45	30.50	31.00	33.50	35.75
R 60	31.00	31.50	33.75	37.00
R 75	31.50	32.00	33.50	36.50
R 90	31.75	32.25	33.75	36.25
R 105	31.25	31.50	33.50	36.00
R 120		30.50	33.25	35.50
R 135				35.00
R 150				
R 165				
180				

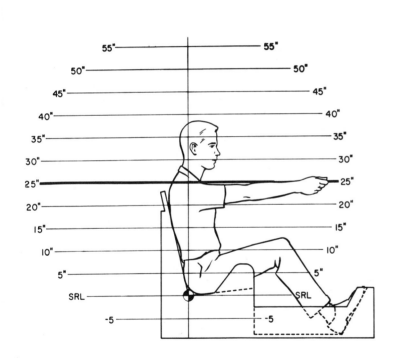
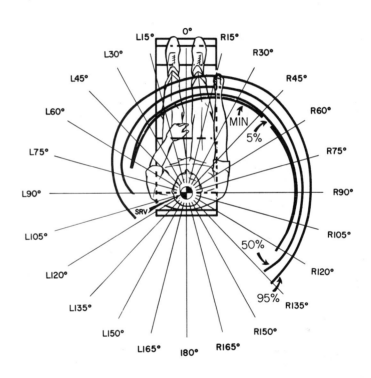

68. Grasping reach, in inches, to a horizontal plane 25 inches above the seat reference point. Right arm.

Table 53. Grasping Reach, in Inches, to a Horizontal Plane 30 Inches Above the Seat Reference Point (Kennedy, 1964). (right arm)

Angle to Left or Right	Minimum	Percentiles 5	50	95
L 165				18.75
L 150				19.25
L 135				20.00
L 120				18.75
L 105				19.00
L 90			16.75	20.75
L 75			18.75	22.50
L 60	17.00	17.25	20.75	24.50
L 45	18.25	19.00	22.50	26.50
L 30	19.75	21.50	24.50	28.25
L 15	22.00	23.75	26.75	29.50
0	23.75	25.50	28.50	31.00
R 15	26.00	27.25	29.75	33.00
R 30	27.75	29.00	31.50	34.25
R 45	28.75	30.25	32.25	34.75
R 60	30.00	31.00	32.75	35.75
R 75	30.75	31.25	33.00	35.50
R 90	31.00	31.25	33.25	35.75
R 105	30.75	31.00	33.00	35.25
R 120		30.25	32.50	34.75
R 135				34.50
R 150				
R 165				19.50
180				20.25

Table 54. Grasping Reach, in Inches, to a Horizontal Plane 35 Inches Above the Seat Reference Point (Kennedy, 1964). (right arm)

Angle to Left or Right	Minimum	Percentiles 5	50	95
L 165			14.75	21.00
L 150			13.75	20.00
L 135			13.25	19.00
L 120		10.75	13.25	18.75
L 105		12.25	14.00	18.75
L 90	12.75	13.75	15.50	20.00
L 75	14.25	15.00	17.25	21.00
L 60	15.25	16.00	18.75	21.50
L 45	16.25	17.25	20.50	24.75
L 30	18.00	19.25	22.50	26.25
L 15	19.25	21.00	24.75	27.00
0	20.75	22.25	26.50	28.50
R 15	22.75	24.75	27.75	31.00
R 30	24.50	26.75	29.25	32.75
R 45	26.75	28.25	30.50	33.75
R 60	28.00	29.00	31.00	33.75
R 75	28.75	29.50	31.25	34.00
R 90	29.00	29.75	31.25	33.50
R 105	29.00	29.75	31.50	33.50
R 120	28.50	29.00	31.00	33.50
R 135			28.50	33.50
R 150				31.50
R 165				21.75
180			16.50	22.25

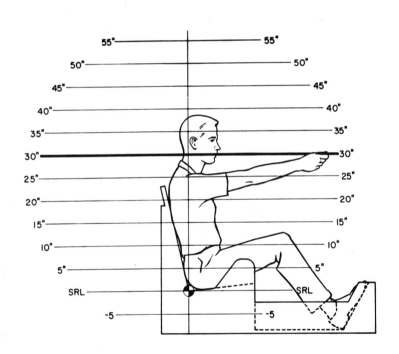

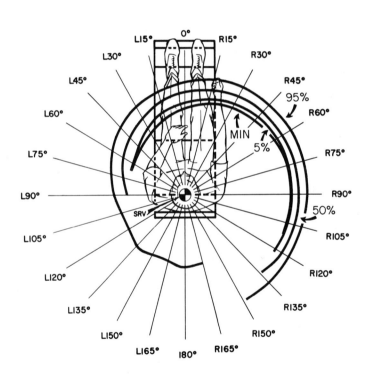

69. Grasping reach, in inches, to a horizontal plane 30 inches above the seat reference point. Right arm.

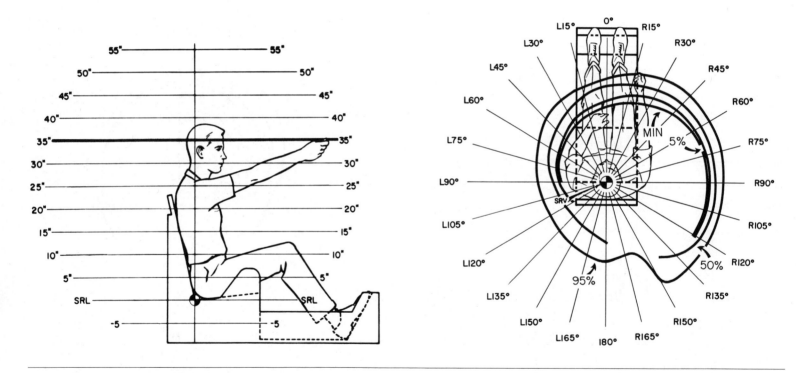

70. Grasping reach, in inches, to a horizontal plane 35 inches above the seat reference point. Right arm.

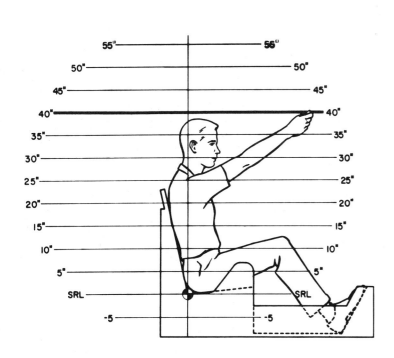

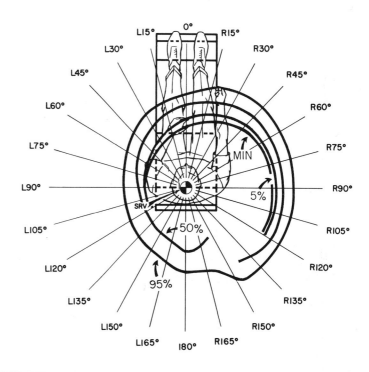

71. Grasping reach, in inches, to a horizontal plane 40 inches above the seat reference point. Right arm.

Table 55. Grasping Reach, in Inches, to a Horizontal Plane 40 Inches Above the Seat Reference Point (Kennedy, 1964).
(right arm)

Angle to Left or Right	Minimum	Percentiles		
		5	50	95
L 165			15.50	21.50
L 150			14.75	20.00
L 135			14.00	19.25
L 120		11.25	13.25	18.50
L 105		11.75	13.25	18.25
L 90	12.00	12.25	13.75	18.25
L 75	12.25	12.50	15.00	18.75
L 60	12.50	13.25	16.25	20.00
L 45	13.00	14.00	17.75	21.50
L 30	13.75	15.50	19.50	23.50
L 15	15.25	17.00	21.25	24.50
0	17.00	19.00	23.00	25.75
R 15	18.75	21.00	24.50	28.50
R 30	21.00	22.75	26.25	30.50
R 45	23.25	24.75	27.75	31.50
R 60	24.25	25.50	28.00	31.25
R 75	25.00	26.00	28.00	31.50
R 90	25.00	26.25	28.25	31.50
R 105	25.75	26.75	28.50	31.75
R 120		26.25	28.75	31.50
R 135			27.00	31.00
R 150				29.25
R 165			16.75	23.75
180			17.75	23.50

Table 56. Grasping Reach, in Inches, to a Horizontal Plane 45 Inches Above the Seat Reference Point (Kennedy, 1964).
(right arm)

Angle to Left or Right	Minimum	Percentiles		
		5	50	95
L 165		10.50	14.00	20.00
L 150	8.50	8.75	12.25	18.25
L 135	7.50	7.75	11.00	16.75
L 120	7.00	7.50	10.50	15.50
L 105	6.75	7.25	10.25	15.00
L 90	6.75	7.25	10.50	15.00
L 75	6.75	7.50	11.00	15.25
L 60	7.00	7.75	12.00	16.25
L 45	7.50	8.50	13.50	18.25
L 30	8.50	9.50	15.00	19.75
L 15	10.00	11.00	16.50	21.25
0	11.25	12.75	18.25	22.75
R 15	13.00	15.50	20.00	24.75
R 30	14.75	17.50	22.00	26.25
R 45	17.25	19.00	23.50	27.00
R 60	19.25	20.50	24.00	27.25
R 75	19.50	20.50	24.00	27.50
R 90	19.75	21.00	24.25	27.75
R 105	20.25	21.50	24.50	28.00
R 120	19.75	21.25	24.50	27.75
R 135	18.75	20.00	23.25	27.75
R 150		15.50	20.75	26.00
R 165		14.75	18.00	22.75
180		12.75	16.50	21.50

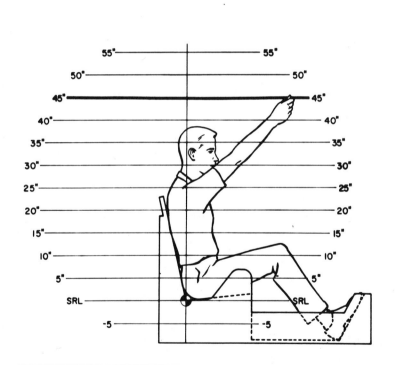

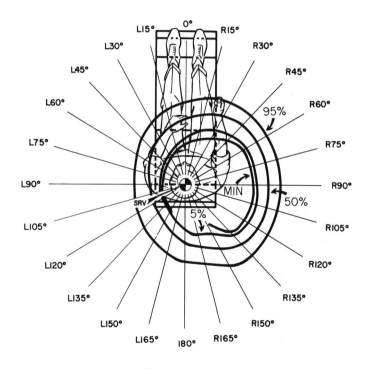

72. Grasping reach, in inches, to a horizontal plane 45 inches above the seat reference point. Right arm.

7. Body Surface Area

The surface area of the whole body, and of its parts, is useful to the designer or engineer in such contexts as: human response to temperature and humidity (since heat exchange and sweating occur at the body surface); tolerance of forces during acceleration or blast; exposure to light or to ionizing radiation; and oxygen consumption. The basal metabolic rate expresses oxygen consumption per square meter of body surface, as a percentage of "expected" uptake for a person of a given sex, height, and weight.

Hutchinson's (1950, 1952) attempt to derive a universal "human shape factor" illustrates the heating engineer's practical concern with body surface area measurement, but his mathematically derived model was accurate to only 7 per cent on ten subjects ranging widely in height, weight, and depth.

For good summaries of techniques for determining body surface area, see Guibert and Taylor (1951) and Sendroy (1961). The standard formula of DuBois, which gives good results, is

$$\text{Surface area} = 71.84 \ W^{0.425} \ H^{0.725}$$

For its practical use, logarithms may be used, and tables and charts are widely available in physiology texts. This formula was based on only nine subjects, with surface areas ranging from 0.8 to 2.3 square meters.

The graphic method of Sendroy and Cecchini (1954) is based on 252 individuals, from fetuses to very large persons, with surface areas ranging from 0.05 to 2.3 square meters. It differs by only ± 0.035 m^2, or 2.0 per cent, on the average, from direct surface area measurement of 107 subjects. Table 57, derived from their graph, is from the *Handbook of Biological Data* (1956); an alternative method, based on revised and extended calculations, is the nomogram of Sendroy and Collison (1960), presented in Fig. 73.

A promising development is the use of photographic anthropometry. Gavan, Washburn, and Lewis (1952) offer many practical suggestions. For surface topography, useful in designing closely fitting gear, Hertzberg, Dupertuis, and Emanuel (1957) have devised a stereophotogrammetric technique based on aerial photography. Though accurate, it is slow and costly. Geoghegan (1953) photographed subjects in a variety of standardized poses, calculating whole body, limb, and segmental areas from measurements of the photographic negative, and estimating volumes. For four subjects of average build, total body area was close to that given by the DuBois formula. Pierson (1962a) has developed a monophotogrammetric method.

Guibert and Taylor (1951) devised an elaborate planigraphic method; more simply, Sendroy and Cec-

Table 57. Surface Area, in Square Meters, for Known Height and Weight.

(Sendroy and Cecchini, 1954, and Handbook of Biological Data, 1956).

Height, in Centimeters

Body Weight Kg	20	30	40	50	60	70	80	90	100	110	120	130	140	150	160	170	180	190	200	210	220	230	240	250	260
5	.18	.20	.23	.26	.29	.33	.37	.42	.48	.55	.62														
10		.35	.36	.38	.41	.44	.48	.52	.57	.64	.69	.76													
15					.54	.57	.60	.63	.67	.72	.77	.83	.89												
20							.68	.72	.76	.80	.85	.91	.97	1.03											
25								.80	.84	.88	.93	.98	1.03	1.09	1.15										
30									.92	.96	1.01	1.05	1.10	1.16	1.22	1.28									
35										1.04	1.08	1.12	1.17	1.23	1.29	1.35	1.42								
40										1.11	1.15	1.20	1.25	1.30	1.36	1.42	1.48	1.55							
45											1.23	1.27	1.32	1.37	1.43	1.48	1.54	1.61							
50											1.30	1.34	1.39	1.44	1.49	1.54	1.60	1.67	1.74						
55											1.37	1.42	1.46	1.50	1.55	1.61	1.67	1.73	1.80						
60											1.44	1.48	1.52	1.57	1.62	1.67	1.73	1.79	1.85	1.92					
65												1.54	1.58	1.63	1.68	1.73	1.79	1.85	1.91	1.97					
70												1.61	1.65	1.70	1.75	1.80	1.85	1.91	1.96	2.02	2.08				
75												1.68	1.72	1.76	1.81	1.86	1.91	1.96	2.02	2.07	2.13				
80												1.74	1.78	1.82	1.86	1.91	1.96	2.02	2.07	2.13	2.18	2.25			
85												1.81	1.84	1.88	1.92	1.97	2.02	2.07	2.13	2.18	2.24	2.31			
90												1.87	1.90	1.94	1.98	2.03	2.08	2.13	2.18	2.24	2.30	2.36			
95												1.97	2.01	2.05	2.09	2.14	2.18	2.24	2.30	2.36	2.42	2.48			
100													2.03	2.07	2.12	2.16	2.20	2.24	2.30	2.35	2.41	2.47	2.54		
105													2.10	2.14	2.18	2.22	2.26	2.31	2.35	2.41	2.47	2.53	2.60		
110													2.17	2.21	2.24	2.28	2.32	2.36	2.41	2.47	2.53	2.58	2.65	2.73	
115													2.23	2.27	2.30	2.33	2.38	2.42	2.47	2.53	2.58	2.64	2.71	2.78	
120														2.33	2.36	2.39	2.43	2.48	2.53	2.58	2.63	2.70	2.77	2.84	2.93
125														2.39	2.42	2.45	2.49	2.53	2.58	2.63	2.69	2.76	2.83	2.90	2.97
130														2.44	2.47	2.51	2.54	2.59	2.63	2.68	2.75	2.82	2.88	2.95	3.02
135														2.50	2.53	2.56	2.60	2.64	2.69	2.74	2.81	2.87	2.93	3.00	3.08
140														2.55	2.58	2.62	2.66	2.70	2.74	2.80	2.87	2.93	2.98	3.06	
145														2.61	2.63	2.67	2.71	2.75	2.80	2.86	2.92	2.98	3.04		
150														2.66	2.69	2.73	2.77	2.81	2.86	2.92	2.97	3.03	3.09		
155														2.72	2.74	2.78	2.83	2.87	2.92	2.97	3.03	3.08			
160														2.77	2.80	2.83	2.88	2.92	2.97	3.02	3.08				
165															2.86	2.89	2.93	2.97	3.02	3.07					
170															2.91	2.94	2.98	3.03	3.07						
175															2.96	2.99	3.03	3.08							
180															3.01	3.04	3.08								
185															3.06	3.09									

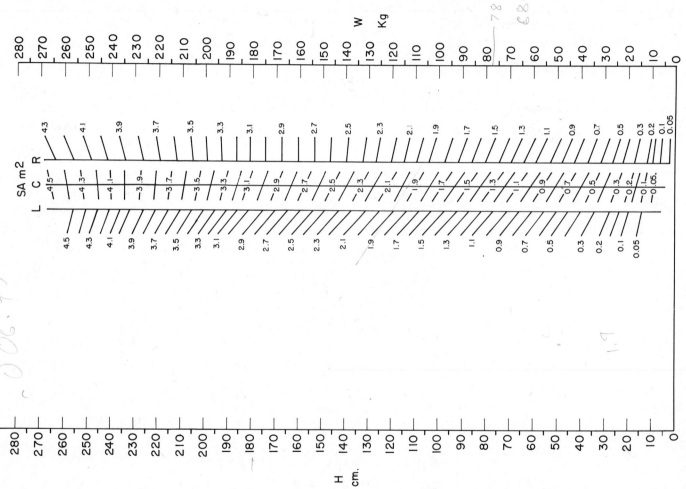

73. Nomogram for the determination of human body surface area (SA) from height (H) and weight (W). Read SA along line most nearly parallel to line joining H and W (Sendroy and Collison, 1960).

chini (1954) utilized "somatotype" photographs (front, side, and back views, at a known magnification factor), following Sheldon *et al.* (1940). They obtained body area by adding front and side view areas, determined planigraphically from the negatives by a photoelectric areameter, and multiplying by two. The total area thus obtained differed on the average by ± 0.044 m^2, or 2.3 per cent, from those derived from their height and weight table (Table 57). The photographic and graphic methods thus gave about equally good results in approximating true surface area.

Since all of the authors cited have been concerned with developing techniques which will be accurate for the individual case, there are no large studies of population groups. In most cases, as summarized in the *Handbook of Growth* (1962, p. 344) surface area is calculated by the DuBois formula for presumably representative subjects of different racial or ethnic groups —even though Stevenson (1928), for example, has shown that the DuBois formula is not suitable for Chinese. The number and representativeness of subjects used for such calculations should be critically assessed before published values are accepted. For description of such groups as flyers, soldiers, or children, it is recommended that the designer utilize Table 57 for total body surface. For areas of parts of the body, he should adapt one of the photographic methods cited, selecting subjects representative of the group with which he is concerned.

8. The Center of Gravity and Moment of Inertia of the Human Body*

INTRODUCTION

The need for biomechanical data regarding the center of gravity (*CG*) and moments of inertia of the human body and body segments arises in several fields of application. Such data are useful in determining the stability and angular acceleration of equipment occupied or operated by persons in various postural attitudes; in the design of seats, particularly aircraft ejection seats and fastening devices; in dummy construction; in assessing the ability to apply torque while in the weightless state, and the consequences of such application; and in the study of human biomechanics. Until recently, only limited data were available, derived largely from studies of Braune and Fischer (1889) and Fischer (1906). Renewed interest since World War II has resulted in a number of investigations of human body *CG* location, distributions of mass, and moments of inertia. The results of some of these researches will be presented in summary form, with a discussion of techniques and their practical application.

*The material in this section was written by Benjamin C. Duggar, formerly at the Harvard School of Public Health and presently with Bio-Dynamics in Cambridge, Massachusetts. It is reprinted with permission of the author and the editor of *Human Factors,* in which an earlier version appeared (4:131–148, June 1962).

Studies in this field provide two types of information: basic data concerning general biological characteristics of the body segments and measured *CG* locations for whole bodies in specified postural attitudes and clothing outfits. Information of the first type has broad application, since it can be used to calculate the *CG* or moment of inertia of the human body in postural attitudes for which no actual measurements have been reported. Data of the latter type are more accurate, though more limited. The biotechnologist must therefore decide whether to apply portions, or even averages, of the body segment data to his problem, whether to use the published measurements for actual persons in positions and wearing equipment similar to those in his own problem, or whether neither alternative is suitable and new measurements must be made for his specific situation. Although the designer faced with such a problem is fortunate in having sufficient data available to allow several distinct approaches to his problem, the paucity of information concerning the accuracy of his estimate is discouraging.

Body segment data have been derived primarily from dissected cadavers, generally of short stature, below average weight, and of middle age or older. Although segment characteristics can be computed from *CG*, volume, and other indirect measurements of living subjects, the accuracy of the computed values remains to be verified. Ultimately such calculated values as seg-

ment *CG*, weight, or moment of inertia about the *CG* for living subjects depend on segment densities based on cadaver measurements, and must ignore the dynamic nature of body mass distribution.

ACCURACY—REQUIREMENTS AND POSSIBILITIES

The establishment of a logical criterion of required accuracy for *CG* or moment of inertia data should be the first step. For example, the most exacting measurements cannot locate the *CG* of a particular man to within 0.3 cm. Indeed, Mosso (1884) reported that the *CG* of a recumbent man shifts with each deep inspiration. Mosso's measurements were conducted by balancing a platform on which the subject assumed a supine position. The platform was stabilized by suspending a 25-kg weight about 80 cm below the platform on a fixed vertical rod. Variations in the body *CG* were balanced by placing containers of liquid at the head or the feet of the supine subject. Although the detail of Mosso's description is insufficient for exact calculation of head and footward excursions of the *CG*, the greatest shift (which occurred during forced inspiration) was probably between 0.5 and 0.7 cm.

Cotton (1932) found that respiratory movements caused such relatively large shifts in the supine *CG* as to preclude determination of the anterior-posterior *CG* location by the "differential tilting" method. Supine subjects were positioned on a horizontal balance table stabilized by vertical springs at the end. The forces on the springs were determined according to Hooke's law from the deflection of the end of the table. Measurements were made with the subject alone, and also with an added weight at the end of the balance table. Cotton calculated the height of the CG of the system above the fulcrum by equating moments about the fulcrum for these two conditions. The difficulty caused by respiration was overcome by placing the subjects perpendicular to the balance table and parallel to the tilt axis, since the CG excursions were nearly symmetrical with respect to the sagittal plane. In this method, and with the subject holding his breath during the determination, repeated measurements had a mean error of 0.7 per cent for antero-posterior CG location.

Swearingen (1953) observed that repeated readings of CG location in the same living subject, using a highly sensitive balance, did not vary more than 0.67 cm. This variation he attributed to slight differences in the subject's position during remeasurement. Swearingen's equipment consisted of five platforms mounted one above the other. The top platform supported the subject in an adjustable seat, the second and third platforms were horizontally adjustable, and the fourth platform was separated from the fifth, or base, by a ball and socket joint in the center and electrical contact points at each corner. Measurement of the CG location for the upright seated posture, for example,

was made by balancing the system with the subject tipped horizontally, then rebalancing with the seated subject tipped approximately 20 degrees from the horizontal. Two planes passing through the CG were thereby established, and the exact location could be measured with respect to any suitable reference point. The extent to which this procedure altered the CG location by displacement of the viscera and body fluids was not reported. The high degree of accuracy reported for measurements using this elaborate equipment applies only to the conditions actually experienced during measuring.

For standing individuals, Hellebrandt, Genevieve, and Tepper (1937) reported that the projection of the CG onto the base of support shows a small but incessant oscillation. They hypothesized that such swaying is essential in maintaining balance. Reynolds and Lovett (1909), attempting to determine the CG in the erect position, also encountered considerable body sway in their subjects. Dempster (1955a) reported that the location of the CG of a pantographic representation of a man closely approximated the location of the actual CG, and that deep inspiration and complete expiration caused the pantographic CG to shift over a range of more than 6 cm. Although the actual body CG would certainly not shift by nearly as much as that of the pantographic representation, the implications are clear.

The human body is a dynamic system, so that its *CG* location with reference to the surroundings or even with reference to anatomical landmarks over time, can be specified only with an inherent uncertainty. If the body is undergoing acceleration or a change in attitude, the resultant shift of forces will further alter the *CG* location. For example, Mosso (1884) observed a noticeable redistribution of the blood and a shift in the *CG* when a subject assumed a horizontal position after standing. Cotton (1932) reported that after exercise the *CG* shifts toward the head as accumulated blood migrates out of the lower limbs. Although the shift of the *CG* under accelerative forces has not been studied, it may be significant for even small loads unless pressure suits or other means of affecting the distribution of the body contents prevent or compensate for this effect.

The moments of inertia of the body and of body segments are also sensitive to redistribution of the body contents. Although no direct measurements have been reported on the constancy of moment of inertia values, it is likely that the differences occasioned by rising from a supine to a standing position, for example, significantly increase the moments of inertia of the lower extremity. Such differences probably alter only slightly the whole-body moment of inertia taken about the *CG* or the soles of the feet. Since the moment of inertia represents the sum of each incremental mass times the square of the distance from the mass to the axis of rotation, it is theoretically possible to shift the *CG* without changing the moment of inertia. Similarly the moment of inertia may be changed with respect to one axis and unchanged with respect to another.

The above observations are intended not to discourage the use of published *CG* and moment of inertia data or the experimental determination of the needed values, but rather to emphasize that the required accuracy and the conditions for which the data are sought limit the possible methods. Stringent requirements may render a wholly satisfactory solution impossible. If the design objectives are such that only the average *CG* location is required for a group of persons, or even for the same person over a period of time, the accuracy of a simple platform and scale measuring system may suffice. A one per cent error in a simple balance system will be of the same order of magnitude as error caused by such natural phenomena as breathing. Moments of inertia may be approximately determined from tabular data or oscillation measurements, but the accuracy of estimation remains to be established.

BODY SEGMENT CHARACTERISTICS

Cadaver Measurements

Studies involving the dissection of cadavers which are

summarized here are by Braune and Fischer (1889 and 1892), Fischer (1906), and Dempster (1955a). Characteristics of body segments amputated for medical reasons are likely to be abnormal and will not be included. (For example, an amputated lower leg studied by Bartholomew [1952] represented a much smaller percentage of total body weight than has been reported in any other study, of cadavers or living persons.) Essential data and identification of the cadavers are summarized in Table 58. The bodies were frozen before dismemberment to prevent loss of body fluid, although Dempster made preparatory cuts from the crease of the crotch overlying the ischial ramus down to the bone before freezing. Cuts were then made passing through the joint centers with the limbs in midrange position (Dempster) or with arms and legs straight (Braune and Fischer). The average segment masses and *CG* locations are summarized in Tables 59, 60, and 61. Location of the *CG* of the frozen segments was determined by suspending the member alternately from two points. The intersection of vertical planes passed through the points of suspension determined the exact *CG*. Variations in distribution of body weight and in segment *CG* locations can be attributed mainly to individual differences in age and body build of the cadavers, and to a smaller extent to differences in body orientation during freezing, dismembering techniques, and the selection of reference points.

Table 58. Characteristics of Thirteen Dissected Cadavers Described by Braune and Fischer (1889), Braune and Fischer (1892), and Dempster (1955a).

Designation	Age (Years)	Supine Height (Cm)	Weight (Kg)	Cause of death	Remarks
Braune and Fischer (1889)					
2[a]	45	170	75.1	Hanging	"Very muscular"
3	50	166	60.75	Hanging	"Very muscular, of normal build "
4		168.8	55.7	Hanging	"Muscular with little fat "
Braune and Fischer (1892)		165.0	58.5		
		150.5	44.057		
Dempster (1955a)					Somatotype[b]
14815	67	168.9	51.26	Unknown	4-5-2 1/2
15059	52	159.8	58.29	Cerebral hemorrhage	3-5-3
15062	75	169.6	58.29	Generalized arterio-sclerosis	4-2-4
15095	83	155.3	49.56	Unknown	4-3-4
15097	73	176.4	72.35	Esophageal carcinoma	4-5-3
15168	61	186.6	71.22	Coronary thrombosis	3-3-4
15250		180.3	60.33	Acute coronary occlusion	3-3-4
15251		158.5	55.79	Chronic myocarditis	4-4-2

[a] Cadaver No. 1 of Braune and Fischer was not dissected.

[b] Somatotype, on a 7-point scale, was estimated from postmortem photographs and without benefit of a weight history. See Sheldon et al., 1940.

Table 59. Average Segment Characteristics of
Three Cadavers Dissected by
Braune and Fischer (1889).

Part	Weight (Kg)	Per Cent of Total Body Weight	Center of Gravity Location with Respect to the Joint Axes for Each Limb, Expressed as Percentage of Limb Length	
			From Above	From Below
Upper arm (each)	2.127	3.3	47.0	53.0
Forearm (each)	1.335	2.1	42.1	57.9
Forearm and hand	1.872[a]	2.9	47.2[b]	52.8[b]
Hand (each)	0.533	0.85		
Upper leg (each)	6.793	10.75	43.9	56.1
Lower leg (each)	3.025	4.8	41.95	58.05
Foot (each)	1.067	1.7	43.4[c]	56.6[c]
Lower leg and foot	4.127[a]	6.5	51.9[d]	48.1[d]
Head, neck and trunk, minus the limbs	33.990	53.0		
Head	4.440	6.95	Average locations not given. See individual cadaver data.	
Torso	29.550	46.3		
Entire body	63.85	100.0		

[a]Combined weights of several segments were slightly larger than
the sum of the individual components due to loss of material
during sawing.
[b]Per cent of length from cubital axis to lower edge of flexed
fingers.
[c]Per cent of length from front to rear of foot.
[d]Per cent of length from knee axis to sole of foot.

Table 60. Average Segment Characteristics of Two
Cadavers Described by Braune and Fischer (1892).

Part	Weight (Kg)	Per Cent of Total Body Weight	Center of Gravity Location with Respect to the Joint Axes for Each Limb, Expressed as Percentage of Limb Length	
			From Above	From Below
Upper arm	1.51	2.93	45.9	54.1
Forearm and hand	1.30	2.54	46.0	54.0
Upper leg	5.78	11.23	43.4	56.7
Lower leg	2.32	4.53	42.4	57.6
Foot	0.95	1.88	41.7	58.3
Lower leg and foot	3.28	6.42	53.4	46.6
Head and trunk	26.12	51.31	58.1	41.9
Entire body	51.28	100.00		

Table 61. Average Segment Characteristics of Eight
Cadavers Dissected by Dempster (1955a).

Part		Weight (Kg)	Per Cent of Total Weight	Center of Gravity Location with Respect to Stated Reference Points. Expressed as Percentage of Total Distance between Reference Points.
Upper arm	R	1.614	2.77	43.6% to gleno-humeral axis; 56.4% to elbow axis
	L	1.536	2.63	
Forearm	R	0.954	1.64	43.0% to elbow axis; 57.0% to wrist axis
	L	0.914	1.57	
Hand	R	0.388	0.67	50.6% to wrist axis; 49.4% to knuckle III
	L	0.383	0.66	
Forearm and hand	R	1.343	2.30	67.7% to elbow axis; 32.3% to ulnar styloid
	L	1.297	2.22	
Entire upper extremity	R	2.973	5.09	51.2% to gleno-humeral axis; 48.8% to ulnar styloid
	L	2.875	4.93	
Upper leg	R	5.756	9.86	43.3% to hip axis; 56.7% to knee axis
	L	5.812	9.95	
Lower leg	R	2.741	4.69	43.3% to knee axis; 56.7% to ankle axis
	L	2.732	4.68	
Foot	R	0.832	1.42	24.9% of foot link dimension to ankle axis (oblique); 43.8% of foot link dimension to heel (oblique); 59.4% of foot link dimension to toe II (oblique) [a]
	L	0.872	1.49	

Table 61. Continued

Part		Weight (Kg)	Per Cent of Total Weight	Center of Gravity Location with Respect to Stated Reference Points. Expressed as Percentage of Total Distance between Reference Points.
Lower leg and foot	R	3.592	6.16	43.4% to knee axis; 56.6% to medial malleolus
	L	3.625	6.21	
Entire lower extremity	R	9.481	16.25	43.4% to hip axis; 56.6% to medial malleolus
	L	9.408	16.13	
Trunk minus limbs		33.626	57.61	60.4% to vertex; 39.6% to hip axis
Both shoulders		6.174	10.58	
Head and neck		4.610	7.90	43.3% to vertex; 56.7% to 7th cervical vertebra [b]
Thorax		6.763	11.58	62.7% to 1st thoracic vertebra [b]; 37.3% to 12th thoracic vertebra [b]
Abdomen plus pelvis		16.395	28.09	59.9% to 1st lumbar vertebra [b]; 40.1% to hip axis
Sum of segment weights		58.363	100.0	

[a] Alternately, a ratio of 42.9 to 57.1 along the heel to toe distance establishes a point above which the center of gravity lies on a line between ankle axis and ball of foot.

[b] Vertebral body.

The use of average segment values from cadavers to locate the whole-body CG or moment of inertia involves considerable inaccuracy. Even the use of values from the cadaver most closely approximating the dimensions of the subject to whom the CG predictions will apply may involve an error of 1 cm or more. Indeed, Braune and Fischer calculated from segment data the CG location of cadaver No. 4 and obtained a value 0.8 cm from the actual measured location, which they regarded as very good agreement, considering the difficulties. Individual measurements for the various cadavers, including moments of inertia and segment density when reported, are presented in Tables 62 to 68. Radii of gyration as per cent of segment length are also included for Braune and Fischer's (1892) and Fischer's (1906) data. Moment of inertia about the segment CG (I_{CG}) is equal to the product of segment mass and radius of gyration squared. Moment of inertia about the proximal joint center (I_0) is related to I_{CG} by the formula:

$$I_0 = I_{CG} + md^2,$$

where m is the segment mass and d is the distance from the joint center to the CG. The moments of inertia of the segments were determined by a free-swinging pendulum system. The segments were suspended from the proximal joint center, the oscillation period measured, and the moment of inertia determined by the relation:

$$I_0 = \frac{mgL}{4\pi^2 f^2},$$

where I_0 = moment of inertia about the point of suspension,

m = mass of the segment (weight/g),

L = distance from the CG to the suspension point,

f = frequency of oscillation, and

g = acceleration of gravity (980 cm/sec²);

and

$$I_{CG} = I_0 - mL^2,$$

where I_{CG} = moment of inertia about the CG.

NOTES TO TABLE 62

[a] Per cent of total cadaver height as measured from the floor.

[b] Length of head is vertical projection from below chin to crown. CG location is per cent of distance from atlanto-occipital joint to crown as measured from the joint.

[c] Per cent of distance from the line connecting the centers of the hip joints to the atlanto-occipital joint as measured from this line.

[d] CG location given as per cent of length from humeral head to capitular head (upper arm plus forearm length).

[e] Length measured from cubital axis to lower edge of flexed fingers.

[f] Length measured from center of femoral head to sole of foot.

[g] Length measured from knee axis to sole.

[h] Length measured from knee axis to tibio-talar joint.

Table 62. Individual Segment Characteristics Determined for the
Three Cadavers Dissected by Braune and Fischer (1889).

Weights are in kg, length in cm, and CG location is distance from proximal
joint axis expressed as per cent of segment length.

Part		Cadaver II			Cadaver III			Cadaver IV		
		Weight	Length	CG Location	Weight	Length	CG Location	Weight	Length	CG Location
Entire body		75.10	170.0		60.75	166.0		55.70	168.8	54.8 [a]
Head[b]		5.35	21.0		4.04	20.2		3.93	21.3	45.5
Torso without limbs (includes neck)		36.02			28.85			23.78	59.1	39.0 [c]
Whole arm[d]	R	4.95		53.5	3.55	68.0	49.2	3.52	69.5	55.1
	L	4.79		53.3	3.48	68.2	50.8	3.71	69.5	53.4
Upper arm	R	2.58	31.7		1.99	30.6	43.8	1.73	32.0	50.9
	L	2.56	31.5		1.88	30.2	45.4	2.02	32.0	57.8
Forearm with hand[e]	R	2.37			1.55	37.4	47.5	1.79	37.5	47.2
	L	2.23			1.60	38.0	46.3	1.69	37.5	47.7
Forearm without hand	R	1.70	29.5		1.05	26.3	41.4	1.30	27.0	42.2
	L	1.60	29.5		1.12	27.1	40.6	1.24	27.0	44.1
Entire leg[f]	R	12.12	93.2	41.9	10.65	92.7	40.7	10.11	88.0	40.3
	L	11.89	92.6	41.7	10.25	91.6	42.1	10.65	88.0	37.5
Upper leg	R	7.65	44.0	43.2	6.69	42.0	46.9	6.15	40.0	42.5
	L	7.30	43.3	44.6	6.22	41.0	47.6	6.75	40.0	38.8
Lower leg with foot[g]	R	4.47	49.2	50.0	3.95	50.7	53.1	3.96	48.0	52.1
	L	4.50	59.3	51.7	3.98	50.6	51.4	3.90	48.0	53.1
Lower leg without foot[h]	R	3.21	41.4	42.0	2.87	43.0	43.5	2.97	41.5	41.0
	L	3.32	41.8	41.6	2.88	42.9	41.3	2.90	41.5	42.2
Foot	R	1.10	28.5	40.4	1.06	26.5	43.0	0.99	26.5	45.3
	L	1.16	28.3	42.4	1.09	26.9	43.9	1.00	26.5	45.3

Since Dempster did not report data on the individual segment *CG* locations, estimates of accuracy of these measures are based only on the data from Braune and Fischer's cadavers. Segment *CG* locations seem to be less variable than calculated segment mass. For example, the standard deviation for upper leg *CG* location, as a percentage of mean joint-to-joint length, was 2.3 per cent (for five subjects), while the standard deviation for upper leg as a percentage of total body weight exceeded 20 per cent of the mean value (thirteen subjects). Consequently the variability of *CG* locations can be neglected in calculations based on cadaver segment weights.

Barter (1957) has taken the combined data from the above studies (with the exception of the 58.5-kg man described by Braune and Fischer, 1892) and calculated regression equations of body segment mass as a function of body weight. These equations, together with the standard deviations of the residuals about the regression lines, are presented in Table 69. The standard deviations do not incorporate the uncertainty of the regression value itself, based on only twelve cadavers, so that the total error of prediction will be greater for body weights differing from the average cadaver weight, 59.4 kg. Extrapolation of these regression equations to include persons weighing more than the heaviest cadaver, 75.3 kg, is unreliable.

Although cadaver measurements of *CG* locations and body segment weights have been an accepted standard for more than 50 years, the heights, weights, and ages of such cadavers are not representative of the present-day working man. Application of such measurements to problems requiring high accuracy in the living should be made with caution. In the case of persons closely matching one of the cadavers in body build, height, weight, and age these data should provide calculated whole-body *CG* locations as close as one centimeter from the "true" location determined by the procedure outlined below. Moments of inertia of cadaver segments appear to be more variable than weight or *CG* values and may be greatly in error if applied to persons of different constitution or having different segment length or mass.

Table 63. Individual Segment Characteristics of Two Cadavers.

Weights are in kg, length in cm, and CG location is
distance from proximal (above) joint axis expressed
as per cent of segment length.
Described by Braune and Fischer (1892)
and Fischer (1906).

Part	1st Cadaver			2nd Cadaver		
	Weight	Length	CG	Weight	Length	CG
Upper arm						
R	1.71	28.5	47.4	1.24	25.5	44.6
L	1.85	29.5	46.1	1.25	27.1	45.5
Forearm and hand						
R	1.44	37.0	45.9	1.12	36.0[a]	44.4
L	1.42	36.0	45.8	1.20	35.5[a]	47.9
Entire upper extremity						
R	3.15	65.5	42.9	2.36	59.0	42.7
L	3.27	65.5	42.2	2.47	58.5	46.4
Upper leg						
R	6.45	40.0	44.25	4.86	35.9	43.8
L	6.99	40.0	41.75	4.81	36.6	43.4
Lower leg						
R	2.68	39.0	42.3	2.07	37.9	42.6
L	2.66	39.0	41.0	1.89	37.1	43.9
Foot						
R	.99	20.0	30.0	0.91	20.0[b]	31.9
L	1.00	20.0	30.0	0.91	20.0[b]	34.8
Lower leg and foot						
R	3.68	47.0	53.2	2.98	43.9	56.4
L	3.67	47.0	53.2	2.80	43.1	50.9
Entire lower extremity						
R	10.13	87.0	40.1	7.84	78.9	41.5
L	10.66	87.0	38.3	7.64	79.2	42.0
Trunk and head	28.45	78.0	57.7	23.79	72.8	58.6
Head				3.88	16.0[c]	74.6
Trunk				19.91	56.7[d]	60.8
Entire body	55.66			44.06	150.5	

[a] Distances in the Table are with respect to the cubital axis and the first interphalangeal joint. Right lower arm was 24 cm long, left 25 cm. The distances of the first interphalangeal joint from the hand joint were 12 cm for the right hand, and 10.5 cm for the left.
[b] Foot distances are with respect to the toes and the axis of the upper ankle joint.
[c] Head distances are with respect to the top of the head and the middle of the atlanto-occipital joint.
[d] Trunk distances are with respect to the center of the atlanto-occipital joint and the line connecting the two hip joint centers.

Table 64. Moment of Inertia About the Center of Gravity and Radius of
Gyration Data for Two Cadavers Described by Braune
and Fischer (1892) and Fischer (1906).

Radius of gyration as per cent of segment length (P/l), or
in centimeters (P), moment of inertia (I_{CG}) in gm cm^2 x 10^6.

		1st Cadaver		2nd Cadaver			
		Axis Perpendicular to the Limb Longitudinal Axis		Axis Perpendicular to the Limb Longitudinal Axis		Axis Parallel to the Limb Longitudinal Axis	
		P/l	I_{CG}	P/l	I_{CG}	P	I_{CG}
Head and trunk		0.23	9.17	0.29	10.56		
Trunk				.29	5.57		
Head				.43	0.18		
Entire arm	R	.29	1.04	.31	0.80		
	L	.28	1.11	.30	0.76		
Upper arm	R	.27	0.10	.31	0.08	2.8	0.010
	L	.27	0.12	.29	0.08	2.7	.009
Lower arm and hand	R	.26	0.14	.29	0.12	2.8	.008
	L	.28	0.15	.32	0.15	2.7	.009
Entire leg	R	.30	6.78	.32	4.94		
	L	.30	7.26	.32	4.80		
Upper leg	R	.26	0.70	.31	0.59	4.6	.100
	L	.27	0.81	.31	0.63	4.6	.100
Lower leg and foot	R	.32	0.83	.33	0.62		
	L	.32	0.83	.35	0.64		
Lower leg	R	.25	0.25	.24	0.17	3.1	.020
	L	.26	0.26	.26	0.18	3.0	.018
Foot	R	.30	0.04	.30	0.03	6.2	.035
	L	.31	0.04	.30	0.03	6.2	.035

Table 65. Individual Segment Mass for Eight Cadavers Studied by Dempster (1955a).

Weights are in grams.

Cadaver Number	Body Weight	Trunk minus Limbs	Trunk minus Shoulders	Both Shoulders	Head and Neck	Thorax	Abdomen plus Pelvis
14815	51364	31363	26818	4310	--	--	--
15059	58409	32955	26705	6535	3797	4803	18182
15062	58409	(34558)	(27670)	6888	5227	6136	16364
15095	49886	29300	24431	5743	4348	5341	14515
15097	72500	40568	33409	8039	5337	8754	19187
15168	71364	38369	33377	7229	4850	9053	17237
15250	60455	31558	25909	5708	4371	6620	(14918)
15251	55909	30341	25341	4942	4340	6637	(14364)

Cadaver Number	Entire Lower Extremity	Thigh	Leg and Foot	Leg	Foot	Entire Upper Extremity	Arm	Forearm and Hand	Forearm	Hand
				Left Side						
14815	6255	3495	2602	1961	725	2720	1157	1290	850	445
15059	9855	6482	3384	2629	760	2770	1541	1256	934	325
15062	8290	5520	2835	2080	754	2485	1373	1080	747	332
15095	8313	5285	3041	2218	814	2132	1133	1003	703	317
15097	11907	7093	4846	3860	967	3899	2199	1691	1191	500
15168	11111	6258	4812	3552	1209	3453	1909	1515	1104	417
15250	11337	7700	4045	2991	949	3080	1663	1400	1002	390
15251	(8092)	4660	3432	2564	796	2459	1315	1140	780	339
				Right Side						
14814	6176	3385	2613	1963	655	2641	1212	1342	865	457
15059	9580	6115	3472	2674	800	3277	1920	1340	995	352
15062	8303	5370	2907	2165	746	2695	1528	1134	815	311
15095	7715	4770	2878	2205	767	2125	1123	1024	710	317
15097	11920	7155	4825	3899	924	3947	2171	1777	1250	517
15168	11904	6902	4765	3606	1095	3673	1970	1699	1265	452
15250	11791	7215	3955	2954	865	3035	1614	1414	1021	400
15251	(8092)	5135	3322	2459	808	2394	1372	1017	713	295

Table 66. Specific Gravity of Body Segments [a] (From Dempster, 1955a).

Body Segment		14815	15059	15062	15095	15097	15168	15250	15251	Mean
Trunk minus limbs		1.04	1.05	1.05	1.04	1.03	1.02	1.02	1.00	1.03
Trunk minus shoulders		1.05	1.06	1.00	1.05	1.06	1.03	1.00	1.00	1.03
Shoulders		1.00	1.04	1.02	1.05	1.03	1.04	1.05	1.05	1.04
Head and neck		--	1.12	1.13	1.12	1.10	1.10	1.10	1.11	1.11
Thorax		--	0.81	0.94	0.92	1.01	0.91	0.95	0.90	0.92
Abdomino-pelvic		--	1.00	1.00	1.00	1.02	1.03	1.03	1.00	1.01
Entire lower extremity	R	1.06	1.07	1.07	1.07	1.04	1.06	1.04	1.06	1.06
	L	1.06	1.07	1.05	1.07	1.05	1.06	1.05	0.06	1.06
Thigh	R	1.04	1.06	1.04	1.05	1.05	1.05	1.04	1.05	1.05
	L	1.04	1.06	1.03	1.05	1.04	1.05	1.05	1.05	1.05
Leg and foot	R	1.05	1.11	1.08	1.11	1.07	1.09	1.06	1.07	1.08
	L	1.09	1.11	1.08	1.12	1.07	1.10	1.08	1.08	1.09
Leg	R	1.10	1.12	1.08	1.11	1.08	1.09	1.07	1.09	1.09
	L	1.11	1.12	1.08	1.11	1.08	1.10	1.09	1.09	1.09
Foot	R	1.02	1.17	1.08	1.17	1.12	1.11	1.01	1.07	1.09
	L	1.02	1.17	1.07	1.17	1.14	1.14	1.05	1.09	1.10
Entire upper extremity	R	1.10	1.25	1.07	1.10	1.07	1.10	1.10	1.09	1.11
	L	1.11	1.09	1.08	1.11	1.08	1.10	1.10	1.09	1.10
Arm	R	1.09	1.07	1.06	1.01	1.07	1.09	1.10	1.09	1.07
	L	1.09	1.07	1.06	1.01	1.07	1.08	1.10	1.08	1.07
Forearm and hand	R	1.11	1.18	1.10	1.01	1.12	1.13	1.11	1.13	1.11
	L	1.09	1.14	1.10	1.16	1.11	1.12	1.12	1.13	1.12
Forearm	R	1.14	1.12	1.11	1.20	1.08	1.12	1.14	1.10	1.13
	L	1.14	1.15	1.11	1.18	1.09	1.14	1.14	1.05	1.12
Hand	R	1.14	1.41	1.05	1.33	1.09	1.15	1.13	1.08	1.17
	L	1.14	1.27	1.05	1.28	1.12	1.09	1.10	1.09	1.14

[a] Calculated from mass data on cadaver parts and volumetric data derived from immersion in water.

Table 67. Moments of Inertia About the Center of Gravity
of Body Segments, gm cm^2 x 10^6. (From Dempster, 1955a)

Cada-ver Number	Entire Upper Extremity Right	Entire Upper Extremity Left	Arm Right	Arm Left	Forearm and hand Right	Forearm and hand Left	Forearm Right	Forearm Left	Hand Right	Hand Left	Entire Lower Extremity Right	Entire Lower Extremity Left	Thigh Right	Thigh Left
14815	0.90	1.10	0.190	0.122	0.220	0.187	0.058	0.059	0.007	0.005	4.60	4.90	0.21	0.26
15059	.78	.78	.130	.118	.137	---	.041	.051	.005	.004	6.70	5.70	.70	.64
15062	.99	.96	.102	.115	.180	.155	.055	.043	.004	.005	5.67	6.05	.76	.82
15095	.58	.79	.062	.079	.128	.128	.039	.035	.003	.005	5.20	4.60	.69	.65
15097	1.10	.98	.220	.222	.298	.287	.072	.055	.011	.009	9.20	9.10	1.27	1.14
15168	1.40	1.42	.145	.191	.197	.188	.061	.072	.003	.002	8.60	10.00	3.12	3.29
15250	1.57	1.35	.166	.155	.232	.218	.068	.074	.004	.003	9.50	9.80	1.44	1.22
15251	0.91	1.02	.122	.112	.152	.146	.054	.050	.003	.003	5.20	5.60	0.61	0.61

	Leg and Foot Right	Leg and Foot Left	Leg Right	Leg Left	Foot Right	Foot Left	Trunk Minus Limbs	Trunk Minus Shoulders	Shoulders Right	Shoulders Left	Head and Neck	Thorax	Abdomino-pelvic Region
14815	0.81	0.82	0.307	0.321	0.018	0.021	---	---	---	---	---	---	---
15059	.85	.73	.340	.330	.025	.025	15.5	14.0	0.378	0.355	0.22	0.45	---
15062	.86	.55	.298	.308	.028	.029	23.7	15.9	.324	.500	---	---	---
15095	.75	.73	.275	.260	.013	.026	13.4	13.0	.421	.417	---	---	---
15097	1.65	1.56	.620	.650	.040	.037	22.1	21.0	.700	.800	.31	1.19	3.24
15168	1.64	1.66	.620	.560	.038	.043	24.3	23.1	.800	.520	.23	2.18	9.70
15250	1.40	1.29	.620	.560	.035	.035	14.9	16.4	.425	.425	.32	0.96	2.44
15251	0.96	0.94	.360	.340	.032	.033	14.9	14.0	.420	.480	.39	0.99	1.96

Table 68. Moments of Inertia About the Proximal
Joint Center of Body Segments, gm cm^2 x 10^6.
(from Dempster, 1955a)

Cadaver Number	Entire Upper Extremity Right	Left	Arm Right	Left	Forearm and hand Right	Left	Forearm Right	Left	Hand Right	Left	Entire Lower Extremity Right	Left	Thigh Right	Left
14815	3.50	3.82	0.375	0.334	0.670	0.675	0.183	0.188	0.027	0.028	12.6	13.3	2.08	2.16
15059	2.23	2.60	.278	.372	.443	---	.155	.165	.016	.011	13.4	13.7	2.30	2.33
15062	2.59	2.66	.319	.252	.449	.456	.136	.121	.013	.015	13.6	14.0	2.30	2.25
15095	2.04	2.14	.212	.239	.397	.406	.124	.123	.013	.014	13.2	12.3	2.08	1.78
15097	5.07	4.91	.657	.629	.852	.845	.249	.279	.031	.032	26.0	24.6	3.54	3.44
15168	4.14	4.36	.541	.566	.652	.715	.212	.236	.027	.025	25.5	26.8	5.26	5.53
15250	4.51	4.52	.506	.593	.697	.655	.224	.221	.016	.015	24.5	26.7	4.28	4.35
15251	2.71	2.53	.320	.323	.455	.466	.145	.153	.011	.009	15.0	15.5	1.99	2.27

Cadaver	Leg and Foot Right	Left	Leg Right	Left	Foot Right	Left	Trunk Minus Limbs[a]	Trunk Minus Shoulders[a]	Shoulders[b] Right	Left	Head and Neck[c]	Thorax[d]	Abdomino-Pelvic Region[1]
14815	2.29	2.46	0.895	0.928	.076	.057	---	---	---	---	---	---	---
15059	2.64	2.67	1.130	1.140	.047	.047	42.7	30.4	0.607	0.539	1.31	1.51	---
15062	2.48	2.00	.882	.770	.051	.053	60.5	50.4	.735	.886	---	---	---
15095	2.58	2.12	.902	.805	.013	.046	39.6	32.8	.886	.937	---	---	---
15097	5.09	5.23	2.660	2.530	.086	.063	64.5	50.4	1.530	1.800	1.87	4.73	8.72
15168	5.02	4.89	1.810	1.770	.136	.115	78.0	62.8	1.300	1.090	1.43	6.56	12.60
15250	4.20	3.95	1.970	1.950	.057	.064	60.3	37.7	.827	.799	1.44	3.48	3.96
15251	2.84	2.84	1.180	1.150	.065	.057	50.9	34.6	.961	1.080	1.04	3.12	3.77

[a] Suspension from hip joints.

[b] Suspension from sternoclavicular joint.

[c] Suspension from seventh cervical vertebral body.

[d] Suspension from twelfth vertebral body.

Table 69. Regression Equations for Computing the Mass (in kg)
of Body Segments (From Barter, 1957).

Body Segment	Regression Equation	Standard Deviation of the Residuals
Head, neck and trunk	= 0.47 x Total body wt. + 5.4	(± 2.9)
Total upper extremities	= 0.13 x Total body wt. - 1.4	(± 1.0)
Both Upper arms	= 0.08 x Total body wt. - 1.3	(± 0.5)
Forearms plus hands [a]	= 0.06 x Total body wt. - 0.6	(± 0.5)
Both forearms [a]	= 0.04 x Total body wt. - 0.2	(± 0.5)
Both hands	= 0.01 x Total body wt. + 0.3	(± 0.2)
Total lower extremities	= 0.31 x Total body wt. + 1.2	(± 2.2)
Both upper legs	= 0.18 x Total body wt. + 1.5	(± 1.6)
Both lower legs plus feet	= 0.13 x Total body wt. - 0.2	(± 0.9)
Both lower legs	= 0.11 x Total body wt. - 0.9	(± 0.7)
Both feet	= 0.02 x Total body wt. + 0.7	(± 0.3)

[a] N = 11, all others N = 12.

Measurements on Living Subjects

Several methods of determining body segment characteristics of living subjects have been tried with some success. One technique for determining limb weight is to measure the shift in whole-body CG when only the limb in question is moved. From a knowledge of the limb CG and the measured whole-body shift, it is possible to determine limb weight. Similarly the limb CG can be determined if limb weight is known. A somewhat analogous procedure can be used to determine the moment of inertia of a limb, but these procedures are seldom used, since a highly sensitive balance is required, and the accuracy is unknown. The principal sources of error in these methods include the normal shifts of whole-body CG as demonstrated by Mosso, as well as the necessity of relying on cadaver data for limb weight or CG location.

Dempster's measurements of segment volumes for a variety of body builds are presented in Table 70. Also included in this table are estimated weights of body segments based on volume measurements and corrected by the average density of the corresponding segments (from Table 66). Subjects used in the segment volume determinations were a selected group of male university students. Volume was measured by immersion of limbs to predetermined levels, displaced fluid being equal to the segment volume. Dempster attributed the significant differences between cadaver and the living

Table 70. Ratio of Mean Volume of the Limb Segments to Body Volume (Dempster, 1955a).

Segment	Rotund (Per Cent)	Muscular (Per Cent)	Thin (Per Cent)	Median (Per Cent)
Entire upper extremity	5.28	5.60	5.20	5.65
Arm	3.32	3.35	2.99	3.46
Forearm plus hand	1.95	2.24	2.23	2.15
Forearm	1.52	1.70	1.63	1.61
Hand	0.42	0.53	0.58	0.54
Entire lower extremity	20.27	18.49	19.08	19.55
Thigh	14.78	12.85	12.90	13.65
Leg plus foot	5.52	5.61	6.27	5.97
Leg	4.50	4.35	4.81	4.65
Foot	1.10	1.30	1.46	1.25

Percentage of Body Weight of Segments Calculated from Volume and Specific Gravity Data[a] (Dempster, 1955a).

Entire lower extremity	20.47	18.69	19.29	19.74
Thigh	14.79	12.86	12.92	13.65
Leg and foot	5.68	5.78	6.46	6.14
Leg	4.67	4.52	5.00	4.83
Foot	1.15	1.36	1.53	1.31
Entire upper extremity	5.53	5.87	5.45	5.92
Arm	3.38	3.42	3.05	3.53
Forearm and hand	2.08	2.39	2.38	2.29
Forearm	1.62	1.82	1.74	1.72
Hand	0.46	0.59	0.64	0.60
Trunk minus limbs	47.99	50.89	50.52	48.67

[a]Volumes measured for young living subjects, specific gravity measured for dissected segments of cadavers.

subjects' volumes—namely, that upper arms, thighs, and buttocks were a much larger proportion of total volume in the living—to anthropometric differences between young living subjects and the older cadavers.

In a study of maximum horizontal push and pull forces, Dempster (1955a) used segment volume, an assumed density of 1.00, and cadaver segment *CG* locations to calculate the whole-body *CG* location of three subjects in each of twelve positions. The calculated *CG* was then compared to the *CG* determined by balancing moments about the back edge of a seat. In no instance did the *CG* location found by these alternative methods differ by more than 1 cm.

The relevant topics of body composition, density, and specific gravity will be treated in chapter 4. In the present context, it is pertinent to note that specific gravity or density, as determined by underwater weighing, can be predicted to within 2 per cent from the somatotype component of endomorphy, or fat (Sheldon, Stevens, and Tucker, 1940; Brozek, 1955), and as well or better from the triceps and subscapular skinfolds (Chen, Damon, and Elliot, 1963; Damon and Goldman, 1964), which are much easier to obtain. The prediction formulas for body density based on these two skinfolds will be found in chapter 4.

The moment of inertia of a limb or distal limb segment about its proximal joint can be measured directly by the "quick-release" method. This technique requires a means of measuring force and acceleration. The subject places the test limb against a force-measuring restraining lever and exerts maximum force. The experimenter measures this force, then releases the restraining lever and records the limb acceleration. The force measurement is converted to torque, and the moment of inertia is computed by dividing torque by acceleration. Fenn (1938) reported several measurements obtained in this manner and compared them with calculated values based on Braune and Fischer's data (Table 71). Agreement between the methods was within 7 per cent for all measurements. Bresler and Frankel (1950) used the quick-release method to measure the moment of inertia of the lower leg and shoe, then compared these results with calculated values based on "Fischer's Coefficients" and with calculated values obtained by Weinbach's (1938) method. "Fischer's Coefficients" are the percentage segment weights from Braune and Fischer's (1889) cadaver number IV, the average *CG* locations from cadavers II, III, and IV, and the radius of gyration as a per cent of segment length from Fischer's (1906) cadaver. The average difference between the experimental and "Fischer's Coefficients" values was about 10 per cent (Table 72). Weinbach's method was reported to give values about 10 per cent less than those calculated by the coefficients.

Weinbach's (1938) method for estimating the *CG*

Table 71. Moments of Inertia About the Proximal Joint Centers
Measured by the "Quick Release" Method and
Calculated from Braune and Fischer's Data.

Fenn (1938)	I_o (Gm Cm^2 x 10^6)	
	Measured	Calculated
Forearm (W.B.L.)	0.59	0.56
Leg (M.N.)	5.35	5.74
Leg (W.B.L.)	3.77	3.79
Leg (A.P.)	4.44	4.80
Artificial limb and muscles	2.96	2.99

Bresler and Frankel (1950)	I_o (Gm Cm^2 x 10^6)	
	Measured	Calculated
Lower leg and shoe,		
Subject 1	3.80	3.72
Subject 2	2.72	3.60
Subject 3	3.26	3.50

Table 72. Static Moments and Moments of Inertia of the Lower Leg
and Foot About the Proximal Joint Centers.

Measured by the "Oscillation" Method as Versus
Calculated from Fischer's Coefficients
(from Bartholomew, 1952)

Subject	Height (Cm)	Weight (Kg)	Static Moment (Kg Cm)		I_o (Gm Cm^2 x 10^6)	
			Measured	Calculated	Measured	Calculated
A	175.3	65.8	108.5	118.8	1.19	4.18
B	185.4	68.0	115.2	139.6	2.50	5.37
C	188.0	72.6	138.3	149.9	2.47	6.12
D	172.7	86.2	128.3	151.4	2.41	5.20
E	193.0	77.1	141.0	152.1	3.39	5.71
Amputated leg	160.0	74.8	58.8 [a]	104.0	0.52 [a]	0.95

[a]Measurements were made on the amputated leg 2-3 hours after
amputation. No control measurements before amputation were
reported. Shank weight was 1.86 kg (2.5% of body weight,
foot weight was 0.86 kg (1.2% of body weight).

and moments of inertia of the body or body segments requires "volume contour maps," constructed by direct measurement of volume increment along the length of a body segment, or by approximation from front and side view photographs. The *CG* is obtained by plotting the product of the incremental volume and per cent of total segment length, measuring the area beneath this curve, dividing by the area beneath the volume contour map, and multiplying by total segment length. Moment of inertia is approximated from a third contour map representing the product of incremental volume and length squared, the area of which represents moment of inertia about the end of the segment. For other methods of obtaining volumes and contours from photographs, see pp. 154–157.

Bartholomew (1952) compared results obtained by the pendulum oscillation method with those calculated from "Fischer's Coefficients." Measurements were restricted to the shank and foot of five living subjects, and to one freshly amputated leg. The static moment about the joint center (product of weight and distance of *CG* from joint) was first measured by a simple lever support; then, with the subject relaxed, the free oscillation period of the limb was measured. Bartholomew's measured static moments and moments of inertia, together with corresponding data calculated from the coefficients, are tabulated in Table 72. Agreement was good for the static moment values, but poor for the moment of inertia values for the living subjects. Since agreement between measured values for the amputated leg and the calculated values was also poor, Bartholomew concluded that using "Fischer's Coefficients" was less accurate than the lever support and pendulum oscillation method. However, since the amputated leg weighed less than 2.5 per cent of total body weight (in contrast to 5.5 to 7 per cent reported for cadavers or living persons), there is reason to consider it abnormal. The use of Braune and Fischer's data to predict characteristics of this leg may therefore have been inappropriate.

Duggar (1963) compared moments of inertia of the arm obtained by the pendulum oscillation method with moments of inertia calculated from Braune and Fischer's (1892) radius of gyration data, actual measured arm length, and Dempster's (1955a) volume and specific gravity data. Moments of inertia obtained by the pendulum measurements for ten subjects averaged only 74 per cent of the calculated values. Duggar concluded that the use of the pendulum oscillation method may lead to serious error if the muscles and joints exert a significant damping effect.

Whole-body Measurements

The literature is replete with examples of *CG* determinations on living subjects, but few determinations of whole-body moment of inertia have been reported.

Many of the former studies, directed toward specific objectives, have produced data of limited general value. An early investigation of whole-body CG was that of Reynolds and Lovett (1909). Their method consisted of simply equating moments about a fulcrum and solving for an unknown distance. A diagram of this method is shown in Figure 74. If the scale is zeroed after the platform is placed between points A and C, the simple equation

$$\frac{AC}{AB} = \frac{\text{weight of man}}{\text{weight at scale}},$$

can be solved to give the distance from A of the vertical plane B which passes through the whole-body CG. Similarly, with the subject supine, another plane passing through the CG can be established. The intersection of these planes will determine the CG. However, the results depend on the postures assumed by the subject, are influenced by respiratory movements and body sway in the erect stance, and reflect shifts in the body contents with changes in posture.

Cotton (1932) used the principle of differential tilting (previously described) to avoid inaccuracies resulting from shifts of the viscera and body fluids. His apparatus was sensitive enough to permit accurate measurement with only slight angular change of the body. Cotton reported that for supine subjects, fully clothed and wearing shoes, the mean height of the CG above

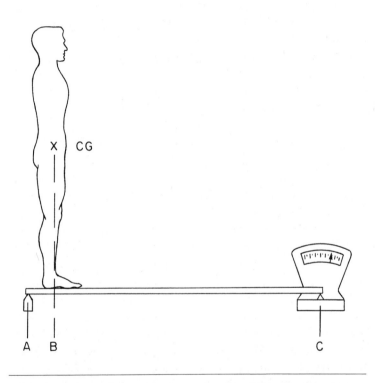

74. The platform resting on fulcrum at A and scale at C is used to determine the location of B, the vertical plane passing through the body CG.

the table was 10.9 cm for men, and 10.1 cm for women.

Åkerblom (1948) studied movements of the vertical projection of the *CG* line in the sagittal plane by use of a "statograph," consisting of a platform, the displacement of which was related to the *CG* of the load, and means for recording this displacement. Åkerblom also took radiographs of the subjects to relate the *CG* line to various skeletal locations. Using fully clothed standing subjects, Åkerblom found that in the sagittal plane the average individual *CG* line varied between 1 and 5 cm in front of the ankle joints, with continual oscillations of from 1.5 to 5 cm around these average values. In the frontal plane the oscillations of the *CG* line were from 1.5 to 3.5 cm. When the subjects were standing comfortably, the *CG* line was 1.6 cm, on the average, in front of the knee joint.

Extensive measurements on whole-body *CG* locations have been reported by Swearingen (1953), who tested five subjects, representing a wide range of body sizes and weights, in 67 positions. An additional twenty-seven subjects were used to check the ranges established by the original five in two of the positions. Ninety-six per cent of the centers of gravity of the additional twenty-seven subjects fell within the experimentally determined range for the sitting position, and 89 per cent within the range for the standing position. A selected few of Swearingen's findings are summarized in Table 73, and anthropometric data for his five sub-jects in Table 74. The apparatus used by Swearingen (described above) required that the subject be rotated through an angle of about 20 degrees between measurements, which undoubtedly caused some displacement of the movable viscera and body fluids. In addition, some measurements of the *CG* location for vertical stances were derived from measurements with the subject horizontal. The resultant shift of the *CG* and loss of accuracy are undetermined.

It is of interest to note that Braune and Fischer's measurements of whole-body *CG* for their cadaver Number 4 in a standing position (92.5 cm from the floor) fell outside Swearingen's "range" (97.2 ± 3.8 cm). A calculated *CG* for this cadaver sitting with seat 90 degrees to back, and legs 90 degrees to thighs, is 21 cm from the back of the chair and 21 cm above the seat. Swearingen's measured range for a similar seated position was a 2.2-cm circle about a point located 21.3 cm from the back, 23.8 cm from the seat. Consequently the cadaver data also fall outside Swearingen's seated range. These discrepancies are particularly important, since the data from Braune and Fischer's cadaver Number 4 have been widely quoted, as by Lay and Fisher (1940, Fig. 23), Steindler (1955, p. 112), and Dreyfuss (1960). Number 4 had 14.1 per cent of total body weight concentrated in the lower legs and feet. Such a large percentage doubtless resulted in an unusually low whole-body *CG*.

Table 73. Locations of the Centers of Gravity of Man (5 Subjects).
(from Swearingen, 1953)

Table 73. Continued

Body Positions and Reference Points	Location of Average CG in Cm from Reference Points (Horizontal, Vertical)		Radius (Cm) within Which the CG for All Subjects (N=5) Is Located
Sitting with seat 90° to back, legs 90° to thighs. Distances measured horizontally from the seat back and vertically from the seat bottom			
Both arms down at sides.	21.3	23.8	2.2
Both arms in lap	22.5	24.4	2.2
One arm forward, one hand in lap	23.5	25.7	1.6
Both arms extended straight forward.	24.8	27.3	2.2
Both arms extended over head	22.5	30.5	1.9
One arm over head, one hand in lap	21.9	27.3	1.9
Both arms extended laterally	20.6	27.3	1.9
One arm extended laterally, one hand in lap	21.3	26.0	1.9
Trunk flexed on thighs, arms extended forward	38.4	13.2	2.9
Sitting back erect, seat 90° to back, legs 50° to thighs Distances measured from seat back and bottom			
One hand on stick control, one hand on control at side of seat	19.7	25.1	2.2
Trunk flexed on thighs, arms around knees	34.0	12.4	2.5
Sitting back erect, seat 90° to back, legs 110° to thighs Distances measured from seat back and bottom			
One hand on stick control, one hand on control at side of seat	23.0	25.1	2.2
Both hands on overhead control	25.1	26.8	2.9
Sitting back erect, seat 108° to back, legs 180° to thighs. Distances measured from seat back and bottom			
One hand on stick control, one hand on control at side of seat	25.1	20.6	2.5
Both hands on overhead control	26.8	22.7	2.2
Seated in commercial airline seat in full reclining position. Vertical distances measured from the ischium parallel to the plane of the subject's back, horizontal distances perpendicular from this plane			
Trunk 115°, knees 145°, hands in lap	23.8	19.7	2.2
Trunk 115°, knees 145°, both arms extended forward	27.3	20.6	2.2
Knees 145°; head and arms on forward seat	31.1	26.5	1.6
Trunk 115°, knees 90°, hands in lap	20.0	17.1	2.5
Lateral deviations: effect on CG measurements are horizontal distances from the mid-sagittal line and vertical height from the ischium (in all tests the pelvis was not moved)			
Standing, body straight, arms at sides	0.0	14.9	2.2

Table 73. Continued

Head flexed to side	1.3	14.6	1.9
One arm extended laterally, body straight	1.3	16.2	1.6
One leg abducted	3.8	17.1	1.9
All body parts moved laterally	11.7	18.4	3.5

Vertical shifts of the CG, measured from the ischium, accompanying maximum abduction of arms and legs

Standing, body straight, both arms abducted		17.9	2.2
Both legs abducted, arms at sides		17.5	1.9
Both legs and both arms abducted		21.0	1.9

Vertical distances of the CG, measured from the floor level

Body standing straight		97.2	3.8
Standing, both arms abducted		101.0	3.8
Standing, both legs abducted		92.1	5.7
Both arms and both legs abducted		95.6	5.4

Table 74. Anthropometric Measurements [a] of the Five Subjects. (from Swearingen, 1953)

	J.	B.	M.	N.	T.
Age	39	39	29	60	39
Weight	68.9	68.9	102.	80.3	51.4
Stature	172.7	182.9	177.2	176.5	164.5
Sitting height	88.3	95.2	92.7	94.0	85.1
Trunk height	58.4	62.2	61.0	57.2	58.4
Eye level	76.2	81.9	78.7	79.4	72.4
Buttock knee	58.4	62.2	62.2	59.7	57.2
Patella height	53.3	56.5	55.9	52.7	50.2
Abdominal girth	76.8	73.7	96.5	88.9	66.0
Thigh circumference	47.6	45.7	61.0	52.0	39.4
Chest depth	20.3	21.0	27.3	24.1	17.1
Abdominal depth	19.7	19.0	25.4	24.1	16.5

[a] Age in years; weight in kg; all others in cm.

Whole-body measurements have also been used to locate the *CG* of man-equipment combinations. For example, Hertzberg and Daniels (1950) reported measurements for the *CG* of a fighter plane (F-86) ejection seat and pilot with normal equipment. Their procedure was simple, involving only the suspension of the loaded seat alternately in two positions. Plumb bobs from the points of suspension were projected onto an attached reference surface, and the point of intersection used to locate the *CG*. Such measurements can be made inexpensively and precisely, but should not be generalized to apply to all persons unless a wide range of body size is tested. Based on nine subjects representing the heights and weights of virtually 100 per cent of Air Force flying personnel (heights ranged from 165 to 186 cm, weights from 54.4 to 88.9 kg), the combined man-seat *CG* was found to fall within a 2.8-cm radius of the average location. The *CG* shifted forward of the seat reference point with increasing weight of the subject, on a line approximately 30 degrees to the seat pan. Variations in buttock position on the seat could shift the *CG* location as much as 1.25 cm.

The range of *CG* locations measured by Swearingen (1953) for seated subjects (2.25 cm horizontal and vertical range) is reasonably close to that given by Hertzberg and Daniels, namely 2.8 cm horizontal and vertical range, although the latter study included about 27 kg of equipment (flying clothes, helmet, seat-style sus-tenance kit, and back-pack parachute) plus the ejection seat itself. The effect of personal equipment and clothing would probably be to increase the variation in *CG* locations, since different habits of wearing this gear induce variations in seated position and posture, and since the added weight of helmet and boots at the extremities influences the whole-body *CG* location. On the other hand, the inclusion of the ejection seat, with a fixed *CG*, provides a stabilizing effect on the man-seat *CG* location.

Santschi, DuBois, and Omoto (1963) reported individual *CG*'s and moments of inertia, as well as fifty anthropometric measures, for sixty-six living subjects whose heights and weights were representative of those reported for male Air Force flying personnel (Hertzberg, Daniels, and Churchill, 1954). Oscillation frequencies of a compound pendulum suspended alternately from two parallel axes were used to compute the *CG* locations and moments of inertia about all three axes (x-axis represented by the intersection of sagittal and transverse planes through the *CG*, y-axis represented by the intersection of the frontal and transverse planes, and z-axis represented by the intersection of the sagittal and frontal planes). Mean values and standard deviations are given in Table 75, together with multiple correlations of the moments of inertia with height and weight. The high correlation coefficients indicate that, for subjects similar to those studied, useful predictions

Table 75. Mean Whole-Body Centers of Gravity and Moments of Inertia, Together with Correlation Coefficients and Regression Equations Relating Stature and Weight to Moment of Inertia (N = 66, from Santschi et al., 1963).

Position	Axis	Center of Gravity [a] (Cm) Mean	S.D.	Moment of Inertia (Gm Cm²x10⁶) Mean	S.D.	$R_{1.sw}$	Moment of Inertia Regression Equations [b] (Gm Cm² x 10⁶) S.E.	
Standing (arms at sides)	X	8.9	0.51	130.0	21.8	.98	4.73	-262.0 +1.68S +1.28W
	Y	12.2	0.99	116.0	20.6	.96	5.96	-240.0 +1.53S +1.15W
	Z	78.8	3.68	12.8	2.5	.93	0.95	-0.683 -0.044S +0.279W
Standing (arms over head)	X	8.9	0.56	172.0	29.5	.98	6.36	-371.0 +2.39S +1.63W
	Y	12.2	0.99	155.0	28.6	.96	7.79	-376.0 +2.38S +1.47W
	Z	72.7	3.38	12.6	2.1	.86	0.98	1.6 -0.038S +0.234W
Spread Eagle	X	8.4	0.48	171.0	30.6	.98	5.54	-399.0 +2.51S +1.69W
	Y	12.2	0.99	129.0	24.1	.96	7.06	-305.0 +1.91S +1.29W
	Z	72.4	4.82	41.4	8.9	.93	3.19	-114.0 +0.677S +0.484W
Sitting (elbows at 90°)	X	20.1	0.91	69.1	10.6	.92	4.53	-104.0 +0.637S +0.804W
	Y	12.2	0.99	75.4	13.1	.92	5.10	-153.0 +1.01S +0.669W
	Z	67.3	2.89	37.9	6.6	.97	1.64	-59.6 +0.34S +0.502W
Sitting (forearms down)	X	19.6	0.86	70.5	11.0	.91	4.50	-89.0 +0.574S +0.771W
	Y	12.2	0.99	77.0	13.6	.92	5.28	-144.0 +0.913S +0.802W
	Z	68.1	2.95	38.2	6.7	.97	1.54	-60.8 +0.341S +0.514W
Sitting (thighs elevated)	X	18.3	0.94	44.2	6.8	.89	3.16	-38.2 +0.242S +0.529W
	Y	12.2	0.99	43.0	6.6	.77	4.14	-25.1 +0.193S +0.449W
	Z	58.7	1.98	29.7	5.8	.92	2.26	-34.4 +0.146S +0.509W
Mercury Position	X	20.1	0.86	74.4	10.6	.93	4.24	-107.0 +0.699S +0.768W
	Y	12.2	0.99	85.1	15.8	.94	5.61	-198.0 +1.27S +0.794W
	Z	68.8	2.89	38.7	6.3	.96	1.85	-50.9 +0.297S +0.492W
Relaxed (weightless)	X	18.5	0.84	104.0	15.0	.96	4.20	-120.0 +0.788S +1.13W
	Y	12.2	0.99	99.8	15.0	.94	5.13	-157.0 +1.08S +0.879W
	Z	69.9	3.66	40.6	6.1	.96	1.74	-53.4 +0.346S +0.440W

[a] Location of CGs are with respect to the back plane, anterior superior spine of the ilium, and top of the head.
[b] S is stature in centimeters; W is weight in kilograms.

of moments of inertia can be made from height and weight alone.

The theoretical accuracy of the compound pendulum used by Santschi *et al.* was reported to be between 2 and 8 per cent for moments of inertia, depending on the body position and axis, and about 0.5 per cent for the *CG* location. The amplitudes of the pendulum oscillations were small ($\pm 1^0$ arc), the lengths of the pendulum arms long. Consequently angular changes in body position during measurement were small enough to avoid significant shifts in body contents. Periods of oscillations were long with respect to respiration rate, so that errors due to this source should be negligible. Measurements about the x- and y-axes were made with the subject upright, whereas measurements about the z-axis were made with the subject supine. The extent to which shifts in the gravity vector might affect the moments of inertia was not reported.

Weinbach (1938) reported a number of "relative" whole-body moment of inertia values obtained by his volume contour technique, using 100 as the total body height. To convert these relative values to actual moments of inertia requires knowledge of body heights, which were not reported. Weinbach observed that the relative moment of inertia decreases with age, while the relative whole-body *CG* location is fairly stable as the subject ages. Other investigators have derived whole-body moments of inertia from Braune and Fisch-

Table 76. Whole-Body Moments of Inertia
about the Center of Gravity.
(from Noskoff, 1942)

Calculations based on a man with a total height of 178 cm,
and total weight of 77 kg

Estimated weight distribution as per cent of total weight:

Head --------------	6.18	Hands (2) --------	1.65
Trunk -------------	43.82	Upper legs (2) ---	21.76
Upper arms (2) ----	6.76	Lower legs (2) ---	13.54
Lower arms (2) ----	6.29		

Whole-body moments of inertia (gm cm^2 x 10^6) about the
CG. I_h is for lateral rotation of the body about the
horizontal axis. I_o is for posterio-dorsal rotation
about the lateral axis. I_v is for rotation about the
vertical axis.

	I_h	I_o	I_v
Upright standing position, arms at sides	169.6	164.6	9.2
Seated pilot position, both hands on wheel	64.1	93.6	46.6
Seated radio operator position, arms on table	61.2	86.1	39.5
Standing, machine gun operating position	133.2	130.8	21.5
Seated bombardier, crouched forward	46.3	63.8	31.7
Seated infantry man with 30 pounds of equipment and cartridges, arms on thighs	71.1	85.2	40.2

er's data, from treating the human body as a set of geometrical shapes, or from measurements on manikins. The accuracy of such procedures remains to be determined.

Noskoff (1942) calculated whole-body moments of inertia about the center of gravity axes for six positions, assuming a total body weight of 77 kg and height of 178 cm. The weight distribution used and the calculated moments of inertia are listed in Table 76. The accuracy of Noskoff's estimates has not been verified, but the segmental weight distribution used is decidedly different from that obtained by Braune and Fischer or by Dempster. Noskoff reported that some of the weights of the body segments were obtained from Steindler (1935), and that segment weights not mentioned by Steindler were estimated. Steindler himself (1935: page 26, Table 1) credits the segment weights and moments of inertia values as "from Jules Amar," 1920.

APPLICATION OF DATA

The selection of the appropriate method for determining the human *CG* or moment of inertia depends on three interacting factors: the accuracy required, the characteristics of the population to which the data are to apply, and the body positions for which the *CG* or moment of inertia is to be located. These considerations indicate the approach to the problem and limitations of the solution. A rational approach to practical problems involving the determination of *CG* or moments of inertia is outlined in the following paragraphs.

(1) Establish the accuracy required. The importance of establishing criteria of accuracy on a sound functional basis cannot be overemphasized. Considerations of accuracy may not only dictate the appropriate method of determining *CG* or moment of inertia, but—when stringent—may limit the population to which the results apply, as well as the equipment and postures which this population may assume. In other cases the accuracy required may be so slight as to permit the use of "guesstimates" based on published data.

(2) Determine the characteristics of the population. The average location of the whole-body *CG* or moment of inertia of truck drivers may differ significantly from that of clerical workers or basketball players. The range of values containing any given percentage of a group may likewise vary from one group to another. In some cases the range may be so great that the population must be divided into anthropometrically distinct subgroups. At present no extensive survey of whole-body *CG* locations or moments of inertia related to body build or somatotype has been published. Swearingen's (1953) data show that unusual distributions of weight between the trunk and legs cause unusual shifts in the *CG*, but the data are inadequate to permit generalization. Consequently, the accurate use of tabulated *CG* data requires that the measured subjects match the using population anthropometrically.

The importance of anthropometric matching can be illustrated by calculating the *CG* location of a seated individual (upper legs horizontal; back, lower legs, and arms at 90 degrees) using segment masses obtained from two different sources. Segment weights for a thin subject weighing 56 kg, calculated from Barter's regression values (Table 69) and from Dempster's volume-equivalent segment weights (Table 70) are listed in Table 77, together with the vertical and horizontal distances of segment *CG's* from a reference point. The *CG* locations, with respect to the reference point, are calculated by separately summing the products of segment weights with horizontal and vertical distances, and then dividing by total body weight. The two calculated *CG* locations differ by nearly 2.1 cm because of the wide variation in age and body build of the subjects.

The volume measurements, corrected for density, in the lower portion of Table 70 are particularly useful in estimating segmental weights for specific constitutional types. The "median" body build corresponds to the most prevalent group of body builds among Air Force fighter pilots as described by Hertzberg *et al.* (1954).

If moment of inertia values are to be calculated from segment data, the weights should be calculated from data most closely resembling the person to whom the

Table 77. Comparison of Calculated Locations of the Whole-Body Center of Gravity.
Weight Distribution from Barter (1957) and Dempster (1955a)

CG calculated for a seated 56 kg man, upper legs horizontal, lower legs and arms at 90°.
Reference point at floor level and directly under head and trunk CG.

	Horizontal (X) and Vertical (Z) Distance of CG Locations from Reference Point (Cm)		Segment Weight (W) in Kg		Static Moments			
Segment	X	Z	W_1 from Table 69	W_2 from Table 70	W_1X	W_2X	W_1Z	W_2Z
Feet (2)	46.5	3.0	1.82	1.71	84.63	79.52	5.46	5.13
Lower legs (2)	40.0	31.0	5.26	5.60	210.40	224.00	163.06	173.60
Upper legs (2)	17.0	48.0	11.58	14.47	196.86	245.99	555.84	694.56
Hands (2)	23.0	65.0	.86	.72	19.78	16.56	55.90	46.80
Lower arms (2)	8.5	65.0	2.04	1.95	17.34	16.58	132.60	126.75
Upper arms (2)	0.0	80.7	3.18	3.42	0.0	0.0	256.63	275.99
Head and Trunk	0.0	80.0	31.72	28.29	0.0	0.0	2537.60	2263.20
			56.46	56.16	529.03	582.65	3707.09	3586.03

Whole-body CG coordinates with respect to reference point:

$$X_1 = \frac{\Sigma W_1 X}{\Sigma W_1} = \frac{529.03}{56.46} = 9.37 \text{ cm}, \qquad Z_1 = \frac{\Sigma W_1 Z}{\Sigma W_1} = \frac{3707.09}{56.46} = 65.66 \text{ cm}$$

$$X_2 = \frac{\Sigma W_2 X}{\Sigma W_2} = \frac{582.65}{56.16} = 10.37 \text{ cm}, \qquad Z_2 = \frac{\Sigma W_2 Z}{\Sigma W_2} = \frac{3586.03}{56.16} = 63.85 \text{ cm}$$

Difference between calculated CG locations:

$$X_1 - X_2 = 1.0 \text{ cm}, \qquad Z_1 - Z_2 = 1.81 \text{ cm}$$

$$\sqrt{(1.0)^2 + (1.81)^2} = 2.07 \text{ cm difference}$$

results will apply. Segment lengths for persons of vary-ing constitution are not available. Therefore, for maxi-mum accuracy, segment lengths must be obtained from the actual using population (or from photographs). Since the moment of inertia is proportional to the square of the length of the segment, accuracy of length measurement is particularly important. If volume measurements for the subject in question can be ob-tained, or if nude front and side view photographs are available, Weinbach's volume contour method can be used.

(3) Determine the relevant body positions or range of positions. Swearingen's (1953) data indicate the extent to which body posture affects CG location and range, with respect both to fixed points in the environ-ment and to anatomical landmarks of the subject's body. Positions requiring bending of the spinal column, particularly maximum flexion, increased the variability of CG location. Calculations based on segment data will also be subject to error with postural changes, since trunk reference points will be more difficult to locate, and since displacement of trunk contents will alter the distribution of mass.

Shifts of the CG from a measured location caused by limb movements can be accurately calculated from seg-ment data. For maximum accuracy, photographs of the subject which clearly indicate the limb joint locations should be obtained, with reference points marked on the subject before photography. Horizontal and verti-cal referents as well as a scale should appear in the photographs. Locations of the segment CG's can be approximated by using data from the Braune and Fischer cadaver (Tables 62, 63) which most closely resembles the subject in body build, or by using Demp-ster's average values (Table 61). If the subject's build closely resembles that of any of the cadavers, the cadaver's segmental percentage weight should be used.

A shift in the whole-body moment of inertia because of change in posture or movement of the limbs will be equivalent to the algebraic sum of the individual seg-ment changes in moment of inertia about the axis of rotation. The contribution of each segment to total moment of inertia is determined by the equation:

$$I_{\text{Total}} = (I_{CG} + mx^2),$$

where $I_{CG} =$ moment of inertia about segment CG,
       $m =$ mass of the segment,
       $x =$ distance of the segment CG from the axis of rotation.

Consequently the change in I_{Total} is the difference between the sum of mx^2 before and after change in posture.

SUMMARY

Tabular data concerning the range of locations of the center of gravity (*CG*) of subjects in a variety of positions may be used to specify the location of an individual *CG* within a radius of 1.75 to 9 cm for a large range of body builds. Locations for positions not included in the tables can be calculated by use of body segment data. Available data on mass distribution of body segments have been derived from dissected cadavers and from volume measurements on living subjects. Significant differences exist between the two sets of data because of differences in age and body build between the two groups of subjects. Application of the appropriate data will depend on the anthropometric characteristics of the population under study.

Location of the *CG* of body segments is based solely on cadaver data, but errors in these locations when applied to populations differing anthropometrically should not be significant in comparison to the segment mass errors. Calculations based on appropriate segment data and applied to particular subjects whose bone length and posture can be measured should be accurate to within 1 cm.

The *CG* of living persons can be determined by use of a platform or chair suspended separately from two different sets of locations. Accuracy of such measurements can be made to within 0.7 cm under ideal laboratory conditions, but will be subject to larger error if the subject wears bulky equipment or maintains the attitude for long periods of time, with consequent position shifting. Since the *CG* varies dynamically with respiration, acceleration, and even with the heart beat, static determination of *CG* has certain limitations. Within these limitations, it can provide useful data.

Whole-body moments of inertia for various positions and axes of revolution can be calculated from segment data, or may be determined by direct measurement using the pendulum oscillation method. With the subject in the desired postural attitude and resting on a chair or platform suspended so as to permit free oscillation, a highly accurate moment of inertia may be determined. Accuracy of whole-body moment of inertia calculated from segment data has not been determined, but the known errors in calculated segment mass and *CG* location indicate that the principle source of inaccuracy will be in the contribution of the trunk and head to moment of inertia. Calculations based on volume contours have not been evaluated for accuracy, but should be limited only by the accuracy of volume measurement corrected for segment density.

For practical purposes, the *CG* and moment of inertia must be used in the light of the accuracy required, anthropometric characteristics of the population under consideration, and equipment and body positions anticipated.

Biomechanics and Equipment Design —— III

1. Introduction

Biomechanics includes the range, strength, and speed of human movements, as well as body composition and response to such physical forces as acceleration and vibration. Biomechanical data are useful in control and tool design, in workspace and task layout, and in the protection of personnel against mechanical force. Biomechanical abilities have been studied chiefly among small groups of white males, American or European. Comparable data for females, other racial groups, or for specific ages or occupations are few or nonexistent. This chapter is not intended to be a comprehensive survey of all existing data in the field, but a summary of those aspects of biomechanics that are most immediately applicable to equipment design.

Where biomechanical measurements were obtained under different conditions from the relevant work situation, in respect to the placement of operator or control, the type of control, or forces exerted by or on the operator, solutions can usually be obtained by extrapolation from existing data, with the addition of a safety factor. With a unique design, or where precise limits of biomechanical abilities are needed, a special study may be required.

2. Range of Motion at the Joints of the Body

APPLICATIONS OF JOINT MOVEMENT DATA

Data on the range of joint movement help the designer plan the placement and excursion of controls as well as the operator's movements in performing tasks. To maintain efficiency, required body movements should be kept well within the limits of comfort.

LINKS AND JOINTS

Body joints are formed wherever two or more bones articulate. Immovable joints, like those between the skull bones, are not relevant here. The movable joints, strengthened by means of ligaments (tough, fibrous bands), are of several different types, the three most important being hinge joints (fingers, knees), pivot joints (elbow—also a hinge joint), and ball-and-socket joints (shoulder and hip). The range of joint motion is determined by the bony configuration, by the attached muscles, tendons, and ligaments, and by the amount of surrounding tissue, all of which vary to some extent from person to person and from joint to joint.

Bones articulated at movable joints may be regarded as the rigid levers of the body's mechanical system. These levers or movable body segments may be likened to the engineer's "links," which are intermediate rods or pieces transmitting force or motion. Body links, however, are not identical to mechanical links; they are functional rather than structural dimensions and may be defined as the spanning distance between centers of rotation of adjacent joints (Fig. 75). A body link may change its length to a variable extent. The humeral link, or upper arm between the shoulder and elbow joints, for example, may vary in length by $1\frac{1}{4}$ inches or more depending on arm position.

The human body is, in the foregoing terminology, an open chain system of links rotating around joint centers. The end members of these open chain links, the hands and feet, can occupy a wide variety of positions in space as a result of the cumulative ranges of the intervening joints (Dempster, 1955a and 1955b).

FACTORS INFLUENCING THE RANGE OF
JOINT MOVEMENT

The range of joint motion varies from person to per-

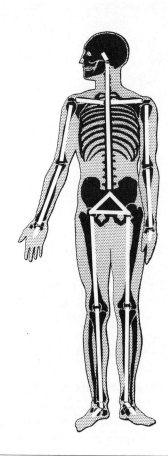

75. Body links and joints—a simplified kinematic system. The detailed individual links of the vertebrae, fingers, toes, etc., are not shown (after Dempster, 1955a).

son anatomically, as mentioned above, and also because of other factors which the designer may need to consider when dealing with specialized populations.

Age: Joint mobility decreases only slightly between age 20 and age 60, barring arthritis or other disease. Between the 1st and 7th decades joint mobility declines about 10 per cent (West, 1945), but no significant differences between youth and normal middle age have been found (Salter and Darcus, 1953; Hewitt, 1928). The prevalence of arthritis (degenerative, senile, or osteoarthritis) increases so markedly beyond age 45 that any older population will have a considerably decreased average joint mobility (Smyth *et al.,* 1959).

Sex: Women exceed men in the range of movement at all joints but the knee. Differences vary from minor increases to as much as 14 degrees at the wrist (Sinelnikoff and Grigorowitsch, 1931). See Table 82.

Race: There may be racial differences in joint mobility, but no data are currently available.

Body build: Slender men and women have the widest range of joint movements, fat ones the smallest. Average and muscular body builds, in that descending order, are intermediate (Barter, Emanuel, and Truett, 1957). These differences may reach practical significance, especially those between the thin and the fat groups, where variations of more than 10 degrees in a given movement are not uncommon (Sinelnikoff and Grigorowitsch, 1931).

Exercise: Any joint of the body tends to become restricted in movement if it is not used regularly within the limits of its normal range (Keegan, 1964). Conversely, physical exercise may increase the range of motion of a joint. However, excessive exercise can result in the so-called "muscle-bound" condition, which increases bulk and limits joint excursion.

Occupation: Some specialized tasks involve the repetition of certain body movements. As a result, the range of movement at the affected joints will tend to increase.

Fatigue: Severe fatigue will restrict the effective range of joint motion by decreasing not only motivation but muscle strength as well.

Disease: Arthritis, poliomyelitis, or other diseases affecting the joints, muscles, or nervous system can severely restrict body movements or completely immobilize a joint.

Motivation: Motivation influences the limits of joint motion by determining the effort exerted to attain the maximum amount of movement.

Right versus left side: There is normally so little variation that the two sides can be considered identical (Gilliland, 1921). In arm rotation, for example, group differences between left and right ranged from 0 to 5 degrees (Salter and Darcus, 1953).

Body position: The range of movement of one part of the body is affected by the position or movement of neighboring parts; thus, hand rotation can be considerably increased if shoulder movements are added to those at the elbow. Wrist flexion is greater with the hand pronated than supinated. Prone position movements may not coincide with those made in the erect position.

Zero gravity: Whereas all normal body movements may be possible under zero gravity (Dzendolet and Rievley, 1959), the effect of altered proprioception on joint movement is problematical.

Clothing and personal equipment: Light clothing has little effect on joint movement, but bulky clothing such as cold-weather or flying gear considerably reduces the range of motion (Saul and Jaffee, 1955; Nicoloff, 1957). The Army Arctic uniform markedly restricts movements at the neck, shoulder, arm, waist, and crotch—shoulder flexion, for example, is reduced by over 20 degrees (Dusek, 1958). It should be remembered that the joint motion values presented below are for nude or lightly clothed subjects.

MEASURING JOINT MOTION

Techniques

Joint motion is measured at the angle formed by the long axes of two adjoining body segments (link lines), or, in some cases, at the angle formed by one body segment and a vertical or horizontal plane. The total range of movement is measured between the two extreme positions of the joint, by a pendulum or protractor goniometer or by an "electrogoniometer" (Karpovich, Herden, and Asa, 1960).

Discrepant values reported by various investigators arise from differing amounts of "forcing" by the subject, as in active versus passive joint movement; from variations in body position (such as prone versus supine, or with forearm pronated versus supinated); and from differences in the joint component moved. In wrist flexion and extension, for example, the forearm may be held rigid and the hand moved, or the hand held rigid against a flat surface and the forearm moved. Readings are higher in the latter case, though controlled comparative studies are lacking.

Subjects

The joint movement data in Tables 78–82 are, for the most part, those obtained by Dempster (1955b) and reanalyzed by Barter, Emanuel, and Truett (1957), on thirty-nine college students selected to represent the major physical types in the military forces. These subjects include ten "thin," eleven "muscular," seven "rotund," and eleven "median" individuals. Although the group includes larger proportions of extreme types than occur in the general male population, the average joint movement values are probably not markedly affected. Sinelnikoff and Grigorowitsch (1931) measured one hundred "healthy" European male workers and one hundred "healthy" European females for the only satisfactory study of women. Glanville and Kreezer's (1937) data come from only ten "healthy" males. The functional wrist movement data of Daniels and Hertzberg (1952) were obtained on a U.S. Air Force population of seventy-nine men (sixty-six for some measurements).

The Data

The ranges, in angular degrees, of each of the types of *voluntary* movement possible at the joints of the body are presented below in Tables 78 through 82. The eight types of movement (Figs. 76, 77, and 78) are:

Flexion: bending, or decreasing the angle between the parts of the body.

Extension: straightening, or increasing the angle between the parts of the body.

Adduction: moving toward the midline of the body.

Abduction: moving away from the midline of the body.

Medial Rotation: turning toward the midline of the body.

Lateral Rotation: turning away from the midline of the body.

Pronation: rotating the forearm so that the palm faces downward.

Supination: rotating the forearm so that the palm faces upward.

Table 78. Range of Movement at the Joints of the Arm and Hand [a].

Joint	Type of Movement [b]	Range of Movement (degrees)			
		Mean	S.D.	5th Percentile [c]	95th Percentile [c]
Wrist	Flexion [d]	90	12	70	110
	Extension [d]	99	13	78	120
	Total [d]	189	21	154	224
	Adduction [d]	27	9	12	42
	Abduction [d]	47	7	35	59
	Total [d]	74	13	53	95
Forearm	Supination [e]	113	22	77	149
	Pronation [e]	77	24	38	116
	Total [e]	190	30	141	239
Elbow	Flexion	142	10	126	158
Shoulder	Flexion	188	12	168	208
	Extension	61	14	38	84
	Total	249	19	218	280
	Adduction	48	9	33	63
	Abduction	134	17	106	162
	Total	182	20	149	215
	Rotation:medial	97	22	61	133
	Rotation:lateral	34	13	13	55
	Total	131	24	92	170

[a] Barter, Emanuel, and Truett, 1957: 39 male subjects representing varied types of body build.

[b] See Figure 76.

[c] Computed from the Standard Deviation.

[d] These are "forced" movements in that the hand is physically restrained and the forearm then rotated about the wrist joint. Normal movements, in which the hand is rotated about the wrist, would have less excursion.

[e] Elbow at 90° angle.

Table 79. Range of Wrist Flexion and Extension While
Grasping a Control.

(Daniels and Hertzberg, 1952)

Subjects and Test Conditions: 79 male subjects, average age 28
years, representing varied body builds grasped a vertical handgrip located approximately 19 inches forward and 13 1/2 inches
above the Seat Reference Point. Sixty-six subjects were used to
determine the extreme limits. Flexion and extension of the right
wrist were measured from the "neutral" or resting position of the
handgrip selected as most comfortable by each subject. This position
averaged 19° to the left of a midsagittal (fore-and-aft) plane.

Type of Movement	Range of Movement (Degrees)			
	Mean	S.D.	Percentile[a] 5th	Percentile[a] 95th
Flexion--to left of neutral position				
Comfortable, usable limits	46.0	15.7	20	72
Extreme possible limits	91.0	16.6	64	118
Extension--to right of neutral position				
Comfortable, usable limits	33.6	13.7	11	56
Extreme possible limits	71.8	16.0	46	98
Total movement, extension - flexion				
Comfortable, usable limits	76.6	23.3	38	115
Extreme possible limits	164.2	22.6	127	201

[a] Computed from S.D.

Table 80. Range of Movement at the Joints
of the Leg and Foot[a].

Joint	Type of Movement[b]	Range of Movement (Degrees)			
		Mean	S.D.	5th Percentile[c]	95th Percentile[c]
Ankle	Flexion	35	7	23	47
	Extension	38	12	18	58
	Total	73	14	50	96
	Adduction (or inversion)	24	9	9	39
	Abduction (or aversion)	23	7	11	35
	Total	47	13	26	68
Knee	Flexion(standing)	113	13	92	134
	Flexion(kneeling)	159	9	144	174
	Flexion (prone)	125	10	109	141
	Rotation:medial	35	12	15	55
	Rotation:lateral	43	12	23	63
	Total	78	16	52	104
Hip	Flexion	113	13	92	134
	Adduction	31	12	11	51
	Abduction	53	12	33	73
	Total	84	14	61	107
	Rotation:medial (seated)	31	9	16	46
	Rotation:lateral (seated)	30	9	15	45
	Total	61	14	38	84
	Rotation:medial (prone)	39	10	23	55
	Rotation:lateral (prone)	34	10	18	50
	Total	73	16	47	99

[a] Barter, Emanuel, and Truett, 1957: 39 male subjects representing varied
types of body build.
[b] See Figure 77.
[c] Computed from standard deviation

Table 81. Range of Movement at the Neck.

Type of Movement [a]	Range of Movement (Degrees)	
	Mean	S.D.
Flexion (ventral) [b]	60	12
Flexion (ventral) [c]	67	9
Flexion (dorsal) [b]	61	27
Flexion (dorsal) [c]	77	10
Flexion (right or left) [b]	41	7
Rotation (right or left) [b]	79	14
Rotation (right) [c]	73	5
Rotation (left) [c]	74	4

[a] See Figure 78.

[b] Glanville and Kreezer, 1937: 10 male subjects.

[c] Buck et al., 1959: 100 subjects, 47 males, 53 females.

Table 82. Difference in Range of Joint Motion in Men and Women [a].

Joint	Type of Movement	Mean Difference (Degrees) [b]
Wrist	Flexion-extension	+14
	Adduction-abduction	+11
Elbow	Flexion-extension	+ 8
Shoulder	Abduction (rearward)	+ 2
Ankle	Flexion-extension	+ 4
Knee	Flexion-extension	0
Hip	Flexion	+ 3

[a] Sinelnikoff and Grigorowitsch, 1931: 100 male and 100 female subjects.

[b] "Plus" (+) denotes greater range in women.

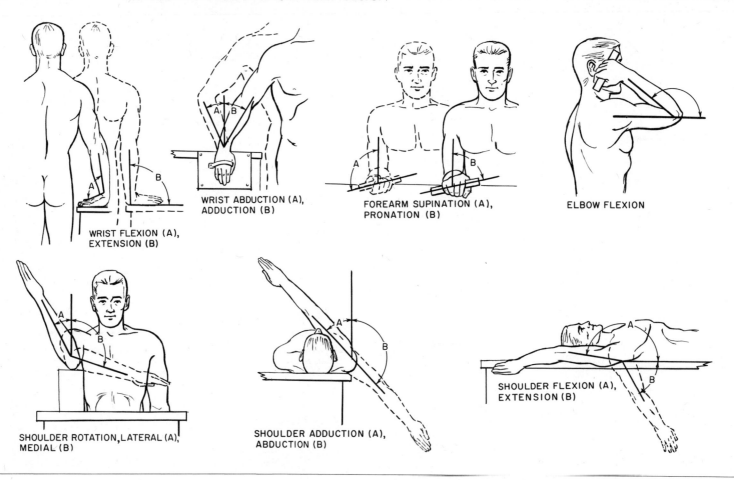

WRIST FLEXION (A),
EXTENSION (B)

WRIST ABDUCTION (A),
ADDUCTION (B)

FOREARM SUPINATION (A),
PRONATION (B)

ELBOW FLEXION

SHOULDER ROTATION, LATERAL (A),
MEDIAL (B)

SHOULDER ADDUCTION (A),
ABDUCTION (B)

SHOULDER FLEXION (A),
EXTENSION (B)

76. Methods of measuring range of motion at the joints of the arm and hand (after Barter et al., 1957).

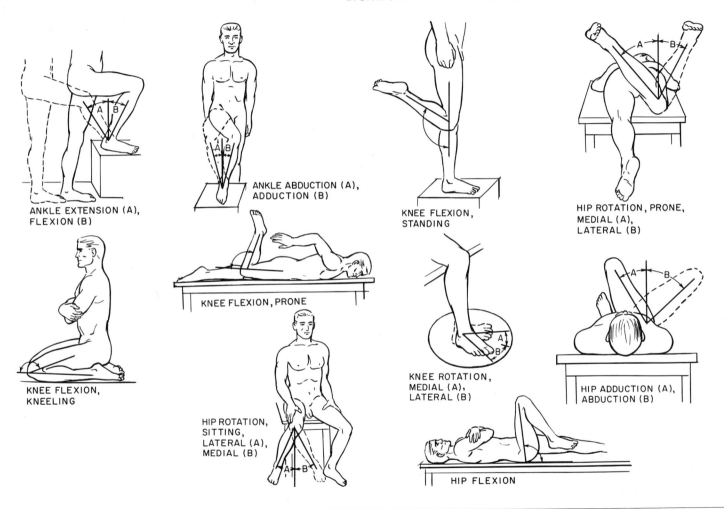

ANKLE EXTENSION (A),
FLEXION (B)

ANKLE ABDUCTION (A),
ADDUCTION (B)

KNEE FLEXION,
STANDING

HIP ROTATION, PRONE,
MEDIAL (A),
LATERAL (B)

KNEE FLEXION, PRONE

KNEE FLEXION,
KNEELING

HIP ROTATION,
SITTING,
LATERAL (A),
MEDIAL (B)

KNEE ROTATION,
MEDIAL (A),
LATERAL (B)

HIP ADDUCTION (A),
ABDUCTION (B)

HIP FLEXION

77. Methods of measuring range of motion at the joints of the leg and foot (after Barter et al., 1957).

DYNAMIC JOINT MOVEMENT

The joint movement ranges in Tables 78–82 are static data taken in artificially standardized ways. For a dynamic study of joint movement, several photographic techniques have been devised, including those of Eberhart and Inman (1951), Taylor and Blaschke's (1951) kinematic analysis, Dempster's (1955b) multiple-exposure films, and Drillis' (1959) "gliding cyclograms." All deserve careful study by the design engineer faced

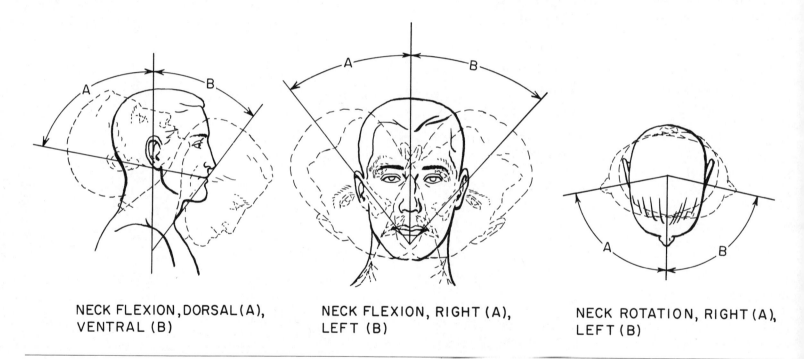

NECK FLEXION, DORSAL (A), VENTRAL (B)

NECK FLEXION, RIGHT (A), LEFT (B)

NECK ROTATION, RIGHT (A), LEFT (B)

78. Methods of measuring range of motion at the neck.

with a dynamic problem for which the present data fall short.

Head Movements in Vertical Sighting

An example of dynamic anthropometry involving body movements is Brues's (1946) study of head and eye movements in sagittal-plane sighting, undertaken in connection with gunsight and sighting panel design. Fig. 79 shows the location of the head and eyes at various angles of gaze, based on twenty-one Air Force men. These data were applied to equipment design in the reference cited (Brues, 1946), which is recommended as an example of applied anthropometry. When looking directly upward, the average subject tips his head back until the back of his head is 2¼ inches behind the plane of the back rest. An additional inch should be allowed for larger heads. The operator also shortens his sitting height by ½ inch when looking directly overhead. Eye movement is virtually a circular arc from 90 degrees above to 45 degrees below the horizontal—the arc is perfect from 67.5 degrees above to 25 degrees below horizontal. The ear hole does not move in a circular arc, as commonly supposed, but follows a paraboloid curve. And finally, the pivot point or center of the arc described by the eye in sagittal-plane sighting is not at the ear hole, but is roughly 2 inches below and ½ inch behind it.

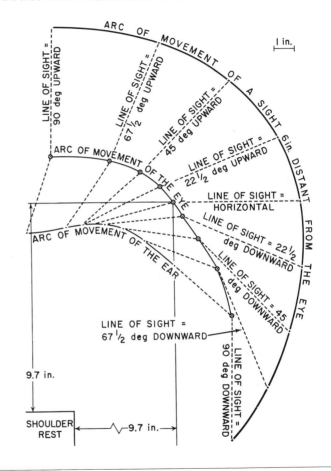

79. The location of the head and eyes at various angles of gaze (after Brues, 1946).

3. Muscle Strength

APPLICATION OF MUSCLE STRENGTH DATA

The voluntary muscles of the body stabilize the joints, maintaining desired body positions; rotate the body segments around the joints, resulting in movement and locomotion; and transmit force to objects outside the body such as controls or tools. This applied force is the "fundamental human output in all control systems" (Birmingham and Taylor, 1954).

Data on voluntary muscle strength are used by the design engineer to determine maximum and optimum control resistances, the forces required in various manual tasks, and the arrangement of weights for safe, efficient lifting or carrying.

"Maximum" control resistances should be based on the strength exertable by the weakest potential operators. If such persons can be accommodated, all others as well will be able to use the control. "Maximum" control resistances should not be exceeded, particularly with critical controls. It should also be remembered that men may have to operate power-assisted controls or other devices under emergency conditions when power fails.

Optimum or "operational" resistance levels should not require the application of maximum strength from any worker. Operational levels, affecting comfort and efficiency, should be low enough to prevent fatigue or discomfort, but still high enough to prevent inadvertent operation of the control and to provide kinesthetic cues to control movement, if desired. The precise levels between which the designer must work or at which he should aim remain to be determined for most controls. The data presented below refer to maximum strength levels, not comfort or efficiency.

As with control resistance, schedules for other physical tasks or for lifting or carrying must be set with regard to the greatest strength of the weakest workers. Performance at or near the limits of physical capacity should not be required, since fatigue will develop rapidly, and actual injury (muscle sprain or rupture, strained or torn ligaments) may occur.

Levels of human strength are also relevant to equipment design under special conditions such as space travel where, because of restrictions on space and weight, conventional sources of power may be logistically impractical or too expensive. If there is a human passenger aboard, he may "have to act as a sensor, controller, and to a modest extent, power source as well, to pay his way" (Krendel, 1960).

HOW MUSCLES WORK

Muscles can be defined as "machines burning carbonaceous fuels at low temperatures" (Haggard, 1946) which convert chemical energy to mechanical energy. The source of this chemical energy is the oxidation of food-stuffs, primarily carbohydrates, glycogens and fatty acids, to carbon dioxide and water. With favorable conditions, muscles have an efficiency—that is, work output / chemical energy used—of about 20 to 25 per cent (Wilkie, 1960).

The force developed by a muscle depends mainly upon two factors: (1) Muscle tension, which "reaches a maximum when the length is greatest and there is, momentarily, no change in length" (Elftman, 1941). The tension, or contractile force, decreases as the muscle shortens and as its rate of shortening increases (Arkin, 1941; Elftman, 1941; Darcus, 1951). (2) The mechanical advantage of the body's lever system. The long bones are the lever arms, the joints the fulcra. Power is applied at the points of muscle attachment.

Both of these factors vary with changes at the joints. At the elbow, for example, the force of *extension* is greatest when the elbow is flexed. This position permits the triceps to attain its greatest length, but does not afford the optimum mechanical advantage. In elbow *flexion,* on the contrary, the force is greatest toward the midpoint of the full range of movement (Darcus, 1954). Here the increasing mechanical advantage of the forearm relative to the upper arm more than compensates for the reduction in strength caused by the shortening of the biceps (Haggard, 1946).

Although human muscles in maximum contraction can exert large forces—often exceeding 1,000 pounds in the line of the tendon—such forces cannot be directly utilized since all muscles work at mechanical *dis*advantage (high ratio of resistance to power) which permits rapid movement (Hugh-Jones, 1947). A rough estimate of muscle strength is "about 4 kg per cm^2 cross section of the muscle" (Hettinger, 1961).

The power output utilizable as external, mechanical power varies with the individual, the task, and the duration of the output. For a task like cycling, where the mechanical system is simple, the gears readily adjusted, and the output easily measured, the range of maximum power output developed in various studies using average as well as champion cyclists, who develop about 20 to 30 per cent more power, was as follows (Wilkie, 1960): for single movements of less than 1 second duration—less than 6 hp; for brief exercise of 0.1 to 5.0 minutes duration—0.5 to 2.0 hp; for steady work of 5 to 150 minutes duration—0.4 to 0.5 hp; and for "all day" work—±0.2 hp.

Strength and Endurance

Strength is defined as the peak force, or maximum possible exertion achieved during an instant of time. Endurance can be defined as the ability to maintain a submaximal force over a given period of time, whether seconds, minutes, or hours. In general, there is a positive relationship between strength and endurance so defined, since strong persons can usually maintain a given submaximal force for longer periods of time than weaker persons. Caldwell (1961) reported a correlation between strength (for a single maximum response) and endurance (in a constant, submaximal, holding response) of 0.88. However, when individual differences in strength were eliminated, the correlation between strength and endurance dropped markedly. That is, when endurance was measured in terms of a subject's ability to maintain for as long as possible a stated percentage of his maximum strength, the correlation between strength and endurance fell to 0.13 (Caldwell, 1963). This low level of correlation means that a weak person can maintain, say, 50 per cent of his maximum level almost as long as a more powerful person can maintain 50 per cent of his much greater maximum strength.

FACTORS INFLUENCING MUSCLE STRENGTH

Many factors—biological, psychological, environmental, and occupational—affect muscle strength. Awareness of these factors will help the designer evaluate the physical abilities of his potential operators. Tasks requiring strength easily exerted by young soldiers, for example, will exceed the capacity of many other selected groups or the general male population.

The following personal characteristics, by no means an exhaustive list, are relevant to equipment design.

Biological Factors

Age: Strength increases rapidly in the teens, more slowly in the early 20's, reaches a maximum by the middle to late 20's, remains at this level for 5 to 10 years, and thereafter declines slowly but continuously (Hunsicker, 1955; Hunsicker and Greey, 1957). From the various studies relating strength to age summarized by Fisher and Birren (1946), the following rough estimates may be made: By the age of 40 muscle strength is approximately 90 to 95 per cent of the maximum in the late 20's. By age 50 it is about 85 per cent, and by 60 about 80 per cent of this maximum. Müller (1959) found muscle strength at 65 to be about 75 per cent of that exerted in youth. But not all physical abilities decline with age at the same rate; hand grip, for example (which forms a large part of age-strength data), seems to be relatively stronger in later years than other types of muscular performance. The strength of the back muscles drops faster with age than that of either the hands or the arms (Simonson, 1947). In old age, strength may be a small fraction of one's earlier peak.

A distinction must be drawn between age differences among groups and age changes in the individual. Virtually all current data on aging during adult life come from "cross-sectional" or one-time surveys among persons of various ages. Since men and women in their 50's and 60's are shorter and weaker than younger persons, the customary interpretation is that stature and strength decrease in mid-adult life. But only repeated observation of the same subjects throughout life can establish the presence, extent, and time of onset of age changes. The few "longitudinal" studies in existence show no such decrements in the same men between the late teens or early 20's and the 50's or 60's (Damon and Stoudt, 1963). Age differences may reflect the life experience of different cohorts or differential survival as well as true age changes.

The equipment designer is not concerned with this basic biological distinction. His interest is in the distribution of physical characteristics in the population at a point in time, that is, in age differences.

Sex: Women are in general about two thirds as strong as men, the amount varying for different muscle groups. In the forearm flexors women have only about 55 per cent of men's strength, whereas for the flexors and extensors of the hip, and flexors of the lower leg, their proportional strength increases to about 80 per cent. Other muscle groups are intermediate. Hettinger (1961) believes that the strength of a muscle per unit cross section is virtually the same in both men and women, and that muscle size largely determines strength. After correcting for differences in gross body size, other investigators found that women have about 77 per cent of the strength of men (Asmussen and Heebøll-Nielsen, 1962). In any event, the response of muscles to training appears to be less in women than in men. With the same kinds of muscle training, strength increased faster and to a greater extent among men; thus, basic sex differences in muscle strength were more marked after training (Hettinger, 1961).

Race: Racial differences in muscle strength result from variations in gross size, body build, and in such culturally conditioned factors as occupation, type of physical activity, and diet. Data on racial differences in strength are at present inadequate to permit any general conclusions.

Body build: This is closely related to strength. In a group of young, healthy college men of various body builds, strength varied tenfold, with the 95th percentile 4 to 5 times stronger than the 5th (Hunsicker, 1955). As might be expected, the mesomorphic (muscular or "athletic-looking") person is often the strongest. However, for tasks involving exertion of a force over extended periods of time, a less powerfully built person may be more efficient (Karpovich, 1953). Cullumbine *et al.* (1950) found that physique did not appear to be related to ability to perform moderate exercise. For rapidly fatiguing, severe exercise, however, slender subjects performed best, followed by normal and stocky subjects, and, finally, obese subjects. For moderate exercise carried to fatigue, those of normal build were superior, followed closely by the slender. The worst performers were the stocky and the obese. Among 158 college men studying physical education, mesomorphy was related to both motor skills and to strength, but at fairly low levels, with $r = 0.2$-0.4 (Sills, 1950). In another study, mesomorphy correlated with grip strength very slightly ($r = 0.17$-0.20) among young white soldiers, but appreciably ($r = 0.41$-0.48) among Negroes (Damon *et al.*, 1962).

A general size factor, common to all dimensions, extends also to strength. A person who is large in one body dimension tends to be above average in all body measurements as well as in limb strength. Stature and weight correlated with handgrip ($r = 0.42$-0.48), with elbow flexion force ($r = 0.45$-0.47), and with elbow extension force ($r = 0.44$-0.68) (Roberts, Provins, and Morton, 1959). Caldwell (1963) found that the maximum strength of eighteen male and eighteen female college students, as measured by pull on an isometric dynamometer handle, was closely related to body size, as indicated by the following correlations: stature, $r = 0.76$; weight, 0.74; upper arm length, 0.60; forearm length, 0.75; upper arm circumference, 0.71; forearm circumference, 0.88. All correlations were statistically significant at $p < 0.01$. However, size was unrelated to endurance as measured by the length of time the subject could exert 50 per cent of his maximum strength.

Among twenty-seven medical students, Rasch and Pierson (1963) found lower correlations between isometric strength, as measured by flexion and extension forces of the elbow, and weight and height of 0.44 and 0.21, respectively. With isotonic strength as measured by various barbell exercises, weight had a correlation of 0.45 and height 0.11.

Among persons of equal size, differences in strength may be caused by variations in the amount and quality of muscle tissue, body shape and proportions, and innervation or nerve stimulus (Martin, 1921).

Body position: This is an important factor influencing strength. For example, in an upward movement of the arm, strength at the weakest position of elbow flexion was only 44 per cent of that at the strongest position (Hunsicker, 1955). Similarly, higher torques could be applied by the hand with the elbow at 90 degrees of flexion than at 150 degrees (next best) or 30 degrees (Salter and Darcus, 1952). Fortunately, when large forces must be overcome, people assume the position from which maximum strength can be exerted. This position is not necessarily the one in which lesser forces can be most easily maintained (Darcus, 1954).

Handedness: In right-handed persons (roughly 90 per cent of all), the left hand averaged about 10 per cent weaker than the right (Hunsicker, 1955). Schochrin (1934) found the left leg also to be 10 per cent weaker. In general, the preferred side is the stronger.

Fatigue: This greatly reduces strength. For example,

after maximum force had been exerted on a control stick or foot pedal for 5 minutes, strength was reduced to about one third of its maximum value—even after a 5-minute rest period (Hertel, 1930). Similarly, the average grip strength for a one-minute period was only 60 per cent of the maximum (Tuttle, Janney, and Thompson, 1950). Maximum strength is usually attained at the beginning of an exertion. After pulling on a dynamometer at hourly intervals for 42 hours, pull strength declined, though an occasional subject could marshal sufficient strength in a single exertion at the very end of the period to equal or better his earlier efforts (Hunsicker, 1957). With a maximum pull (forearm flexion) once a minute for 30 minutes, there was an irregular but gradual downward trend in strength amounting to 10 per cent at the end of the period (Ikai and Steinhaus, 1961). In general, only about 15 per cent of maximum strength can be exerted over long periods throughout the day without muscle fatigue (Hettinger, 1961).

Fatigue can be avoided or postponed by comfortable working positions, periodic posture changes permitting muscle stretching and flexion, decreased intensity or duration of the activity (or both), training or practice, and rest periods. As an example of the importance of rest periods in alleviating fatigue, Yochelson reported that, following a standard exercise, there was a 40 per cent recovery from fatigue after a one-minute rest, a

65 per cent recovery after 2 minutes, 85 per cent after 4 minutes, and 95 per cent after 8 minutes. Recovery from a single maximum strength effort probably takes less than 10 minutes (Hunsicker and Greey, 1957).

Exercise: This can increase strength and endurance within the limits imposed by a person's innate physical potential. Differences in strength as measured before and after training can be considerable (Reedy, 1953). Increases in strength after training of as much as 100 per cent of the pretraining level have been recorded—though such large increments are rare. More commonly, increases in the range of 30 to 50 per cent above the beginning levels are observed in "normal" subjects after 12 weeks of training (Hettinger, 1961).

Recent experiments have shown that a greater increase in muscle strength can be obtained by exercising with isometric muscle contraction (no body movement or change in muscle length) than with isotonic contraction, which involves both body movement and change in muscle length. For most practical purposes, Hettinger (1961) found that the greatest improvement in the strength of any muscle group could be obtained with a single muscle contraction against a resistance for 1 to 2 seconds a day. There was no additional increase in strength with more than a single maximum contraction per day, so that long, strenuous periods of exercise were unnecessary for training. With one maximum muscle contraction every other day, the increase in strength fell to 80 per cent of that obtainable with daily sessions; with two sessions per week the comparable figure was 60 per cent, and with a single weekly session only 40 per cent.

Maximum muscle contractions were not essential for training since only 40 to 50 per cent of this level can also produce the maximum effect. Under such conditions, however, the duration of the contraction must be increased, since the force and duration of the contraction were interrelated. For example, when the force was only two-thirds of maximum strength, it had to be maintained for 4 to 6 seconds to achieve the same training effect as a maximum effort for 1 to 2 seconds. When the strongest daily muscle contractions fell below 20 per cent of the maximum exertable, regardless of duration, strength decreased. When 20 to 30 per cent of the maximum muscle strength was used, strength remained stable. This is apparently the range of normal daily activity where repetition of the same efforts produce neither an increase nor decrease in strength (Hettinger, 1961).

Similar training methods may have different results on different people. Of 102 subjects, 12 per cent showed a very good training effect (strength increase of 3 to 5 per cent per week for 8 weeks), and 10 per cent showed little or no training effect; others were intermediate. In general, "athletic" or mesomorphic types improved the most. There were also differences in the rate of strength increase in different muscles. Those used most commonly in daily activities will show the least increase—presumably because they are operating closer to the level of maximum strength—whereas muscles used less will show greater training effect. Trainability is said to be greatest in the summer and fall, and least in the winter. These differences have been experimentally related to seasonal differences in ultraviolet radiation, either natural or artificial (Hettinger, 1961).

Rasch and Pierson (1963) found that improvements in isotonic strength as a result of exercising with barbells were not accompanied by significant increases in isometric strength, and they suggest that since the converse may also hold, isometric exercises may not be of value in preparing for activities which are primarily isotonic in nature.

Strength endurance can be improved even more than strength in single exertions. In one experiment ergographic measurements of the endurance of the forearm flexors showed a 500 per cent increase after exercise. In another experiment, within two months after stopping daily exercise, the capacity of the legs to perform work was reduced to one-third of the maximum during the training period (Steinhaus, 1933).

Exercising one arm or leg significantly increases the strength of the other limb as well ("cross-education"). In extension of the knee, exercise increased the amount of weight that could be lifted by 34.7 per cent, but the unexercised leg also increased by 34.1 per cent. In grip strength the parallel increase was less marked, an 11.4 per cent increase for the exercised hand but only 1.5 per cent for the unexercised hand (Hellebrandt, Parrish, and Houtz, 1947). It should be noted that the larger increases for the unexercised limbs occurred after isotonic muscle contractions (involving change in the length of the muscle) and hence may be caused primarily by the increased blood supply resulting from the exercise. With isometric muscle contractions (no change in muscle length), no increases in the strength of the unexercised limb were found by one investigator (Hettinger, 1961).

Health: Muscle strength may be reduced by illness, acute or chronic.

Diet: Hunger or inadequate diets decrease strength (Hunsicker, 1955). During one experimental study, the handgrip squeeze of thirty-two subjects averaged 128 pounds before they were placed on a "famine" diet of 1570 calories a day. After 12 weeks of this diet, the squeeze had dropped to 104 pounds, and after 24 weeks, to 92 pounds. Comparable figures for backlift were 364, 290, and 255 pounds (Keys *et al.,* 1950).

Drugs: Under the influence of amphetamine sulfate ("benzedrine"), 14 mg per 70 kg of body weight, performance was significantly improved for three different classes of athletes. Eighty-five per cent of weight throwers, 73 per cent of runners, and from 67 to 93 per cent of swimmers performed better with amphetamine sulfate than with a placebo. Weight throwers obtained the greatest amount of improvement—about 3 to 4 per cent (Smith and Beecher, 1959). In a test of maximum forearm flexion, three 10-mg tablets of amphetamine sulfate produced a statistically significant average increase in strength of 13.5 per cent over the control level. With the same test, epinephrine (adrenalin) (0.5 ml of a 0.1 per cent solution) and alcohol (15 to 20 ml of 95 per cent ethyl alcohol in water) produced insignificant increases (Ikai and Steinhaus, 1961).

Diurnal variation: Strength is generally greatest in midmorning, with a lesser peak in the afternoon (Martin, 1921). The exact time varies with the individual and with other external factors (see below). In one test, thirty-five subjects exerted an average squeeze on a rubber bulb of about 150 mm of mercury at 6:00 a. m., 200 mm at 8:00 a.m., and a maximum of about 330 mm between 11:00 a.m. and noon. The squeeze thereafter decreased slightly to 300 mm at 10:00 p.m. This pattern was not the result of the relative immobility of the joint during sleep, or the time of day at which measurements were begun, but it closely paralleled the rise and fall of body temperature—even when body temperature was artificially increased or decreased (Wright, 1959). In another test, the maximum hand squeeze of twenty college students on a dynamometer was about 15 pounds (or 13 per cent) less immediately upon arising, than it was in the early afternoon (Jeanneret and Webb, 1963).

Environmental Factors

Temperature: Heat of 85°F or more when coupled with high humidity impairs human performance (Mackworth, 1950). Single efforts or brief bursts of moderate activity lasting seconds or minutes are little affected, but endurance over longer periods is greatly reduced.

Cold temperatures will not affect muscle strength if adequate protective clothing is worn, but such clothing adds weight and may impede body movement, requiring additional muscle power to perform a given action.

Altitude: Strength decreases with altitude, since anoxia decreases the physiological efficiency of muscle (Dill, 1938). Individual variability and acclimatization determine the strength decrement. Full acclimatization may take from weeks to months, depending on the altitude. Persons who are fully acclimatized can maintain their sea level static strength (as measured by handgrip) up to about 13,000 feet. Above this altitude strength decreases slowly but consistently to about 16,500 feet, then remains fairly constant to about 23,000 feet, where it declines more rapidly (Ruff and Strughold, 1942).

Altitude affects endurance more than brief muscular exertions. Endurance time for moderate to heavy muscular activity begins to decline at altitudes as low as 6,500 feet (Ruff and Strughold, 1942). At around 20,000 feet, muscular endurance, even for the fully acclimatized, is only about one half of that at sea level (Pugh and Ward, 1953). Above 23,000 feet heavy exertions of only the briefest duration, seconds to minutes, have been deemed possible (Ruff and Strughold, 1942), but the strenuous physical activity of mountain climbing has been continued without supplementary oxygen up to 28,000 feet or higher (Pugh and Ward, 1953).

Unacclimatized persons suffer more severely from the effects of altitude than the acclimatized. Above 10,000 feet or so, depending on individual variation, decrements in both muscle strength and endurance become increasingly apparent, culminating in partial muscular paralysis and unconsciousness at 23,000 to 25,000 feet (Schneider, 1939).

Acceleration: Unfortunately there are insufficient data on the muscular forces that can be exerted under various types of acceleration. Accelerations up to 5 G do not markedly influence strength (Brown & Lechner, 1956) unless large forces, or moderate forces over long periods, are required. Accelerations above 5 G impair strength progressively. "Highly effective arm movements cannot be executed above 6 G, although wrist and finger movements can be executed up to at least 12 G" (Bondurant, Clarke, Blanchard, Miller, Hessberg, and Hiatt, 1958). Muscular forces acting against the direction of acceleration are decreased, those exerted with the acceleration are increased, and those exerted perpendicular to the acceleration are affected least. The force of push movements on a control stick during positive (headward) acceleration increased slightly, 183.6 to 187.2 pounds, from 1 G to 3 G, and thereafter decreased slightly, 187.2 to 183.1 pounds, up to and including 5 G. Push-right movements, on the other hand, increased from 1 G to 5 G by 72.1 to 80.8 pounds, while for push-left the increase was from 65.2 to 73.4 pounds (Lombard, Canfield, Warren, and Drury, 1948; Canfield, Comrey, and Wilson, 1948). Further data are needed.

High accelerative forces are encountered in advanced aircraft and rockets. Precise manipulation under G forces requires training, with control boosters where the operator's strength falls short of that required. Fatalities have been ascribed to pilots' physical inability to operate certain controls or ejection devices under high G forces, which is just when ejection devices are most needed (Moseley, 1957). "Optimum" control loadings under normal conditions may prove hazardous in maneuvers involving high accelerative forces (Brown and Lechner, 1956).

The effects of zero gravity on human strength are still little known, but during prolonged weightlessness, reduction in muscle tonus and in strength would be expected, since the skeletal muscles are no longer active in supporting body weight (Loftus and Hammer, 1961). Preliminary experiments with weightless subjects in aircraft flying parabolic trajectories indicate that the maximum torques applied during the sudden twist of a fixed handle were about 67 per cent of the torques that were applied under normal gravity (Whitsett, 1963).

Psychological Factors

Motivation and emotional state. These cause considerable variation in muscle strength. Two persons of identical muscular ability, or the same person at different times, may differ markedly in strength, depending on the effort exerted.

It is well known that the body has reserves of strength which are not directly under the control of the will and are available only under stress (Hettinger, 1961). Fear, panic, rage, and excitement can temporarily increase strength, though skill and accuracy may suffer (Karpovich, 1953). When maximum pull with forearm flexion was measured, there was an increase, over "normal" pull, of 26 per cent under hypnosis, and an increase of 23 per cent under posthypnotic suggestion. Less marked but still statistically significant increases were observed when the maximum effort was preceded by a pistol shot, and when the subject was instructed to shout during the pull. These results led the experimenters to state that "in every voluntarily executed, all-out, maximal effort, psychologic rather

Table 83. Backlift: Vertical Pull.

Test Conditions: Backlift is generally measured by a two-handled vertical pull on a horizontal bar or handle at optimum lifting height, about 28 inches above floor level. The legs are straight and the back is slightly bent, then straightened for the lift.

| | | | | Maximum Lift - Pounds | | | |
| | | | | | | Percentile | |
Subjects	Source	Number	Age	Mean	S.D.	5	95
Men							
USAF aircrew students after 3-5 months of physical training	Clarke, 1945	914	18-27	520	90[a]	375	665
Dutchmen	Reijs, 1921	335[b]	21-35	337[c]			
English students	Cathcart et al., 1935	1,704	16+	367	58.9	271	643
Employed Englishmen	Ibid.	10,304	14-65+	363	67.7	251	474
Unemployed Englishmen	Ibid.	1,250	14-65+	315	60.8	214	415
Women							
Dutch women	Reijs, 1921	270[b]	21-35	202[c]			
English college women	Cathcart et al., 1927	460	18-23	216	34.4	160	272
English factory workers, employed	Ibid.	3,076	14-55	183	38.8	119	247
English factory workers, unemployed	Ibid.	413	19-55	165	39.2	101	229

[a]Derived from percentile distributions.

[b]Approximate.

[c]Median of yearly means.

Table 84. Leglift: Vertical Pull.

(Clarke, 1945)

Test Conditions: Subject grasps a horizontal bar at optimum lifting height, about 28 inches above floor level. The back is held vertical, and the legs are slightly bent, then straightened for the lift.

Subjects	N	Age	Maximum Lift (Pounds)		Percentiles	
			Mean	S.D.	5th	95th
USAF student aircrewmen	914	18-27	1,480	290[a]	1,010	1,950

[a]Derived from percentile distributions.

than physiologic factors determine the limits of performance" (Ikai and Steinhaus, 1961).

Occupational Factors

Occupation: White-collar workers are significantly weaker in muscle strength as measured by handgrip and lumbar pull, by some 10 to 20 per cent, than manual workers, whether skilled, semiskilled, or unskilled (Schochrin, 1934; Cathcart, Hughes, and Chalmers, 1935; Nemethi, 1952).

Clothing and personal equipment: Maximum push and one-minute push endurance were reduced by 12 to 14 per cent with a harness simulating restrictive clothing (Fox, 1957). Heavy items increase body weight, requiring additional muscular energy for any movement. Tight or bulky items restrict the range of body movement, possibly precluding the positions for exerting maximum power.

Workspace equipment: Backrests increase pushing strength; footrests increase pulling strength. In one study the addition of a backrest increased the push exertable on a dynamometer handle by 20 to 200 per cent, depending on the elbow angle used (Caldwell, 1962). Restraining harnesses or seat belts generally increase pulling strength, but by restricting body movements may also eliminate the most efficient positions for other types of muscular effort, such as lifting with the leg and back muscles.

MEASURING MUSCLE STRENGTH

Techniques

The strength of any muscular activity such as squeezing, pushing, or pulling, can be measured in pounds of pressure on a dynamometer. Most studies have been made with spring steel dynamometers in which the deformation of the spring was directly proportional to the force exerted upon it. Increasingly, electrical strain gage dynamometers are being used. These measure variations in the electrical resistance of a wire as its diameter changes under an applied force and can record the total strength curve over a period of time, as well as the strength of a single exertion. In addition, strain gage dynamometers measure isometric muscular contraction—*i.e.,* changing tension at constant length—which spring dynamometers cannot do.

Two basic types of muscle strength can be distinguished: the strength of a single maximum effort and strength over a time period, or endurance, as measured either continuously or from a series of efforts.

Subjects

Many muscle strength studies include few subjects, and those not always representative of the larger populations from which they were drawn. Adequate series should ideally contain at least fifty subjects, though it is often necessary to accept smaller series. The compo-

80. Functional lifting test (after Emanuel et al., 1956).

Table 85. Functional Lifting Ability--Maximum Lift.

(Emanuel, Chaffee, and Wing, 1956)

Subjects and Test Conditions: Nineteen young men, 16 of them college students, aged 17-28, lifted a weighted ammunition case (25 1/2 x 10 3/4 x 6 inches) from the floor to various heights from 1 to 5 feet. The case was grasped by handles attached near the top of either end, and a leglift was used insofar as possible. Subjects were told to lift the greatest weight possible without a feeling of injury. See Figure 80.

Task	Maximum Weight Lifted - Pounds				
	Mean	S.D.[a]	Percentiles 5th	95th	Range
Lift case 1 foot	231	47	142	301	135-320
2 feet	193	40	139	259	130-275
3 feet	119	31	77	172	75-190
4 feet	81	19	55	112	50-120
5 feet	58	16	36	83	30- 85

[a]Nonnormal distribution; use with caution.

sition of any group of test subjects should approximate as closely as possible the population for whom the equipment is being designed, with regard to body build, age, sex, and other relevant factors. (See pages 14, 32, and 42.)

Uniformity of the Data

The data from various strength studies are not always directly comparable, because of differences in measuring technique, subjects' body position, type and amount of movement permitted, motivation, kind of dynamometer used, and the number and kinds of subjects. Such studies can, however, be utilized in equipment design if the technique and subjects are clearly specified.

THE DATA

In Tables 83 to 102 the mean, standard deviation, 5th and 95th percentiles, and the range are given for each strength measurement wherever possible. Brief descriptions of the number and type of subjects used, as well as the experimental conditions, are presented for each series (Figs. 80 to 88).

Table 86. Functional Lifting Ability - Without Strain.
(Switzer, 1962)

Subjects and Test Conditions: Subjects were 75 male students aged 17-32 (mean, 19.5) years selected to represent short, medium, and tall U.S. Air Force personnel. Each subject lifted a metal box 12 x 12 x 6 inches, equipped with handles, and weighted with lead shot, to platforms at 18, 42, and 62.5 inches from the floor. The container was lifted by straightening the legs, with the back and arms straight. The subject then used his arms to place the container on the appropriate platform. Each subject was told to exert not a maximum lift, but a "reasonable" lift that could be continued for some time.

Subjects (Height in Inches)	Number	Platform Height (Inches)	Lift (Pounds) Mean	S.D.
Short 63.6-66.6	33	18	124	20.8
		42	73	11.6
		62.5	53	8.6
Medium 68.9-69.7	15	18	138	22.6
		42	92	12.5
		62.5	65	10.8
Tall 71.6-74.9	27	18	146	30.9
		42	96	14.2
		62.5	67	8.5

Table 87. Pedal Pressures Exertable by Extension of the Hip and Knee.

Test Conditions: In the descriptions of the seat and pedal adjustments in the following studies, it should be noted that:

1) backrest angle is formed by the surface of the backrest and the vertical
ii) seat angle is formed by the surface of the seat and the horizontal
iii) thigh angle is formed by the long axis of the femur (upper leg) and the horizontal
iv) knee angle is formed by the long axes of the femur (upper leg) and tibia (lower leg), and
v) ankle angle is formed by the long axis of the tibia (lower leg) and the flat surface of the pedal upon which the foot rests.

Subjects, Seat and Pedal Adjustment	Source	No. of Subjects	Maximum Pressure - Pounds				
			Mean	S.D.	Percentiles		Range
					5th	95th	
USAF heavy bomber student pilots. B-24 aircraft seat, seat belt, and rudder pedals. Average of both legs. Knee angle 111°+5°; backrest angle 13°; seat angle 9°	Elbel, 1949	590	565	96	407[a]	723[a]	
US Armored Forces personnel. Adjustable seat, backrest, seat belt, and pedals. One leg. Position of maximum force: pedal 32.2 inches in front of seat back, 2.4 inches above seat surface	Martin and Johnson, 1952	166	724				350-1,250
German pilots and engineers Junkers A-35 cockpit, "normal" seat and pedal positions. Right leg.	Hertel, 1930	11	480	66			385-605
Same: subjects fatigued, tested 3 minutes after a 5-minute maximum effort			147	24			110-176

Table 87 continued.

Subjects, Seat and Pedal Adjustment	Source	No. of Subjects	Maximum Pressure - Pounds				
			Mean	S.D.	Percentiles		Range
					5th	95th	
Englishmen. Flat padded seat with backrest. Pressure applied only to point of "throw-off," or displacement of eye position. Right leg.	Le Gros Clark and Weddell, 1944	5-10					
thigh horizontal, knee angle 90°			63				48-91
same, no backrest			43				37-49
thigh horizontal, knee angle 113°			89				85-100
thigh horizontal, knee angle 135°			156				150-162
seat angle 10°, knee angle 80°			77				64-90
seat angle 10°, knee angle 90°			59				51-70
British Armored Corps drivers. Flat seat, backrest. Pedal in same vertical plane as hip joint. Right leg. Thigh-seat angle 15°, knee angle 160°	Hugh-Jones, 1947	32	691	187[b]	383[a]	997[a]	
same adjustment, London schoolboys		16	689	125[b]	484[a]	898[a]	
"Powerfully built" Englishmen. Flat seat, backrest. Pedal in same vertical plane as hip joint. Right leg.	Hugh-Jones, 1947	6					
thigh angle 15°, knee 160°			845				
thigh angle 17°, knee 151°			684				

Continued on next page.

Table 87 continued.

Subjects, Seat and Pedal Adjustment	Source	No. of Subjects	Maximum Pressure - Pounds				
			Mean	S.D.	Percentiles		Range
					5th	95th	
thigh angle 5°, knee 164°			559				
thigh angle 15°, knee 169°			530				
thigh angle -10°, knee 165°			346				
thigh angle 16°, knee 129°			319				
thigh angle 10°, knee angle 136°			270				
thigh angle -15°, knee 149°			227				
thigh angle 17°, knee 117°			212				
thigh angle 33°, knee 106°			184				
thigh angle 8°, knee 93°			87				
thigh angle -6°, knee 94°			73				
Male British engineering students. Flat seat, backrest at or around waist height. Right leg.	Rees and Graham, 1952						
Pedal above seat, thigh angle with seat 15°, knee angle 160°		20	383.6	67.5[b]	272[a]	495[a]	247-488
Pedal below seat, thigh angle with seat 0°, knee angle 160°		20	319.2	76.9[b]	192[a]	447[a]	170-483

[a]Computed from mean and standard deviation.

[b]Computed from $SE_m = \dfrac{SD}{\sqrt{N}}$

Table 88. Forces Exertable by Flexion and Extension of the Knee.
(Schochrin, 1934)

Test Conditions: Subject seated, with lower leg vertical and hanging free. He pulls his lower leg under the seat in flexion and straightens it in extension. Force measured at a point 2 inches above the ankle. Mean of both legs.

Subjects and Task	Number	Age	Mean Maximum Force (Pounds)
Healthy Russian men	1,197	19-57	
Extension			102
Flexion			71
Healthy Russian women	378	19-57	
Extension			73
Flexion			50

Table 89. Pedal Pressure Exertable by Extension of the Ankle.

(Hertzberg, 1954)

Subjects and Test Conditions: 94-100 U.S. Air Force males, age not stated. Maximum pressures exertable on brake pedals in a standard aircraft cockpit with varying adjustments. Backrest angle constant at 13°.

Cockpit Adjustment	Maximum Pressure--Pounds		Percentiles[a]	
	Mean	S.D.	5th	95th
"Average" Cockpit - 39" vertically between eye level and heel rest lines. Pedal surface 30° from vertical, and:				
35 1/4" forward of the SRP	137.4	67.8	26	249
38 1/2" forward of the SRP	178.2	72.0	60	297
Pedal surface 10° from vertical, and:				
35 1/4" forward of the SRP	110.9	58.6	15	207
38 1/2" forward of the SRP	134.7	59.4	37	232
Pedal surface 50° from vertical, and:				
35 1/4" forward of the SRP	91.3	42.4	22	161
38 1/2" forward of the SRP	131.7	49.4	50	213
"Low" Cockpit - 37" vertically between eye level and heel rest lines. Pedal surface 30° from vertical, and:				
35 1/4" forward of the SRP	136.9	68.3	25	249
38 1/2" forward of the SRP	183.7	72.7	64	303
Pedal surface 10° from vertical, and:				
35 1/4" forward of the SRP	120.2	64.8	14	227
38 1/2" forward of the SRP	148.6	57.7	54	244
Pedal surface 50° from vertical, and:				
35 1/4" forward of the SRP	88.3	38.9	24	152
38 1/2" forward of the SRP	132.6	51.7	48	218

Continued on next page.

Table 89. Continued.

Cockpit Adjustment	Maximum Pressure--Pounds		Percentiles[a]	
	Mean	S.D.	5th	95th
"High" Cockpit - 41" vertically between eye level and heel rest lines.				
Pedal surface 30° from vertical, and:				
35 1/4" forward of the SRP	142.5	67.5	32	254
38 1/2" forward of the SRP	174.7	76.1	50	300
Pedal surface 10° from vertical, and:				
35 1/4" forward of the SRP	118.1	60.8	18	218
38 1/2" forward of the SRP	121.8	53.1	35	209
Pedal surface 50° from vertical, and:				
35 1/4" forward of the SRP	98.3	46.0	23	174
38 1/2" forward of the SRP	138.3	53.5	50	226

[a]Computed from mean and standard deviation; use with caution.

Table 90. Pedal Pressure Endurance.

(Elbel, 1949)

Subjects and Test Conditions: 75 U.S. Army Air Forces pilots. B-24 aircraft pilot seat, seat belt, and rudder pedals. Knee angle $111° \pm 5°$, backrest angle $13°$, seat angle $9°$.

Pedal Pressure Exerted (Pounds)	Maximum Endurance Time - Seconds		Percentiles[a]	
	Mean	S.D.	5th	95th
200	218.5	122.8	17	421
300	114.6	62.9	11	218
400	68.4	31.7	16	121
500	39.6	21.2	5	75
600	20.1	12.8	—	41
700	11.7	7.8	—	25

[a]Computed from mean and standard deviation; use with caution.

Table 91. Pedal Pressures Maintainable for Two Minutes without Eye Displacement.
(Le Gros Clark and Weddell, 1944)

Subjects and Test Conditions: Five British anatomists and medical students. Flat, padded seat with backrest. Maximum pressure applied to a pedal for 2 minutes up to the point of "throw-off" or displacement of the eye.

Seat Adjustment	Maximum Pressure for 2-Minute Period (Pounds)	
	Mean	Range
Thigh horizontal, knee 135°	53	48-58
Thigh horizontal, knee 113°	47	44-54

Table 92. Hand Strength: Dynamometer Squeeze.

Subjects and Hand	Source	N	Age Mean or Range	Maximum Squeeze -- Pounds				
				Mean	S.D.	Percentiles		
						5th	95th	Range
Men								
U.S. Army Air Forces aircrew students		914	18-25					
Right hand	Clarke, 1945			134	18	105	164	90-203
Left hand				124	16	96	154	65-190
U.S. Navy personnel Mean of 2 hands	Fisher and Birren, 1946	169	22	119	14.4	95[a]	143[a]	79-165
American industrial workers Preferred hand	ibid.	552	34	117	15.4	92[a]	143[a]	73-172
Normal men	Tuttle et al., 1950	200	20-30					
Right hand				108	21	74[a]	142[a]	-
Left hand				95	18	65[a]	124[a]	-
Same subjects - maximum strength exerted over a one-minute period.	ibid.	200	20-30					
Right hand				62	12	42[a]	82[a]	-
Left hand				55	10	39[a]	71[a]	-
U.S. Air Force personnel and civilians	Barter et al., n.d.	99	31					
Right hand				104	27.3	59[a]	148[a]	-
Left hand				94	23.7	56[a]	134[a]	-
U.S. commercial bus and truck drivers	Damon and McFarland, 1955	268	37					
Right hand				121	18.1	91[a]	151[a]	80-190
Left hand				113	16.4	86[a]	140[a]	70-170
U.S. champion truck drivers	ibid.	101	34					
Right hand				129	17.8	100[a]	159[a]	90-170
Left hand				118	18.6	87[a]	148[a]	70-180

Continued on next page.

Table 92. Continued.

Subjects and Hand	Source	N	Age Mean or Range	Maximum Squeeze -- Pounds				
				Mean	S.D.	Percentiles 5th	95th	Range
Spanish-American War veterans	Damon and Stoudt, 1963		81					
Right hand		118		64	17.3	35[a]	92[a]	
Left hand				59	18.1	29[a]	89[a]	
Rubber industry workers	Damon and Stoudt, 1958		40					
Right hand		162		124	21.2	89[a]	159[a]	
Left hand				122	22.2	86[a]	159[a]	
U.S. Army personnel	Damon et al., 1962		24					
Right hand		431		137	--	106	172	
Left hand				132	--	99	168	
Employed Englishmen Mean of 2 hands	Cathcart et al., 1935	10,473	14-65	109	19	77[a]	140[a]	
Unemployed Englishmen Mean of 2 hands	ibid.	1,304	14-65	95	18	66[a]	123[a]	
English students Mean of 2 hands	ibid.	1,722	16+	114	17	86[a]	142[a]	

Table 92. Continued.

Subjects and Hand	Source	N	Age Mean or Range	Maximum Squeeze -- Pounds				
				Mean	S.D.	Percentiles 5th	95th	Range
Women								
U.S. Naval women Mean of 2 hands	Fisher and Birren, 1946	161	22	73	8.8	58[a]	87[a]	46-119
U.S. industrial workers Preferred hand	ibid.	96	32	74	10.3	57[a]	91[a]	46-119
English college students Stronger hand	Cathcart et al., 1927	460	18-23	63	11.7	43[a]	82[a]	
English factory workers Stronger hand	ibid.	3,076	14-55	58	10.8	40[a]	76[a]	
English unemployed factory women Stronger hand	ibid.	413	19-55	55	11.0	37[a]	73[a]	

[a]Computed from mean and standard deviation.

Table 93. Hand and Finger Strength in Grasping.

(Taylor, 1954)

Subjects and Test Conditions: 15 "normal" men applied maximum pressures in four different kinds of prehension as illustrated in Figure 81.

Type of Prehension	Maximum Force--Pounds		
	Mean	S.D.	Range
Palmar	21.5	5.4	15.2 - 31.5
Tip	21.0	4.8	12.5 - 28.0
Lateral	23.2	4.8	16.0 - 30.0
Grasp	90.0	16.0	64 - 120

Used by permission of the publisher from Klopsteg and Wilson's Human Limbs and Their Substitutes, Copyright 1954, McGraw-Hill Book Company.

Table 94. Finger Strength: Flexion of the Finger - Palm Joint.

(Barter, Fry, and Truett, 1956)

Subjects and Test Conditions: 100 "normal, healthy" Air Force and civilian men, aged 19-45 years, placed the right hand palm up on a flat surface and exerted maximum force against a strain gauge bar placed over the fingertip. The observer pressed down on the palm just behind the finger-palm (metacarpo-phalangeal) joint to insure that the joint was not used as a fulcrum.

Finger Moved	Maximum Force (Pounds)	
	Mean	S.D.
2nd (index)	13.25	2.75
3rd (middle)	14.18	4.30
4th (ring)	10.80	3.78
5th (little)	7.16	2.53

TYPES OF HAND PREHENSION

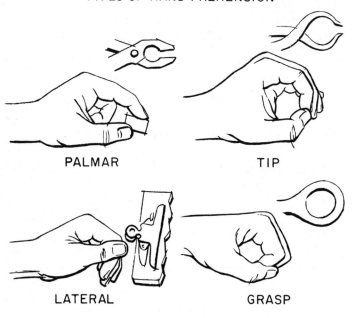

81. Types of hand prehension: palmar, tip, lateral, and grasp
(after Taylor, 1954).

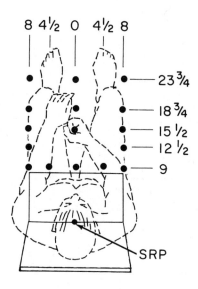

82. The various control stick locations used in testing arm strength.
SRP is the seat reference point.

Table 95. Forces (Pounds) Exertable on an Aircraft Control Stick at Each of Sixteen Different Positions.

(Hertzberg, n.d.)

Subjects and Test Conditions: 33 U.S. Air Force men were used for measurements of the right hand, 15 for measurements of the left or both hands. Each subject sat in a standard cockpit mockup and exerted maximum force on the handle of a control stick 13 1/2 inches above the Seat Reference Point at various forward and lateral locations. See Figure 82.

Movement	Control Location Distance from SRP in Inches		Mean	Right Hand Percentiles		Left Hand Mean	Both Hands Mean
	Forward	Lateral		5th	95th		
Pull	23 3/4	8 left	90	51	129	96	173
		0	102	62	138	89	173
		8 right	103	58	133	81	173
	18 3/4	8 left	74	45	108	89	146
		0	86	56	127	81	160
		8 right	99	58	126	74	160
	15 1/2	8 left	64	39	98	81	133
		0	83	54	113	74	146
		8 right	89	55	119	59	146
	12 1/2	8 left	53	33	77	74	120
		8 right	80	49	108	52	120
	9	8 left	40	26	67	66	93
		4 1/2 left	45	28	66	66	106
		0	57	34	86	52	106
		4 1/2 right	62	39	88	52	106
		8 right	58	39	86	37	106
Push	23 3/4	8 left	64	29	104	93	88
		0	106	54	141	107	110
		8 right	100	56	147	64	99
	18 3/4	8 left	72	36	114	129	121
		0	124	64	177	122	154
		8 right	125	70	198	79	154
	15 1/2	8 left	60	23	118	122	121
		0	86	43	160	93	165
		8 right	100	53	164	57	143
	12 1/2	8 left	36	18	68	79	110
		8 right	74	43	102	43	110
	9	8 left	29	12	44	64	77

Table 95. Continued

Movement	Control Location Distance from SRP in Inches		Mean	Right Hand Percentiles		Left Hand Mean	Both Hands Mean
	Forward	Lateral		5th	95th		
Push	9	4 1/2 left	33	18	54	57	88
		0	46	26	67	50	99
		4 1/2 right	58	34	82	43	99
		8 right	65	37	95	36	99
Move right	23 3/4	8 left	31	19	48	26	33
		0	20	13	30	31	33
		8 right	22	12	51	31	33
	18 3/4	8 left	36	22	61	31	50
		0	25	15	35	44	50
		8 right	24	14	50	44	50
	15 1/2	8 left	43	25	63	35	50
		0	28	20	39	48	61
		8 right	22	13	49	57	55
	12 1/2	8 left	48	31	70	39	61
		8 right	24	16	46	53	61
	9	8 left	55	34	74	48	83
		4 1/2 left	48	31	64	53	72
		0	38	23	49	66	66
		4 1/2 right	27	15	51	57	72
		8 right	22	12	43	53	66
Move left	23 3/4	8 left	21	11	49	24	24
		0	29	14	46	29	47
		8 right	37	20	66	53	65
	18 3/4	8 left	30	16	56	29	35
		0	32	8	53	37	53
		8 right	39	22	70	57	70
	15 1/2	8 left	35	20	58	29	47
		0	38	24	52	41	59
		8 right	40	24	70	57	70
	12 1/2	8 left	44	23	70	37	59
		8 right	39	22	59	57	76
	9	8 left	44	24	65	33	56
		4 1/2 left	49	31	67	37	59
		0	47	30	66	45	70
		4 1/2 right	46	26	78	53	76
		8 right	44	26	72	57	88

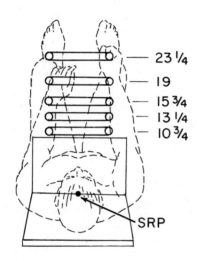

83. The various control wheel locations used in testing arm strength. SRP is the seat reference point.

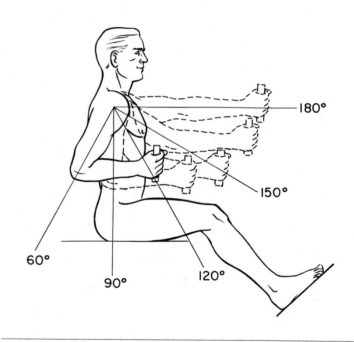

84. The elbow angles at which arm strength at various fore-and-aft positions was tested—subject seated (after Hunsicker, 1955).

Table 96. Forces (Pounds) Exertable on an Aircraft Control Wheel at Each of Sixteen Different Positions.

(Hertzberg, n.d.)

<u>Subjects and Test Conditions</u>: 33 U.S. Air Force men were used for measurements of the right hand, 15 for measurements of the left and both hands. Each subject sat in a standard cockpit mockup and exerted maximum force on a control wheel with grips 18 inches above the Seat Reference Point and some 15 inches apart. Measurements were made at several forward locations and at various degrees of rotation. See Figure 83.

Movement	Control Location from SRP Forward (Inches)	Control Position- Degrees of Turn Right or Left of Neutral		Right Hand			Left Hand	Both Hands
				Mean	5th	95th	Mean	Mean
					Percentiles			
Rotate right	23 1/4	90	left	42	26	82	48	91
		0		60	35	98	39	101
		90	right	40	22	68	35	71
	19	90	left	46	27	94	48	101
		0		63	30	104	43	101
		90	right	41	22	87	43	81
	15 3/4	90	left	53	19	96	43	101
		0		59	27	97	39	101
		90	right	46	20	91	43	91
	13 1/4	90	left	52	21	98	43	111
		90	right	51	19	111	48	101
	10 3/4	90	left	59	27	101	35	101
		45	left	69	24	121	48	132
		0		48	20	96	43	91
		45	right	51	24	118	56	111
		90	right	54	15	112	56	121
Rotate left	23 1/4	90	left	38	21	73	39	71
		0		39	20	86	55	92

Table 96. Continued.

Movement	Control Location from SRP Forward (Inches)	Control Position- Degrees of Turn Right or Left of Neutral		Right Hand			Left Hand	Both Hands
				Mean	5th	95th	Mean	Mean
					Percentiles			
Rotate left	23 1/4	90	right	55	26	109	44	102
	19	90	left	43	22	76	39	82
		0		44	25	95	66	102
		90	right	52	33	104	55	122
	15 3/4	90	left	43	27	82	39	82
		0		46	27	112	61	102
		90	right	50	29	86	61	112
	13 1/4	90	left	44	26	86	44	102
		90	right	45	25	99	66	122
	10 3/4	90	left	47	23	91	55	102
		45	left	54	21	123	50	102
		0		46	26	88	55	92
		45	right	54	31	120	66	133
		90	right	42	21	104	72	122
Pull	23 1/4	90	left	117	73	162	96	182
		0		125	77	182	120	224
		90	right	110	74	186	112	196
	19	90	left	88	60	127	88	154
		0		106	73	169	96	196
		90	right	94	61	149	96	168
	15 3/4	90	left	71	42	144	72	140
		0		94	66	145	80	154
		90	right	80	49	130	80	140
	13 1/4	90	left	67	33	120	56	112
		90	right	60	31	102	80	112
	10 3/4	90	left	55	23	109	40	98
		45	left	67	40	111	56	126
		0		66	44	102	72	126
		45	right	67	39	97	80	126
		90	right	43	18	87	64	98
Push	23 1/4	90	left	131	82	211	103	177
		0		171	105	242	168	265
		90	right	117	49	197	132	191
	19"	90	left	88	37	171	66	162
		0		121	64	235	124	265
		90	right	67	33	140	110	162
	15 3/4	90	left	59	32	139	51	118
		0		90	61	155	88	177
		90	right	53	32	102	73	132
	13 1/4	90	left	54	32	93	44	88
		90	right	51	25	83	59	88
Push	10 3/4	90	left	67	32	125	44	103
		45	left	84	48	149	59	147
		0		86	52	135	73	147
		45	right	67	40	128	88	147
		90	right	52	19	112	59	88

Table 97. Forces Exertable (Pounds) on a Vertical Handgrip from Various Elbow Angles.

(Hunsicker, 1955)

Subjects and Test Conditions: 55 University of Michigan men, aged 17-25 and fairly representative of USAF aircrews, were used. A few extreme physical types were also included, giving a wider range, especially at the weaker end, than normal in military groups. Subjects sat on a flat seat with a backrest and footrest, exerting maximum force on vertical handgrips directly in front of each shoulder. The strength of 6 movements was measured at each of 5 elbow angles. See Figure 84.

Movement	Elbow Angle (Degrees)	Mean	S.D.	Right Arm Percentiles		Range	Mean	S.D.	Left Arm Percentiles		Range
				5th	95th				5th	95th	
Pull	60	63	23	24	74	16-117	64	23	26	110	14-135
	90	88	30	37	135	18-163	80	28	32	122	12-156
	120	104	31	42	154	26-163	94	34	34	152	20-177
	150	122	36	56	189	28-222	112	37	42	168	25-187
	180	120	37	52	171	26-185	116	37	50	172	31-182
Push	60	92	38	34	150	24-174	79	31	22	164	22-180
	90	86	33	36	154	25-178	83	35	22	172	21-190
	120	103	43	36	172	30-220	99	42	26	180	25-203
	150	123	45	42	194	34-210	111	48	30	192	19-211
	180	138	49	50	210	33-215	126	47	42	196	26-215
Move right	60	42	20	17	82	15- 99	50	21	17	83	12- 92
	90	37	18	16	68	13- 94	48	22	16	87	9-112
	120	34	17	15	62	11- 86	45	21	20	89	14-118
	150	33	18	15	64	9-110	47	27	15	113	12-119
	180	34	24	14	62	10-158	43	22	13	92	10-119
Move left	60	52	19	20	87	12- 99	32	17	12	62	9- 77
	90	50	23	18	97	14-113	33	19	10	72	8-102

Table 97. Continued

Movement	Elbow Angle (Degrees)	Mean	S.D.	Right Arm Percentiles		Range	Mean	S.D.	Left Arm Percentiles		Range
				5th	95th				5th	95th	
Move left	120	53	26	22	100	18-122	30	18	10	68	7- 93
	150	54	25	20	104	18-117	29	20	8	66	8-105
	180	50	26	20	104	16-131	30	20	8	64	5-110
Up	60	49	18	20	82	8- 90	44	18	15	82	10- 95
	90	56	22	20	106	5-114	52	22	17	100	10-107
	120	60	24	24	124	10-142	54	25	17	102	10-128
	150	56	28	18	118	10-139	52	27	15	110	11-140
	180	43	22	14	88	9-101	41	23	9	83	4-105
Down	60	51	21	20	89	13- 96	46	18	18	76	12- 88
	90	53	20	26	88	16- 94	49	20	21	92	10-108
	120	58	23	26	98	22-110	51	23	21	102	15-134
	150	47	18	20	80	16- 94	41	16	18	74	12- 80
	180	41	18	17	82	13-116	35	15	13	72	40- 89

Table 98. Forces Exertable (Pounds) on a Vertical Handgrip
from Various Elbow Angles in the Prone Position.

(Hunsicker, 1955)

Subjects and Test Conditions: Same subjects as in Table 97. Lying prone, they exerted maximum force on vertical handgrips directly in front of the shoulders. The strength of 6 movements was measured at each of 5 elbow angles. See Figure 85.

Movement	Elbow Angle (Degrees)	Mean	S.D.	Right Arm Percentiles 5th	95th	Range	Mean	S.D.	Left Arm Percentiles 5th	95th	Range
Pull	60	61	26	21	113	13-130	57	24	17	97	12-113
	90	73	30	24	121	18-136	66	26	23	118	14-127
	120	86	34	31	147	23-168	74	30	22	126	15-150
	150	81	33	29	133	26-141	70	28	21	122	17-148
	180	69	26	31	118	20-127	61	26	18	111	12-123
Push	60	66	26	24	119	17-140	52	21	17	87	11-116
	90	63	23	26	103	17-117	54	22	18	91	12-109
	120	73	28	29	128	20-142	63	27	21	108	15-119
	150	73	30	29	127	21-142	65	26	24	111	21-118
	180	69	26	31	123	25-140	67	28	26	116	21-137
Move right	60	29	12	12	57	9- 67	44	24	11	99	4-101
	90	28	11	13	51	10- 59	40	22	13	92	9-101
	120	28	12	11	58	9- 69	38	23	9	91	8-113
	150	28	14	12	60	11- 75	34	23	8	79	6-110
	180	24	14	9	61	8- 83	31	19	10	67	8- 88
Move left	60	48	22	16	91	15-118	24	12	8	49	4- 77
	90	46	21	16	87	12- 98	22	10	6	45	4- 60
	120	48	25	15	97	12-102	20	9	6	38	5- 46
	150	45	26	15	93	12-131	20	15	5	56	2- 81
	180	37	17	12	71	9- 81	22	19	4	57	2- 83

Table 98. Continued.

Movement	Elbow Angle (Degrees)	Mean	S.D.	Right Arm Percentiles 5th	95th	Range	Mean	S.D.	Left Arm Percentiles 5th	95th	Range
Up	60	44	21	13	85	9-105	35	17	13	71	5- 82
	90	52	22	15	94	8-107	40	18	15	78	8- 84
	120	50	21	13	91	10- 97	40	21	11	81	7-109
	150	41	23	13	83	8-128	31	17	7	62	5- 71
	180	23	12	8	47	7- 63	18	12	5	44	4- 53
Down	60	34	13	13	61	8- 74	30	12	10	51	5- 63
	90	36	13	16	60	8- 70	31	12	12	57	4- 70
	120	35	15	15	61	12- 68	31	14	11	57	4- 64
	150	34	13	15	60	12- 96	28	11	10	48	6- 54
	180	25	10	13	47	10- 50	25	10	7	41	3- 45

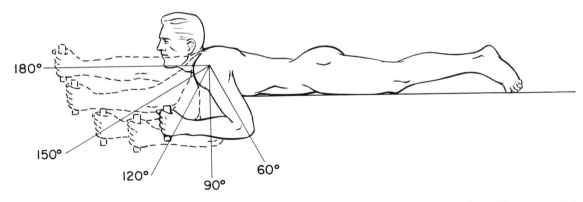

85. The elbow angles at which arm strength at various fore-and-aft positions was tested—subject prone (after Hunsicker, 1955).

Table 99. Forces Exertable (Pounds) on a Horizontal Handgrip
from Various Elbow Angles.

(Hunsicker, 1957)

Subjects and Test Conditions: 30 University of Michigan students, aged 18-27 and approximating
the anthropometric range of U.S.A.F. aircrews, were used as subjects. Each subject sat in a
flat seat with a backrest and footrest, and exerted maximum force on a horizontal handgrip that
was raised or lowered in the same transverse vertical plane as the outer corner of the eye.
The strength of 6 movements was measured at each of 5 elbow angles with the wrist pronated (palm
down) and supinated (palm up). Note: at an elbow angle of 180° the arm is vertical. See
Figure 86.

Direction of Movement	Elbow Angle (Degrees)	Wrist Supinated									
		Right Arm					Left Arm				
		Mean	S.D.	Percentiles		Range	Mean	S.D.	Percentiles		Range
				5th	95th				5th	95th	
Forward	180	32	15	12	58	11- 61	30	10	14	47	11- 58
	150	36	14	17	59	14- 76	38	30	13	69	13-167
	120	43	17	20	71	15- 83	40	18	15	80	14- 81
	90	65	24	25	117	24-121	59	27	25	104	18-140
	60	96	39	34	172	29-128	89	42	35	176	20-195
Back	180	39	19	15	73	14- 77	40	18	17	70	16- 91
	150	37	17	11	66	11- 74	40	15	16	62	14- 83
	120	40	17	11	63	8- 64	40	18	14	66	11- 92
	90	43	19	13	74	10-100	42	21	13	68	11-114
	60	51	25	16	93	11-123	54	23	23	87	19-123
Up	180	113	34	51	165	41-185	111	40	45	173	40-175
	150	103	40	37	161	36-188	104	36	44	164	42-175

Table 99. Continued

Direction of Movement	Elbow Angle (Degrees)	Wrist Supinated									
		Right Arm					Left Arm				
		Mean	S.D.	Percentiles		Range	Mean	S.D.	Percentiles		Range
				5th	95th				5th	95th	
Up	120	88	33	41	143	40-164	94	33	38	152	34-176
	90	63	27	21	107	18-123	75	29	24	131	23-151
	60	45	22	17	78	16-112	49	22	20	89	17- 93
Down	180	87	32	44	135	41-179	78	28	36	124	34-172
	150	93	35	37	150	34-178	84	29	43	136	38-174
	120	92	13	29	148	27-173	84	33	35	136	30-175
	90	80	37	17	143	15-166	80	43	23	160	23-181
	60	59	35	20	132	19-172	58	41	20	138	20-189
Abduction	180	29	12	14	48	12- 67	27	10	8	44	6- 52
	150	32	14	15	60	14- 80	26	10	12	43	10- 47
	120	34	15	17	64	16- 76	28	8	14	45	13- 45
	90	39	24	18	72	17- 77	33	12	16	52	16- 70
	60	44	19	18	73	18- 94	42	20	17	81	16- 93
Adduction	180	28	10	10	44	9- 49	29	9	12	43	10- 58
	150	31	14	12	52	12- 78	32	15	12	62	11- 68
	120	30	11	12	46	10- 50	31	13	14	55	12- 61
	90	31	12	13	48	11- 70	32	12	12	46	8- 70
	60	36	17	13	70	12- 93	38	12	16	64	12- 69

Continued on next page.

Table 99. Continued

Direction of Movement	Elbow Angle (Degrees)	Wrist Pronated									
		Right Arm					Left Arm				
		Mean	S.D.	Percentiles		Range	Mean	S.D.	Percentiles		Range
				5th	95th				5th	95th	
Forward	180	32	12	17	59	15- 60	32	13	12	59	9- 62
	150	40	18	18	66	15- 92	37	18	15	69	13- 82
	120	46	15	23	70	20- 92	43	17	17	71	16- 86
	90	65	24	25	100	22-106	60	28	27	93	20-167
	60	94	36	40	156	30-172	86	35	33	138	32-183
Back	180	28	12	11	48	11- 53	34	15	16	61	12- 71
	150	29	10	12	48	8- 56	32	13	15	52	12- 72
	120	26	10	13	43	9- 48	30	14	12	56	9- 61
	90	32	13	14	54	13- 67	37	18	17	65	14-105
	60	37	16	13	50	11-107	39	18	20	64	19-100
Up	180	95	35	35	156	29-167	101	11	47	171	38-178
	150	99	38	43	165	35-185	100	32	58	159	56-179
	120	91	30	41	138	39-159	91	30	45	145	39-149
	90	69	29	28	112	28-126	77	24	37	123	30-131
	60	49	20	23	79	22- 83	57	22	22	100	21-114
Down	180	87	31	41	143	41-160	76	31	34	138	33-142
	150	90	34	40	154	37-161	79	29	39	136	35-153
	120	92	35	37	161	34-179	75	40	29	148	25-159
	90	83	35	22	142	20-165	75	34	23	136	19-174
	60	81	35	23	158	23-173	74	35	18	139	15-175
Abduction	180	19	7	10	34	9- 44	20	13	10	49	8- 64
	150	21	11	9	39	8- 71	23	16	9	53	8- 83
	120	26	13	9	53	7- 67	22	10	10	39	7- 46
	90	31	15	12	64	11- 67	27	11	11	54	8- 58
	60	41	19	19	72	18- 67	36	15	18	51	15-100
Adduction	180	31	13	16	57	12- 58	28	8	15	41	14- 55
	150	32	7	18	45	15- 56	31	11	17	54	15- 70
	120	34	11	15	47	16- 59	34	8	17	53	16- 55
	90	39	15	16	59	15- 83	38	12	17	60	16- 67
	60	48	18	16	73	15- 84	42	15	20	66	17- 72

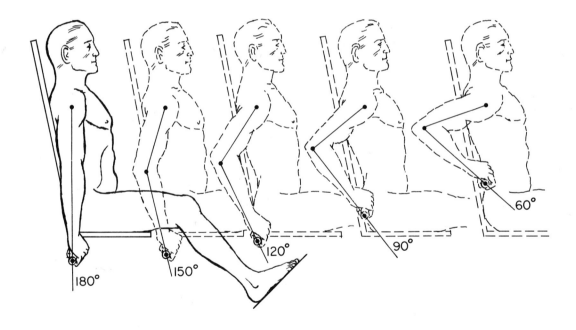

86. The elbow angles at which arm strength at various vertical positions was tested (after Hunsicker, 1957).

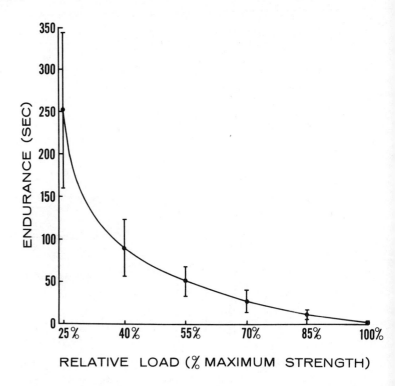

87. The modified prone position (after Brown *et al.*, 1949).

88. Strength endurance. Duration of pull on an isometric dynamometer handle in front of the shoulder, at varying percentages of maximum exertable strength. Subjects were 36 college students, 18 males, 18 females. Abscissa indicates the percentage of each subject's maximum strength that he or she was asked to exert; ordinate indicates time in seconds that these forces could be maintained. Black dots show mean values; vertical lines show ±1 standard deviation (from Caldwell, 1963).

Table 100. Forces Exertable on a Wheel Type of Control at Various Locations from the Modified Prone Position. (Brown et al., 1949, 1950)

Subjects and Test Conditions: a) For push, pull, right, left and rotation right and left (vertical plane), the subjects were 65 University of California students (63 ex-pilots), averaging 24 years of age, 70.7 inches tall, and 160 pounds in weight. Each subject lay on a padded bed tilted 3 - 10° above the horizontal, with a chest support tilted an additional 20°. In this position the subject had a horizontal line of sight. Shoulder straps and foot and ankle supports were used. Forces were exerted on a handlebar assembly with the vertical handgrips spaced 16 inches apart. Control movement was minimal, usually a fraction of an inch and never more than 2 inches. The applied forces were recorded at 9 control positions, 3 horizontal planes at each of 3 forward locations.

b) For up, down, rotation right and left (horizontal plane), the subjects were 37 university men averaging 24 years of age, 26 of whom also participated in part a) above. Test conditions were similar to those described above except that the vertical grips of the handlebar assembly were spaced 17 inches apart, and the maximum control movement was 1 1/2 inches. See Figure 87.

Table 100 continued.

Movement	Location of Control Distance In Front of Shoulder (Inches)	Distance Below Shoulder Level (Inches)	Maximum Force - Pounds Mean	S.D.	Percentiles 5th	95th
Pull	9	0	262	58.2	166	358
		5	259	50.0	177	341
		10	237	40.1	171	303
	13	0	256	47.7	177	335
		5	262	44.2	189	335
		10	240	41.7	171	309
	17	0	257	43.8	185	329
		5	257	41.8	188	326
		10	237	38.7	173	301
Push	9	0	114	22.4	77	151
		5	132	27.4	87	177
		10	134	34.1	78	190
	13	0	127	31.8	75	179
		5	148	40.5	81	215
		10	191	42.4	121	261
	17	0	199	75.4	75	323
		5	231	55.3	140	322
		10	194	46.1	118	270
Move up	9	0	99	21.0	64	134
		5	124	16.9	96	152
		10	153	21.7	117	189
	13	0	84	16.2	57	111
		5	115	18.1	85	145
		10	144	20.2	111	177
	17	0	75	12.2	55	95
		5	103	17.9	74	132
		10	129	21.5	94	164
Move down	9	0	174	30.6	124	224
		5	133	28.4	86	179
		10	143	35.7	84	202
	13	0	155	27.5	110	200
		5	137	28.4	90	184
		10	151	39.2	86	216
	17	0	132	29.6	83	181
		5	124	29.1	76	172
		10	89	28.2	43	135
Move right	9	0	112	28.3	65	159
		5	123	27.9	77	169
		10	113	29.2	65	161
	13	0	106	25.0	65	147
		5	107	24.3	67	147
		10	103	22.7	66	140
	17	0	92	23.0	54	130
		5	95	22.4	58	132
		10	85	22.2	48	122
Move left	9	0	107	33.3	52	162
		5	115	29.7	66	164
		10	111	25.0	70	152
	13	0	103	27.9	57	149
		5	102	24.0	62	142
		10	94	21.2	59	129
	17	0	89	22.1	53	125
		5	90	21.1	55	125
		10	80	19.2	48	112

Continued on next page.

Table 100 continued.

Movement	Location of Control Distance In Front of Shoulder (Inches)	Distance Below Shoulder Level (Inches)	Maximum Force - Pounds		Percentiles	
			Mean	S.D.	5th	95th
Rotate right (in horizon- tal plane)	9	0	176	32.7	122	230
		5	226	39.0	162	290
		10	244	51.7	159	329
	13	0	232	64.0	127	337
		5	282	67.1	172	392
		10	297	54.8	207	387
	17	0	309	73.8	188	430
		5	347	67.7	236	458
		10	309	40.7	242	376
Rotate left (in horizon- tal plane)	9	0	186	41.8	117	255
		5	221	41.8	152	289
		10	234	40.9	167	301
	13	0	222	63.5	117	327
		5	258	60.0	159	357
		10	286	46.4	210	362
	17	0	299	69.2	185	413
		5	336	61.5	235	437
		10	298	57.8	203	393
Rotate right (in verti- cal plane)	9	0	139	26.1	96	182
		5	126	24.6	85	167
		10	127	20.7	93	161
	13	0	127	23.8	88	166
		5	116	21.1	81	151
		10	115	22.4	78	152
	17	0	108	19.7	76	140
		5	101	20.2	68	134
		10	96	22.0	60	132
Rotate left (in verti- cal plane)	9	0	138	29.6	89	187
		5	128	23.2	90	166
		10	131	24.0	91	171
	13	0	129	29.6	80	178
		5	117	22.5	80	154
		10	116	23.3	78	154
	17	0	111	20.9	77	145
		5	100	21.8	64	136
		10	94	21.5	59	129

Table 101. Forces Exertable on a Handgrip in
Pronation and Supination.

(Hunsicker, 1957)

Subjects and Test Conditions: 25 University of Michigan students were selected to approximate the range of body measurements of U.S.A.F. flying personnel. The subjects sat erect, upper arms vertical, elbows at a right angle, and grasped the grip handle of a Kellogg Universal Dynamometer. (See Hunsicker and Donnelly, 1955.) In pronation the hand was turned palm downward; in supination, palm upward.

Movement	Maximum Force (Pounds)				
	Mean	SD	Percentile		Range
			5th	95th	
Pronation - right hand	71	28	29	119	27-120
left	71	31	31	132	29-140
Supination - right hand	64	18	35	93	33-107
left	62	16	30	88	28- 97

Table 102. Forces Exertable on Three Types of Handle.
(McFadden and Swearingen, 1958)

Subjects and Test Conditions: 8 men, aged 29-42, exerted maximum force on a handle for 3-5 seconds under increasing resistance until they were unable to turn the handle 180° in one continuous effort. Three types of handles were used: 1) "L" - a main shaft with a perpendicular bar; 2) "double L" - a main shaft with two bars perpendicular to it and to one another; 3) "T" - a main shaft with two bars perpendicular to it forming a straight line. All three were tested in 5-inch and 11-inch lengths. The force exertable was roughly the same whether the 5- or 11-inch handle was used, but the torque, in inch pounds, was about 2 1/2 times greater with the longer handle. The handles were mounted on a vertical surface 34 1/2 inches above the floor, with 22 inches of clearance to either side. Conditions were intended to simulate the operation of an aircraft cabin door and car door handles. Forces considerably in excess of those attainable in one continuous motion were possible if the subject was allowed to reposition his body and hands.

| Type of Handle | Position of Handle and Direction of Movement | | | Force (Pounds) | | | |
| | | | | Two Hands | | One Hand | |
				Mean	S.D.	Mean	S.D.
T-Handle	0°	or	180°	234	59.0	-	-
	45°		135°	224	57.7	-	-
	90°		90°	225	51.5	-	-
	135°		45°	239	64.1	-	-
Double-L-Handle	0°		90°	203	63.7	-	-
	45°		45°	156	56.2	-	-
	90°		0°	185	47.9	-	-
	135°		315°	166	40.5	-	-
	180°		270°	155	41.0	-	-
	225°		225°	190	51.3	-	-
	270°		180°	186	52.2	-	-
	315°		135°	201	55.0	-	-
L-Handle	0°		180°	156	60.0	150	51.7
	45°		135°	126	46.7	110	34.8
	90°		90°	128	38.6	86	23.8
	135°		45°	151	44.0	93	22.1
	180°		0°	141	43.1	108	20.3
	225°		315°	122	35.9	80	17.5
	270°		270°	138	35.0	90	23.2
	315°		225°	178	50.6	132	35.0

4. Speed of Body Motion

APPLICATION OF SPEED OF MOTION DATA

Information on the speed of human movements enables the design engineer to determine the time required to reach and operate a control, to open and pass through doors and escape hatches, or simply to move a given distance. Such calculations should include both the maximum speed attainable in single exertions or over brief periods, and the speed encountered in normal operation. Critical tasks, like escape through airplane hatches, should be planned with the slowest members of the using population in mind.

REACTION TIME AND MOVEMENT TIME

The time required for a body movement consists of two phases, reaction time and movement time.

Reaction Time

This is the period between perception of a stimulus and the beginning of a bodily response; it can be simple or complex. (1) Simple reaction time is the shorter and involves a predetermined response to a given stimulus, like pressing a button after perceiving a signal. Even the slowest person, barring physical disability, can usually start a physical movement within 0.5 sec-

ond after a signal is given—if the signal is expected and no further decision is necessary. Simple reaction time may be as rapid as 0.2 second or, more rarely, 0.1 second (Woodworth and Schlosberg, 1955). See Table 103. (2) Complex reaction time involves a decision between two or more possible responses to a given stimulus, such as responding to one of ten numbered signal lights by pushing a similarly numbered button. Under such circumstances, response may take 0.6 or 0.7 second (Table 104). Under less standardized conditions which require judgment and discrimination, from 1 to 2 seconds or more may be needed for reaction time (Orlansky, 1948); with inexperienced persons or more difficult decisions, considerably more.

Both simple and complex reaction times are subject to various influences, one of which is the duration of the foreperiod, or interval between a "get-set" signal and the stimulus. Other influences will be discussed below.

Movement Time

This is the period between the first measurable beginning of a body movement and any selected end-point of that movement. There are four basic kinds of body movements: (1) Positioning movements—in which the hand, foot, or other part of the body is placed at a specific location. These consist of primary movements, ending near the desired location, and secondary cor-

Table 103. Simple Reaction Time.

Test Conditions: The subjects pressed (or released) a button immediately upon receiving a stimulus. Time was measured between the onset of the stimulus and the activation of the button. The data were gathered under laboratory conditions and hence the values are minimal; at least 0.5 second more might be expected under working conditions. The values below are derived from various population groups as reported by Baxter, 1942; Forbes, 1945; Orlansky, 1948; Slater-Hammel, 1952; Teichner, 1954; and Woodworth, 1955.

Stimulus	Reaction Time - Seconds	
	Mean	Range
Sound	0.20	0.12-.45
Touch	.20	.12-.45
Light	.30	.15-.50

Table 104. Complex Reaction Time.

Test Conditions: The subjects responded to 1-10 different stimuli, each requiring a distinct physical movement. Each subject reacted by moving a specified finger or pressing a specified button indicated by the stimulus. Time was measured between the onset of the stimulus and the ensuing movement of the finger or button. The data were gathered under laboratory conditions and hence reaction times are minimal. Reaction time under working conditions will be at least 1-2 seconds longer depending on the number and complexity of the choices and the experience of the individual. The values given below synthesize data from Merkel, 1885; McFarland, 1937; Baxter, 1942; and Hick, 1952.

Number of Choices	Approximate Mean Reaction Time (Seconds)
1	0.20
2	.35
3	.40
4	.45
5	.50
6	.55
7	.60
8	.60
9	.65
10	.65

rective movements for precise placement. (2) Repetitive movements—in which a single action is repeated, as in hammering or turning a screw driver. (3) Continuous movements—in which movements are continuously adjusted in response to changing stimuli, as in tracking a moving target or steering a vehicle. (4) Serial movements—in which distinct movements are combined in sequence, as in typewriting or piano playing (McCormick, 1957).

FACTORS INFLUENCING SPEED OF MOTION

The speed of body movement is influenced by the characteristics of the initiating stimulus, the characteristics of the movements themselves, biological and psychological differences among individuals, and environmental variables.

Characteristics of the Initiating Stimulus

While stimulus characteristics do not affect speed once a body movement has started, they do directly influence reaction time and hence total movement time as measured from the onset of the stimulus. The following data are summarized mainly from Teichner's (1954) survey of reaction time studies.

Type: People react more quickly to auditory and tactile than to visual stimuli. (See Table 103.)

Intensity: Reaction time decreases as the intensity of a stimulus increases, though the relationship is not linear. This apparently holds true for all kinds of stimuli.

Duration: When the stimuli are shorter than the reaction time, lengthening very short stimuli shortens reaction time.

Location: Visual reaction time is least when stimuli are located in front of the subject and increases as the stimuli move to the sides. Data are not available on the effect of stimulus location on reactions mediated by the other senses.

Number of sense organs and receptors stimulated: Reaction time is faster when two eyes or ears are used rather than one, and when two senses are stimulated simultaneously, vision and hearing. Reaction time to light decreases as the stimulated area of the retina increases.

The foreperiod: Reaction time is shortened if a preparatory signal is given before the final stimulus. The optimum duration of this foreperiod is between $1\frac{1}{2}$ and 8 seconds.

Characteristics of the Movements

Speed of body movement varies with the type, direction, extent, force, and precision of movement. The influence of each of these factors is summarized briefly below and in the Tables when relevant. Reaction time is unrelated to the speed of the ensuing movement, the correlation between the two being close to zero (Slater-Hammel, 1952).

Type: Continuous movements, whether straight or curved, are faster than movements of the same length involving change in direction.

Direction: For short, accurate, positioning movements the vertical plane is the fastest, followed by the lateral and fore-and-aft planes. Specific directions, from fastest to slowest, are: up and down (no difference), right, pull, left, push. When accuracy is combined with speed, the order is, from most to least accurate: down, pull, up, right and push (no difference), left (Herbert, 1957). The foregoing orders of speed will vary with other factors, such as distance.

Extent: Although short movements take less time than longer movements, all other things being equal, the time taken for longer movements does not increase directly with distance because of acceleration. For example, it may take 0.18 second to move the hand 1 inch, but only 0.22 second to move it twice as far. Similarly, the hand can be moved 12 inches in only twice the time it takes to move it 1 inch (Bailey and Presgrave, 1958).

For repetitive finger movements, as in telegraphy, the extent of the movement has little effect on speed —at least during 5-second periods—between a range of roughly $\frac{1}{25}$ inch to $1\frac{1}{2}$ inches. Arcs of finger movement of about $\frac{4}{5}$ inch were slightly faster than either of these two extreme amplitudes (Bryan, 1892).

Force: Speed decreases consistently, though not linearly, as additional force is required. For example, it took less than twice as long to apply 50 pounds of pressure over a 6-inch distance as to apply only 2 pounds over the same distance (Bailey and Presgrave, 1958). Similarly, decreasing weights moved by the arm from 18 to 9 pounds resulted in only a 25 per cent increase in maximum velocity (Koepke and Whitson, 1940). Weights or resistances of only a few ounces or a pound or two affect speed of movement negligibly.

Precision: Speed decreases as additional precision is required. For example, a visually directed hand positioning movement 10 inches long took about 0.6 second when accuracy to $\frac{1}{2}$ inch was needed, but the same movement took 1.3 seconds, over twice as long, for accuracy to $\frac{1}{32}$ inch (Bailey and Presgrave, 1958).

Biological and Psychological Differences Among Individuals

Among healthy adults, speed of motion is less variable than muscle strength. The most pertinent biological and psychological factors are briefly summarized below. Reaction time has been more thoroughly investigated than speed of movement. The reaction time data discussed below are from Teichner's (1954) survey unless otherwise indicated.

Age: Both reaction time and movement time are fastest at about age 20, and both increase with age thereafter, although this increase is more pronounced for movement time (Pierson and Montoye, 1958). Twenty-four subjects from 62 to 86 years of age reacted to a visual stimulus 11 per cent more slowly than did twenty-four subjects from 19 to 27 years of age (0.275 second versus 0.245 second). For the same groups, the older subjects were 38 per cent slower in a simple guided movement of a control handle to the right or left (0.278 second versus 0.173 second) (Deupree and Simon, 1963). To align various dials and markers, a 70-year-old group took 45 per cent longer than a 22-year-old group. In moving their hands from one dial to another, the older group required 14 per cent more time (Simon, 1960).

Sex: Most of the evidence indicates that men have slightly faster reaction times than women. The difference is about 0.023 second (Seashore and Seashore, 1941). Men also have faster maximum speeds of body locomotion. For example, the world records (in 1964) for the 100-yard dash were 9.2 seconds for men and 10.3 seconds for women. For longer distances the differences are more marked; 44.9 seconds and 53.3 seconds, respectively, for 440 yards. The speed of manipulatory movements has not been established.

Body build: Reaction time is not closely related to body build (Janoff *et al.,* 1950), nor apparently is speed of movement (Pierson, 1962b), although the ability to exert forces quickly and accurately does vary with muscularity. Increased muscularity does not retard speed of movement (Hunsicker, 1955).

Body position: Reaction time increases immediately after any change in body position but quickly returns to normal. Data on speed of movement from different positions are inadequate.

Part of body used: Simple reaction times of the hands are faster than those of the feet by about 0.03 second (Seashore and Seashore, 1941). Right and left sides do not differ; the arms move slightly faster than the legs, the right side faster than the left. The first or index finger is the fastest, followed in order by the second, third, and fourth fingers.

Health: Speed of movement decreases markedly with starvation and with neuromuscular or joint disease (Brozek *et al.,* 1952).

Fatigue: Fatigue induced by lack of sleep for 100 hours or more does not increase reaction time but slows body movement in proportion to the degree of fatigue.

Motivation: This affects both reaction time and speed of movement.

Environmental Variables

The speed of body movement may be influenced by environmental factors, including acceleration, altitude, and temperature.

Acceleration: Body movements are progressively slowed under increasing acceleration, being slowest when the body movement is opposite in direction to the acceleration (Brown and Lechner, 1956). Visual reaction times are also slower under increasing accelerations. In one instance reaction times were increased by an additional 0.05 to 0.20 second between 1 and 4 G, the precise amount varying with the subject and the level of illumination (Brown and Burke, 1958).

Altitude (anoxia): Both simple and complex reaction times were slower at altitude than at sea level (McFarland, 1937). These effects were most marked above 20,000 feet and among unacclimatized persons.

Temperature: There appears to be no change in simple or complex reaction time within an ambient temperature range of −50°F and 117°F. However, at windspeeds of 10 mph and above, ambient temperatures of −15°F and below resulted in a significant increase in reaction time which was probably psychological in nature (Teichner, 1958). Ambient temperatures did not directly affect speed of movement if stiffness due to cold could be avoided.

Clothing and personal equipment: These can reduce speed of movement (McKee, 1957), especially bulky and restrictive cold-weather gear. Gloves, if fitted snugly, have little influence on operation time for most types of hand controls (Bradley, 1957).

MEASURING SPEED OF MOTION

Speed of motion data from different sources may not be directly comparable if they were obtained under different conditions. Reaction time may or may not be included in movement time. The distance, direction, force, and precision of the movement may all vary, as may the number, age, sex, and physical condition of the subjects. When possible, such descriptive information is included in the tabular data presented below.

Warning: Reaction times and movement times take longer in practical situations than in laboratory studies. In one experiment, it took vehicle drivers about 0.5 second longer to respond to a braking stimulus by releasing the accelerator, than it took average subjects to respond to a laboratory stimulus by pressing a button (0.7 second versus the 0.2 second of Table 103) (Bayly-Pike, 1950).

THE DATA

Reaction times are measured in seconds, speed of movement in inches per second or revolutions per minute. Because of the small number of subjects usually involved in speed of motion studies it is not feasible to present the data in percentile tables; hence, mean rates are given. For reaction time, the data in Tables 103 and 104 are summarized from representative studies; for speed of motion, Tables 105 to 115 present the actual experimental data.

Table 105. Speed of Arms and Hands in Various Movements.

Test Conditions: The subjects moved their hands from one position to another as fast as possible consistent with task requirements. Time was measured from the beginning to the end of the physical movement. The number of subjects varied between 5 and 18. Differences in test conditions are noted.

Distance Moved (Inches)	Direction Moved	Force Applied or Weight Moved	Accuracy of Primary Importance?	Velocity (In/sec)
1.0	Right	Negligible	Yes-positioning movements	3.2
1.0	Left			3.6
3.9	Right			7.4
3.9	Left			8.1
15.7	Right			20.6
15.7	Left			21.8
			(Brown and Slater-Hammel, 1949)	
2 3/4	Left	Negligible	Yes-positioning movements	8.9 [a]
2 3/4	Right			9.2 [a]
2 3/4	Up			9.4 [a]
2 3/4	Down			9.4 [a]
2 3/4	In			9.2 [a]
			(Herbert, 1957)	
1	Varied (little difference)	Negligible	No	8.4
6				52.8
11				89.0
16				115.0
			(Peters and Wenborne, 1936)	

Table 105. Continued.

Distance Moved (Inches)	Direction Moved	Force Applied or Weight Moved	Accuracy of Primary Importance?	Velocity
Varied	Varied	6 lbs	No	175
		9		150
		12		140
		15		130
		18		120
		21		110
			(Koepke and Whitson, 1940)	
45°arc	Elbow flexion	Negligible	No	441 degrees/sec.
			(Brozek et al., 1952)	
45° arc	Elbow flexion	Negligible	No	425 degrees/sec.
45° arc	Elbow flexion			448 degrees/sec.
			(Glanville and Kreezer, 1937)	

[a] Includes auditory reaction time, averaging 0.2 second (Table 101).

Table 106. Speed of Arm and Hand Movements with Control Stick.

Test Conditions: The subjects moved an aircraft control stick as fast as possible against varying resistance. These data summarize the results of four studies: Advisory Committee for Aeronautics, 1916; Hertel, 1930; Beeler, 1944; Orlansky, 1948.

Distance Moved (In Inches)	Direction of Movement	Resistance to Movement (Pounds)	Velocity (In./Sec)
± 6	Pull[a]	< 20	75 +
± 6		20 - 100	50 - 75
± 6		100 - 200	25 - 50

[a] Push is about 25% faster (Orlansky, 1948).

Table 107. Speed of Reaching for and Operating Toggle Switches--with Preceding Cue.

(Green and Muckler, 1959)

Test Conditions: The subjects were 10 adult males, 5 being U.S. Air Force pilots on flying status. Each subject, seated in a fighter cockpit mockup, had to reach and operate 9 toggle switches located in different cockpit positions. The subject was first alerted by a 10-second cue light indicating the approximate location of the ensuing stimulus light adjacent to the toggle switch to be activated. Upon perceiving the stimulus light, the subject removed his right or left hand, whichever was closer, from a sidestick control grip located in front of each armrest, and operated the designated switch. Response time was measured from the onset of the stimulus to the activation of the switch.

Hand Used	General Direction of Hand Movement	Toggle Distance From Hand Starting Point (Inches)	Mean Speed of Movement (Seconds)
Right	Forward	6	0.76
Left	Forward	6	.80
Right	Forward	9	.66
Left	Forward	9	.78
Right	Forward	15	.84
Left	Forward	15	.90
Right	Forward	18 1/4	.77
Left	Forward	18 1/4	.78
Either	Forward, center	23	.86

Table 108. Speed of Handwheel Cranking.

<u>Test Conditions</u>: The subjects cranked handwheels of various
radii and loads, for varying periods of time. The location and
orientation of handwheels make little difference in cranking
speed and accuracy as long as they can be operated comfortably.
In the following studies, accuracy was not required.

Number of Subjects	Crank Radius (in Inches)	Crank Load	Cranking Time	RPM
75	4	None	10 Minutes	153
		10 In./Lb	10	114
		30	3	108
		50	2 1/2	83
		70	1 1/2	88
		90	1 Minute	95
	5	None	10 Minutes	150
		10 In./Lb	10	119
		30	7	91
		50	4	96
		70	1 1/2	113
		90	1 Minute	108
	7	None	10 Minutes	141
		10 In./Lb	10	121
		30	10	90
		50	7	87
		70	2	114
		90	1 Minute	93

(Katchmar, 1957)

Table 108. Continued.

Number of Subjects	Crank Radius (in Inches)	Crank Load	Cranking Time	RPM
40	1.2	None	30 Seconds	145
	2.0			135
	2.8			115
	3.5			105
	4.3			95
	5.1			90

(Baines and King, 1950)

Table 109. Speed of Handwheel Cranking in Pointer Alignment.

Test Conditions: The subjects cranked a handwheel which controlled the movement of a pointer. The purpose of the test was to align, as accurately as possible, this pointer with another moving pointer.

Number of Subjects	Crank Radius (Inches)	Crank Load (Pounds)	Cranking Time (Minutes)	RPM
4 to 14	2 1/4 and 4 1/2	2 1/2 and 9	3 to 4	50-200 (180 Optimum)

(Foxboro, 1943a, 1943b; Helson, 1949)

Table 110. Speed of Finger Tapping.

Test Conditions: For telegraphy, the subjects tapped a key 100 times as rapidly as possible; for typewriting, the subjects tapped the keys as many times as possible in 15 seconds.

Number of Subjects	Task	Duration	Taps per Second, Maximum		Taps per Second, Preferred Range
			Mean	Range	
12	Telegraphy	100 tap series	8.5	5-14	1 1/2 - 5

(Miles, 1937)

			Taps per Second							
			Left Hand Fingers				Right Hand Fingers			
			4th	3rd	2nd	1st	1st	2nd	3rd	4th
Unknown	Typewriting	15 seconds	3.2	3.8	4.2	4.4	4.7	4.6	4.1	3.7

(Dvorak et al., 1936)

Table 111. Speed of Leg Flexion.

(Brozek et al., 1952)

Test Conditions: The subject flexed his thigh at the hip once, from the standing position, as rapidly as possible.

Number of Subjects	Type of Movement	Distance Moved	Force Required	Mean Velocity
11	**Leg flexion at hip**	45° **Arc**	Negligible	425 (Deg/sec)

Tables 112 - 115. Speed of Movement - Time and Motion Study Data.

<u>Test Conditions</u>: The following data were determined by stop-watch measurements on a large series of movements of different types and lengths. About 9 subjects were used. These movement times were not the fastest attainable, but "standard" or average times, equivalent to a work pace of "walking 3 miles per hour," or "dealing 52 cards into 4 hands in 30 seconds." The data were compared with time-study data of actual jobs and found to "produce reliable results" (Bailey and Presgrave, 1958). Although the experimental conditions leave much to be desired, the data are unique, to our knowledge, and are presented here not as final values, but as approximations.

Table 112. Time and Motion Data - Speed of Body Motion.
(Bailey and Presgrave, 1958)

Motion	Time - in Seconds
Walking (one pace)	0.60
Sitting	1.32
Standing	1.62
Kneeling on one knee	1.08
Arising from one knee	1.20
Kneeling on both knees	2.64
Arising from both knees	2.88
Bending, hands at or near floor	1.08
Stooping or squatting	1.08
Turning body, moving one foot	0.66
Turning body, moving both feet	1.32
Bending ankle, not over 6-8 inches	0.33
Moving leg at hip or knee, 1-6 inches	.30[a]
Moving eyes and focusing on new point	.48

[a]Plus 0.012 second for each inch above six.

Table 113. Time and Motion Data - Time (in seconds) to Move the Hand Various Distances.

(Bailey and Presgrave, 1958)

Test Conditions: Required precision was to one inch, force less than 2 pounds; one hand only, or both hands if visual direction was not required. See Table 115 for increments to be added under other conditions.

Type of Motion (See Below)	Length of Movement - In Inches																		
	1	2	3	4	5	6	7	8	9	10	12	14	16	18	20	22	24	26	28	30
A	.18	.22	.23	.25	.27	.28	.30	.31	.32	.34	.36	.38	.41	.43	.46	.48	.50	.53	.55	.58
B	.22	.25	.28	.29	.31	.33	.35	.36	.37	.38	.41	.43	.46	.48	.50	.53	.55	.58	.60	.62
BV	.25	.29	.32	.34	.36	.38	.40	.41	.42	.44	.46	.49	.51	.53	.56	.58	.61	.63	.65	.68
C	.29	.33	.36	.38	.41	.43	.44	.46	.47	.49	.52	.54	.56	.59	.61	.64	.67	.69	.71	.74
CV	.32	.37	.40	.43	.46	.47	.49	.51	.52	.54	.57	.59	.62	.65	.67	.70	.72	.74	.77	.79

Type A motions are stopped by impact with a solid object. No muscular effort is used to slow down or stop the action. Examples: downstroke in hammering; slamming a door or drawer.

Type B motions are stopped entirely by muscular control. Examples: upstroke in hammering; opening a door or drawer; tossing an object aside; moving controls without automatic stops. If the action is visually directed it becomes a BV motion.

Type C motions are stopped by muscular control preliminary to a grasping or placing action. Examples: reaching for and grasping a control or other object. Placing an object on a table or desk. If the action is visually directed it becomes a CV motion.

Table 114. Time and Motion Data - Time (in seconds)
to Rotate the Forearm Various Amounts.

(Bailey and Presgrave, 1958)

Test Conditions: Required precision to one inch, force less than 2 pounds. One hand only, or both if visual direction is not required. See Table 115 for increments to be added under other conditions.

Type of Motion (See Table 113)	Amount of Rotation - In Degrees							
	30	45	60	75	90	120	150	180
A	.16	.17	.19	.20	.22	.26	.29	.32
B	.20	.22	.24	.26	.28	.32	.36	.40
BV	.24	.26	.29	.31	.34	.39	.43	.48
C	.34	.36	.38	.41	.43	.49	.53	.58
CV	.44	.46	.49	.51	.53	.59	.63	.68

Table 115. Speed of Movement - Additional Time Required in Hand Movements for Precision, Simultaneous Motion, or Force.

(Bailey and Presgrave, 1958)

Precision - Additional time required, in seconds, for the indicated degrees of precision. Add to the BV or CV values in tables 113 and 114.

Amount of Precision Required (Inches)	Length of Movement (Inches)										
	1	2	4	6	8	10	14	18	22	26	30
1/2	.02	.02	.04	.05	.07	.08	.10	.11	.12	.13	.14
1/4	.08	.10	.13	.15	.17	.19	.23	.27	.31	.33	.35
1/8	.20	.22	.27	.31	.35	.37	.43	.48	.52	.56	.61
1/16	.36	.39	.44	.48	.52	.56	.62	.67	.71	.76	.81
1/32	.54	.58	.64	.68	.72	.76	.81	.86	.90	.94	.99

Simultaneous Motions - Additional time required, in seconds, for simultaneous hand motions, each of which must be visually directed. Add to the BV or CV values in tables 113 and 114.

Amount of Precision Required (Inches)	Separation Distance of Hands (Inches)												
	0	2	4	6	8	10	12	14	16	18	20	22	24
1/2	0	.06	.11	.16	.20	.25	.28	.32	.35	.39	.41	.44	.47
1/8	0	.07	.13	.18	.22	.26	.31	.34	.38	.41	.44	.47	.49
1/16	0	.09	.16	.22	.27	.32	.37	.41	.45	.48	.52	.55	.58
1/32	0	.11	.20	.28	.35	.41	.46	.50	.54	.58	.62	.64	.67

Continued on next page.

Table 115. Continued.

Force - Additional time required, in seconds, for applications
of forces of two or more pounds through various distances, as
in exerting pressure, lifting or carrying. Add to the values
in table 113 and 114.

Amount of Force Required (Pounds)	Length of Movement (Inches)		
	6 or less	12	24
2	.01	.02	.02
4	.04	.04	.04
6	.05	.05	.06
8	.06	.07	.08
10	.08	.08	.10
15	.11	.12	.13
20	.14	.16	.17
30	.19	.21	.23
40	.23	.26	.28
50	.27	.30	.33

Human Body Composition and Tolerance to Physical and Mechanical Force — IV

1. Body Composition

INTRODUCTION

Human body composition can be described in terms of chemical constituents, the cells and tissues formed by these atomic or molecular building blocks, and the organs composed of cells and tissues. A brief presentation of such data can aid the engineer in designing certain kinds of equipment, particularly those concerned with extremes of acceleration, altitude, temperature, vibration, or radiation. Occupants of advanced aircraft or spacecraft may encounter all of these environmental extremes.

Body composition and function have also come to interest the physical anthropologist as he moves beyond his traditional preoccupation with body form to the study of mechanisms, whether to explain human adaptation or to apply his findings to engineering and medicine.

A few examples may be given of the utility of body composition data. (1) *Radiation.* Cosmic or other ionizing radiation may create new chemical compounds in the body. Estimates of the total amount or percentage of various elements in the body may help to predict or assess radiation damage. (2) *Logistics.* Knowledge of the percentage of water and solids in the body is useful in formulating rations. (3) *Mechanics.* The amount and nature of body components play a vital role in determining resonance frequencies and the response of organs and other body components to acceleration, blast, and vibration (see p. 262). (4) *Temperature.* Efficiency and survival under extremes of temperature and humidity are related to amounts of body fat, which acts as an insulator. For example, Baker and Daniels (1956) reported a difference of 1.2°F (0.7°C) in rectal temperature and 5.9°F (3.3°C) in skin temperature among men varying widely in estimated body fat, after sitting almost nude and inactive for 2 hours at 59°F (15°C). The fatter men had higher rectal but lower skin temperatures.

SOURCES AND DATA

Surprisingly few human bodies have been analyzed chemically. This may be because of difficulty in obtaining suitable cadavers or in performing the analyses, but more likely it reflects the biochemist's realization that all body constituents undergo a constant, rapid turnover. Modern tracer studies have shown that even tissues long considered relatively stable, like fat and bone, are quite active metabolically. For example, the half-life of liver fats in small mammals is around 2.7 days, that of "depot" fats—both internal and subcutaneous—5 to 6 days. Thirty per cent of the dietary phosphorus incorporated into bones is eliminated within 20 days. As a result, the dynamics of metabolism have been much more thoroughly investigated than body or tissue composition, which represents a static snapshot of a moving picture.

The number of total body analyses published is not more than twenty, including five to ten adults; Forbes, Cooper, and Mitchell (1953) summarize three modern studies. The familiar figures for the monetary value of total body chemicals, varying from $0.89 to $1.17 or so, rest on single cadavers analyzed in Germany in 1863 (Bischoff, cited by Magnus-Levy, 1910) or in 1910 (Magnus-Levy), and quoted in textbooks ever since (*e.g.*, West and Todd, 1961). Most current estimates are therefore derived from a variety of *indirect* measures involving such techniques as gas or dye dilution, isotopic tracers, densitometry, and anthropometry. All involve assumptions, extrapolations, and appreciable error, as pointed out by Keys and Brozek (1953) in an excellent review. For details of technique and findings in the study of body composition, the reader is referred to the accounts of three Conferences (1956, 1961, and 1963) and to Brozek's (1963) summary.

All body constituents, from the largest (water) to the smallest (iodine and some organic compounds), vary widely from one person to another, both in absolute amount and in percentage of total body weight. For this reason, as well as the paucity of direct analyses, Tables 116 to 120, which follow, are *approximations only*.

Total Body Composition—Chemical Elements

The chemical elements of the body tell us little about the kinds of substances of which the body is made; for this we need their molecular arrangement. Much of the oxygen and hydrogen in the body appears as water, the carbon and nitrogen as protein, the calcium and phosphorus as bone, and so on. Table 116 shows body composition by major chemical elements.

Total Body Composition—Molecular

The water content is high in some tissues, such as whole blood (79 per cent water) and blood plasma, which is blood minus the corpuscles (91 per cent

Table 116. Approximate Elementary
Composition of the Body.
(Dry Weight Basis-Exclusive of Water)
(Williams, 1942)

Element	Per cent
Carbon	50
Oxygen	20
Hydrogen	10
Nitrogen	8.5
Calcium	4
Phosphorus	2.5
Potassium	1
Sulfur	0.8
Sodium	0.4
Chlorine	0.4
Magnesium	0.1
Iron	0.01
Manganese	0.001
Iodine	0.00005
"Trace Elements"	--------

From Williams' A Textbook of Biochemistry, Copyright 1942,
D. Von Nostrand Company, Inc., Princeton, N.J.

Table 117. Human Body Composition -
Approximate Percentages.
(Williams, 1942)

Water	65	
Solids	35	
Protein		15
Fats		15
Minerals		5
Carbohydrate		0.5

Table 118. Subcompartments of Body Water --
Approximate Percentages.
(Edelman and Leibman, 1959)

Intracellular - 55	
Extracellular - 45	
Interstitial and lymph	20
Blood	7.5
Connective tissue, cartilage	7.5
Bone	7.5
Other (glands, gastro-intestinal, spinal fluid, eye)	2.5
Total extracellular	45

Table 119. Subcompartments of Body Fat --
Approximate Percentages.

Subcutaneous	49
Internal "depots" (abdominal, perirenal)	49
"Essential lipids" (bone marrow, nervous system)	2

Table 120. Relative Weight of Body Tissues and Organs.
(Based on 2 men, aged 33 and 46)
(Bischoff, 1863, cited by Magnus-Levy, 1910;
Forbes, Cooper, and Mitchell, 1953)

| Tissue or Organ | Per Cent of Body Weight | |
	Age 33 – Weight, 153 Pounds	Age 46 – Short, Slender
Muscle	41.5	40.0
Fat	18.0	11.4
Skeleton	15.8	17.6
Blood	8.0	--
Skin	6.9	6.3
Brain	2.4	3.0 (Including spinal column)
Liver	2.3	2.3
Alimentary tract	1.8	1.9
Lungs	1.4	3.3 (Containing blood)
Heart	0.5	0.5
Kidneys	0.4	0.5
Spleen	0.2	0.1
Testes	0.04	--

water); and low in others—notably bone (25 per cent water) and adipose tissue (3 per cent water). The composite figure of 65 per cent therefore becomes 72 per cent ±S.D. 3 per cent of water in the fat-free body of normal men (Table 117).

By far the greatest variability among human body components is in fat. One series of 2,017 white soldiers, fairly homogeneous in age, had a coefficient of variation, S.D./mean, of 69 per cent for fat as a percentage of body weight, whereas weight itself had a coefficient of 15 per cent, and height 4 per cent (Newman, 1956). Among thirty-one healthy men from 19 to 42 years of age, Baker and Daniels (1956) found fat percentages (estimated from skinfold measurements) to vary between 2 and 19 per cent of body weight. Among fifteen medical students, Siri (in Lawrence, 1956) reported that fat percentages varied from 5.9 to 49.3 per cent of body weight, while water varied from 37 to 68.7 per cent. Subsequent refinements in technique make fat percentages as low as 2 per cent and as high as 68 per cent unlikely in normal persons. Thus, whereas water in normal persons of similar age and sex can vary almost two-fold, as can calcium also, fat can vary almost twenty-fold. (In disease states such as cardiac or renal failure, or malnutrition, which involve fluid accumulation or "edema," the water percentage can rise to 80 per cent and above.)

Distribution of Water and Fat

The distributions of water and fat in the body, of importance to the designer as well as to the biological scientist, are shown in Tables 118 and 119. The density of human fat is a fairly constant 0.9000 gm/cc (Fidanza, Keys, and Anderson, 1953).

Tissues and Organs

To understand body organization and function, the composition of individual tissues and organs is more important than body composition as a whole. Table 120 presents data on the weight, relative to total body weight, of some important tissues and organs.

Estimating Fat from Body Measurements

Fortunately the simple technique of skinfold measurement permits reasonably close estimation of body fat for young men (Damon and Goldman, 1964). The skinfold calipers used should exert a pressure of 10 gm/mm^2 over an area of 20 to 40 mm^2; the Lange caliper (manufactured by Cambridge Scientific Instruments, Inc., Cambridge, Maryland) is a satisfactory design, as is the Harpenden caliper (Holtain, Limited, Brynberian, near Crymmych, Pembrokeshire, Wales), despite the 90-mm^2 area of its jaw faces. Measurements are made over the right triceps muscle and below the inferior angle of the right scapula.

The triceps skinfold is measured midway between the acromion (shoulder) and olecranon (elbow) points, the precise level being determined with the forearm flexed at 90 degrees. With the subject's arm hanging down freely, the skinfold is lifted parallel to the long axis of the arm. The subscapular skinfold is lifted in a direction parallel to the ribs, with the skinfold angled upward medially and downward laterally at about 45 degrees from the horizontal. At both sites, the skin should be lifted by grasping firmly a fold between the thumb and forefinger about 1 centimeter from the point at which the caliper is to be applied. Readings are made within 3 seconds after application of the caliper.

For athletic young men, the percentage of body fat can be estimated to within 2 per cent of that determined by densitometry, or underwater weighing. First, body density is determined by averaging the values obtained from two equations: (a) Density = 1.0923 − 0.0202 (triceps skinfold, in cm); (b) Density = 1.0896 − 0.0179 (subscapular skinfold, in cm). Then, percentage of body fat = (4.0439/density) − 3.6266. The formulas for density are those of Pascale *et al.* (1956); that for body fat is Grande's (Conference, 1961: p. 128).

Although the same two skinfolds were the best anthropometric predictors of body fat among Chinese men, both young and middle-aged, as well as young women (Chen, Damon, and Elliot, 1963), prediction equations have not yet been validated for groups other than athletic young men.

The Skeleton

For the equipment designer, bone and muscle are the chief solid tissues of the body. Bone and muscle weight, as percentages of total body weight, appear in Table 120; the passive strength of muscle tendon and of bone will be discussed on pp. 275–285, below. Table 121 presents the major components of the skeleton as percentages of total skeletal weight, based on fifty men and fifty women averaging 65 years of age.

AGE, SEX, AND RACE DIFFERENCES

Individual variation in body composition has already been discussed, with ranges of two-fold for water and calcium and twenty-fold for fat noted as occurring among normal persons of the same age, sex, and race. The further influence of age and sex on body composition is indicated in Tables 122–123.

Fat increases with age, and women are fatter than men, relative to total body weight. Since the percentages of fat and water in the body vary inversely, body water decreases with age and is a greater percentage of total body weight in men than in women. With age, both subcutaneous and internal fat increase, the latter more rapidly, so that older persons have relatively less subcutaneous and more internal fat.

Table 121. Portions of Skeleton as Percentages of Total Skeleton Weight. (after Seale, 1959)

| | Number | Per Cent of Total Skeleton Weight | | | | |
		Skull	Spine, Ribs, Sternum	Arms	Legs	Total
Males	50	18	17	19	46	100
Females	50	22	18	17	43	100

Table 122. Body Fat, Age, and Sex Estimated for Persons of "Standard" Weight for Height and Age. (Brozek, 1952; Brozek et al., 1953)

| Age (Years) | Total Body Fat - Estimated (Per Cent) | |
	103 Men	62 Women
25	13	26
35	17	30
45	22	34
55	26	38

Table 123. Body Water, Age, and Sex.
(after Edelman and Leibman, 1959)

		Total Body Water (Per Cent of Body Weight)			
	Males			Females	
Age	Number	Mean[a]		Number	Mean[a]
10-16	11	59		7	57
17-39	248	61		61	50
40-59	127	55		38	47
60-+	20	52		14	46

[a]Standard deviations not given.

As for body water, its percentage fell from 76 per cent among twenty newborn infants to 65 per cent in fifteen infants under 1 year of age, and to 62 per cent in twenty-four children 1 to 10 years of age. Both boys and girls were included. The figures for succeeding ages appear in Table 123.

Racial differences in body fat and water remain to be worked out. Since 361 Negro soldiers showed (Newman, 1956), on the average, considerably less estimated fat than 2,017 white soldiers of the same age (4.6 per cent of body weight as fat, versus 7.4 per cent for whites), one might expect—if body fat and water are inversely related—Negroes to have a higher percentage of body water. However, in a major component of body water, blood volume, seventeen Negro soldiers had only 88 per cent of the total blood, plasma, and red cell volumes of white soldiers matched for age, height, weight, and fatness. See Table 124 (Bass, Iampietro, and Buskirk, 1959). Comstock and Livesay (1963) reported that in Muscogee County, Georgia, Negro males of military age had as much subcutaneous fat as whites, or very slightly less. Negro women below 40 years of age had generally more fat than white women. In a third study which did not support Newman's findings,

Table 124. Whole Blood, Plasma, and Red Cell Volumes of Selected White and Negro Soldiers.
(Bass, Iampietro, and Buskirk, 1959)

	17 Whites		17 Negroes	
	Mean	S.D.	Mean	S.D.
Age, years	21.1		22.3	
Weight, pounds	153		151	
Surface area, meters^2	1.80		1.79	
Fat, per cent of weight	9.7		9.3	
Blood volume, liters	5.71	0.51	5.05	0.38
Plasma volume, liters	3.06	.33	2.72	.26
Red cells, liters	2.65	.22	2.33	.26

Chen, Damon, and Elliot (1963) reported no differences in skinfolds or estimated fat between 65 Negro and 355 white American soldiers, between young Chinese and American women, and between middle-aged Chinese and American men. Young Chinese men, whether students or soldiers, were about two-thirds as fat as young Americans of corresponding occupation.

American Negroes are more mesomorphic than white persons, young soldiers by one point and middle-aged women by $\frac{1}{2}$ point on a 7-point scale (Damon, 1960; Damon *et al.*, 1962). The difference among the young men parallels that for bone weight in Table 125. Since mesomorphy is a rating of bone and muscle and in fact correlates with these tissues (Baker, Hunt, and Sen, 1958), one may infer that young Negro men are about 10 to 15 per cent more muscular than Caucasians. More data are needed.

As for bone, Negroes have significantly heavier skeletons than white persons of comparable age. Among young American soldiers who died in Korean prisoner of war camps, Baker and Newman (1957) reported that nineteen Negro skeletons were 7 per cent (0.65 lb) heavier than eighty white skeletons (dry weight), though the Negroes weighed 5 per cent (7 lb) less than the white men when they entered the Army. The difference occurred in the skull and extremities. In older, dissecting-room subjects, Seale (1959) found that Negroes'

skeletons averaged about 1 pound heavier than white persons' of the corresponding sex. In both races men's skeletons weighed about 2.3 pounds more than women's (Table 125).

Individual variability in skeletal weight, as measured by the coefficient of variability, S.D./mean, was about 15 per cent, as was total body weight, in both young and old groups. This is some 3 times the variability of height and other body and head lengths, $2\frac{1}{2}$ times that of body breadths and circumferences, and twice that of surface measurements.

Table 125. Skeleton Weight by Sex and Race.
(Seale, 1959)

	Number	Mean Age (Years)	Total Skeleton Weight (Dry, Fat-Free) (Pounds)	
			Mean	S.D.
White				
Male	25	66	7.5	1.1
Female	25	66	5.1	1.1
Negro				
Male	25	60	8.5	1.2
Female	25	62	6.2	1.3

2. Human Tolerance to Physical and Mechanical Force

Forces exertable by the human body have been discussed on pp. 198–237. We now consider the body's passive resistance to external force arising from acceleration, blast, blows, or penetrating objects and acting on the total body or on portions of it. In this rapidly developing field, conclusions are soon dated. We therefore present basic material which is likely to be expanded rather than supplanted by later work, and some of which is unpublished or not widely available.

GENERAL CONSIDERATIONS

Comfort, which promotes (but cannot ensure) efficiency, is one segment of a continuum between luxury and death, as shown in Table 126. The borders of each zone can be roughly indicated for many types of environmental stress affecting man, such as acceleration, noise, noxious gases, radiation, temperature, and vibration. Here will be considered chiefly the biomechanics of forces which can damage the human body, rather than psychophysiological performance under physical force—such responses are reviewed by Gauer and Zuidema (1961), the joint armed services *Human Engineering Guide to Equipment Design* (1963), and Di Giovanni and Chambers (1964). There is of course no sharp line between anatomic and physiological damage, so that both will be summarized briefly.

The *tolerance limit* is defined by the designer's purpose. In most human experiments, as on centrifuges and linear accelerators, the limit has been *reversible incapacitation.* For dangerous but vital missions, such as recovery of man from space, the limit may be extended to *survivable injury,* like a broken limb or some weeks in a hospital. Beyond survivable injury is death.

As with all biological data, the limits of human tolerance to mechanical force vary from one person to another and from time to time in the same person. Since human experimentation in this field is hazardous, the experimental data are based on small numbers of subjects, largely military, for acceleration effects, and on dissecting-room material for bone strength. Comparable data on women, children, and even the man on the street are sparse or entirely lacking. The following figures are hence only approximations which cannot pretend to the precision of breaking points of inert materials. The error introduced by human variation is likely to be a factor closer to 2 or 3 than to 10, as shown by Stauffer (1953) for acceleration effects among healthy young men. For breaking strength of body components, the range of human variation can be five-fold or even more.

Table 126. Environmental Effects on Man.

Subjective Reaction	Pleasure			Pain		
	Luxury	Comfort	Discomfort	Tolerable	Intolerable	Unconsci-ousness
Objective Function	Inefficiency	Efficiency		Inefficiency		Incapacitation
Structure				Tissue Damage		
				Reversible	Irreversible	Death

BREAKING STRENGTH OF BODY TISSUES AND ORGANS

The passive resistance to force of the whole body and of various body components provides useful engineering data in several contexts, including the construction of anthropomorphic dummies, the design of boats, vehicles, and aerospace craft; parachutes, supporting harnesses, and protective gear; elevators, ladders, and many other items of equipment involving extremes of acceleration and deceleration. The harmful effects of such forces can be reduced by absorbing the force on structures other than the human body, such as a chassis, restraining harness, or seat, or on energy-absorbing materials (dashboard padding, immersion media); exposing less vulnerable portions of the body; modifying the attitude and duration of impact; and maximizing the body surface exposed. These methods will be discussed more fully later.

The Body as a Whole: Acceleration

The major force encountered by the body as a whole is acceleration,* at the other end of the "gravity spectrum" (Di Giovanni and Chambers, 1964) from weightlessness. Vibration and impact are types of acceleration, impact being deceleration over very short periods of time.

Acceleration, the second derivative of motion with respect to time, may be described in terms of duration, pattern (composition of various linear and angular components), rate of onset and decline, and magnitude. Duration is termed "abrupt" when less than 2

---

*Portions of this discussion, prepared by the authors for the *Human Engineering Guide to Equipment Design* (McGraw-Hill, 1963), are used with permission.

seconds; "brief," from 2 to 10 seconds; "long-term," from 10 to 60 seconds; and "prolonged" when over 60 seconds. There are several systems of nomenclature in use, of which the best one physiologically describes, as a reaction to the physical force, the displacement of the heart with respect to the skeleton, such as head-to-foot ("positive"), foot-to-head ("negative"), or chest-to-back ("transverse"). A force pushing the seated pilot up, as in upward ejection, displaces the heart (and body fluids) downward, and is called by convention "positive." The magnitude of acceleration is expressed in terms of "G," a unit of reactive force opposite in sign to "g," the acceleration due to gravity, which is 32 feet per second^2. A person weighing 150 pounds at rest weighs 300 pounds at 2 g.

Examples of various types of linear acceleration are shown in Table 127. In general, the shorter the time, the greater the acceleration that can be tolerated (Fig. 89). Physical tolerance also depends on the other aspects of acceleration noted above—direction, magnitude, and rate of onset and decline—as well as on several additional considerations, namely, the end points used to determine tolerance, protective devices, body position, environmental conditions, and the subject's age.

Figure 90, from Di Giovanni and Chambers (1964), shows that tolerance is greater for transverse than for head-to-foot acceleration, and is least in the foot-

Table 127. Approximate Duration and Magnitude of Some Accelerations. (Goldman and Von Gierke, 1960)

Vehicle	Activity	Acceleration Duration (Seconds)	Acceleration Magnitude (G)
Elevator	Average acceleration in "fast service"	1-5	0.1-0.2
	Comfort limit for acceleration		0.3
	Emergency deceleration		2.5
Train	Normal acceleration and deceleration	5	0.1-0.2
	Emergency stop (braking from 70 mph)	2.5	0.4
Automobile	Normal stop	5-8	0.25
	Quick stop	3-5	0.45
	Emergency stop	3	0.7
	Crash (potentially survivable)	<0.1	20-100
Aircraft	Normal takeoff	>10	0.5
	Catapult takeoff	1.5	2.5-6
	Crash landing (potentially survivable)		20-100
	Seat ejection	0.25	10-15
	Parachute opening at 40,000 ft.	0.2-0.5	33
	Parachute opening at 6,000 ft.	0.5	8.5
	Parachute landing	0.1-0.2	3-4

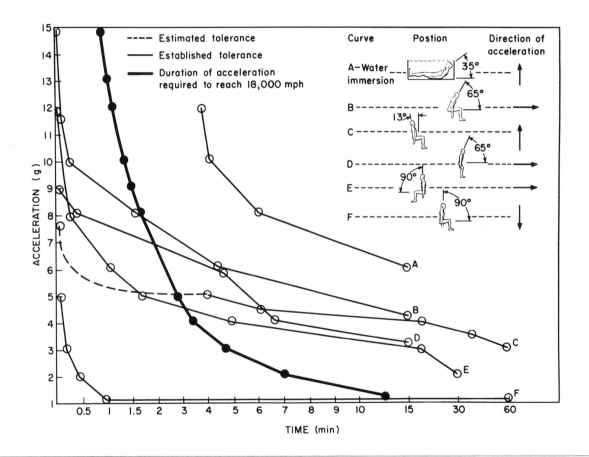

89. Tolerance for acceleration as a function of time and position (Bondurant et al., 1958).

to-head direction. There is little difference between the chest-to-back and back-to-chest directions. The data represent average physiological tolerances, not necessarily maximal, from twenty experiments on young male military volunteers. These physiological end points are, for head-to-foot G, failure of peripheral vision (gray-out), of central vision (blackout), and finally, unconsciousness. For foot-to-head G, the tolerance criteria are severe visual malfunctions, such as blurred vision, pain, headache, "redout," and retinal hemorrhage. For transverse acceleration, the physiolog-

ical end points are extreme chest pain, dyspnea (difficulty in breathing), and visual malfunctions as with foot-to-head G.

Effect of Acceleration on Tissues

Abrupt acceleration (less than 2 sec): In this time period, tissues react by structural damage or failure. The severity of damage depends on such physical properties as tissue elasticity, viscosity, frequency of response, tensile and shearing strength, and compressibility. Stress on solid tissues, like bone, produces little or no displacement until the yield point is reached, when fracture results. (The response of bone to mechanical stress and strain is treated fully below). In viscoelastic tissues like muscle, skin, and internal organs, unit stress produces logarithmic displacement. In body fluids, displacement is linearly proportional to the stress.

The limits of human tolerance to abrupt acceleration are summarized in Figures 91 through 94, inclusive. It should be noted once again that the subjects were healthy young men expecting the force and provided with maximum body support or restraint. Of all the factors which contribute to acceleration tolerance, type of restraint is probably the most important. Protection is achieved by form-fitted contour couches, net couches, inflatable g-suits, water suits, and a large variety of straps, restraints, bindings, and foams (Eiband, 1959; Di Giovanni and Chambers, 1964). Such devices increase g tolerance considerably, but may re-

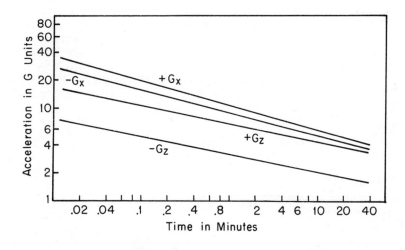

90. Tolerance for transverse (G_x) and vertical (G_z) acceleration (Di Giovanni and Chambers, 1964).

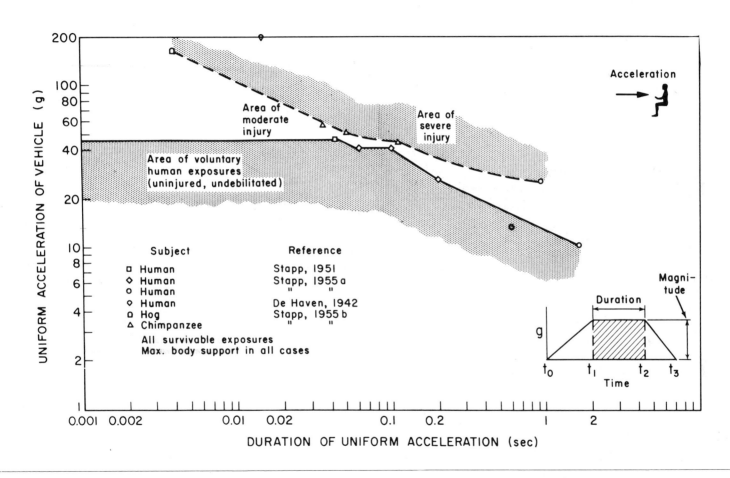

91. Maximum human tolerance limits to transverse acceleration (sternumward G) (Eiband, 1959).

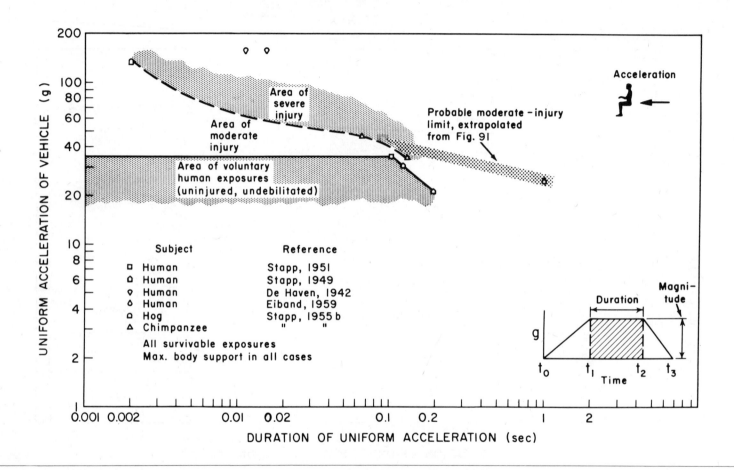

92. Maximum human tolerance limits to transverse acceleration (spineward G) (Eiband, 1959).

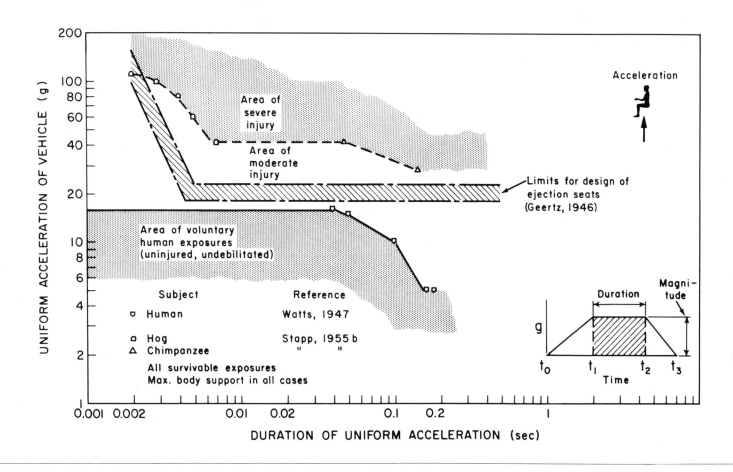

93. Maximum human tolerance limits to positive acceleration (Eiband, 1959).

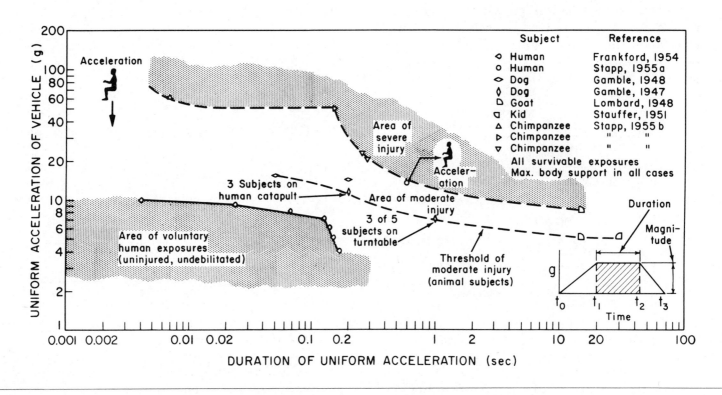

94. Maximum human tolerance limits to negative acceleration (Eiband, 1959).

strict the subject's movement. For a fuller treatment of protective devices, see the *Human Engineering Guide to Equipment Design* (1963) and *Human Factors in Technology* (1963).

Outside the laboratory, survival in abrupt acceleration depends not only on resistance to the actual impact, but also on the ability to recover rapidly enough to release body restraints and leave the vehicle within seconds or, at most, minutes. Fire and water hazards, vibration (as in helicopters), and the enemy in combat situations make immediate exit imperative. Under such conditions, even temporary disability can be disastrous.

Biologically, the limiting factor in human tolerance to headward g (head-to-foot G, as in upward ejection) in the normal, seated posture is spine fracture, particularly just above the lumbar concavity. With optimum alignment, up to 35 g can be tolerated at less than 500 g per second rate of onset, but when the back flexes (bends forward) to the limit of motion, less than 15 g can be tolerated. Animals with spines optimally aligned have withstood 75 g (hogs) and 65 g (chimpanzees) without fracture, at higher than 5,000 g per second onset. Human limits are presumably in this area, but data are lacking. With optimum support, as in immersion baths or form-fitting couches, men might be able to withstand 1,000 g in "zero" time (0.001 sec).

Head-to-foot g (foot-to-head G, as in downward ejection) has not been tested to tolerance limits, but 10 to 12 g for 0.003 to 0.2 second, applied at the rate of 10 to 82 g per second, can be tolerated.

With transverse g, the subject reacts as a boxer to a blow: at a threshold of 30 g with a jolt (change of acceleration with respect to time) of 1,000 g per second, there is a brief period of shock, lowered blood pressure, and weakness. At 40 g and a rate of onset of 1,500 g per second, consciousness is lost.

For seated Air Force subjects restrained only by a lap belt, spine pain at 23 to 27 g for 0.002 to 0.2 second and a rate of onset of 700 to 900 g per second marked the human limit of reversible incapacitation to transverse g (Stapp, 1957).

Free falls: Extensive research in progress at the Civil Aeromedical Research Institute of the United States Federal Aviation Agency (Snyder, 1963) promises to clarify many puzzling features of human tolerance to the extreme impact of free falls. The pioneering work of DeHaven (1942) demonstrated that in rare cases, survival is possible in falls from great heights. His observation that the survivors had landed in the transverse plane on soft material led to the design principle of maximizing the body area impacted, as in backward-facing aircraft seats.

In a first report, Snyder (1963) has identified two sets of factors, physical and biological, which influence survival in free falls. The physical factors include duration of impact, nature of impacted material, extent of secondary impacts, magnitude of force, direction of force (body orientation), and distribution of force. Biological factors consist of age, sex, physical and mental condition, and tissue properties. The initial analysis was based on 168 extensively studied cases from a total of 12,000 survived and 5,000 fatal free falls, about 96 per cent of all reported free falls in the United States over a 2-year period. Only a few preliminary findings can be given from this highly selected material; what is needed, and what should eventually emerge from the study, is a quantitative estimate of the injury and death associated with each combination of physical and biological factors.

The commonest body orientation on impact from free falls was feet-first (61 per cent), next was the head-first position (16 per cent), and third, buttocks-first (8 per cent). Bent knees resulted in less severe injuries. In the material studied, the expected correlations of injury with magnitude of force and with rigidity of surface impacted were not observed, presumably because of the complications introduced by other factors and by the very brief durations of impact. The most important finding was that as the duration of impact is decreased below 0.0006 second, body tissues may not respond as expected, and survival from usually fatal forces may be increased. Such durations are only one-tenth of the briefest time reported in human voluntary deceleration experiments—namely, 0.004 second.

Brief acceleration (*2 to 10 sec*): At about 0.2 second of duration, hydraulic effects begin (Fig. 95). These consist of the displacement of fluids in the cells, changes in the hydrostatic equilibrium between small blood vessels and their surrounding fluids, altered permeability and structural failure of blood vessels, and mechanical displacement or deformation of body structures.

Body orientation determines the physiological effect of the hydraulic displacements. With head-to-foot G, fluids and organs are displaced away from the head and eyes. Functional time limits for tolerance are set by visual blackout and loss of consciousness—the hypoxic effects of Figure 95 (Stapp, 1957). The oxygen-carrying blood does not reach the brain, but pools instead in the lower portions of the body. Vision fails before consciousness because arterial blood pressure must overcome a pressure of about 28 mm Hg in the eyeball, whereas the opposing pressure inside the skull is negligible. Total cessation of blood flow to the brain causes unconsciousness in about 7 seconds (Rossen *et al.*, 1943), and irreversible damage in from 2 to 3 minutes.

With foot-to-head G, fluid is displaced into the head and eyes, causing confusion, pain, visual disturbances, and hemorrhage. About 3 to 5 g can be tolerated for periods up to 30 seconds.

For transverse acceleration in the seated position, tolerance limits, as noted under *Abrupt acceleration*,

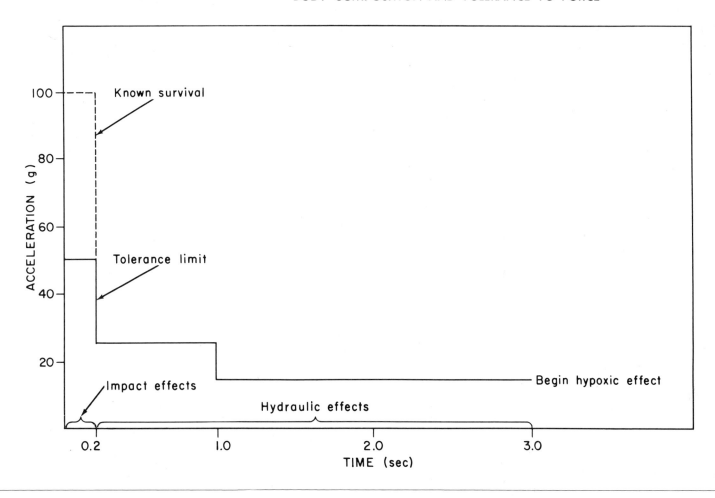

95. Functional limits for tolerance to positive acceleration (Stapp, 1957).

above, are chest pain, dyspnea, and retinal hemorrhage and detachment. Tolerable forces are 3 g for 900 seconds, 10 g for 120 seconds, 15 g from 5 to 50 seconds, and 25 g for 1 second. In the well-supported prone and supine positions, the tolerance limits to transverse acceleration are probably set by the strength of heart or lungs.

Parachute opening shock is a special case of whole-body deceleration for 1 to 2 seconds, with the body in various attitudes. Peak forces, acting for about 0.05 second, vary from 8 to 33 g. Opening shock is greatest at high altitudes. Forces below 20 g are considered to be safe, forces between 20 and 30 g are borderline, and forces over 30 g are dangerous.

Long-term and prolonged acceleration (*10 to 60 and 60+ sec*): The body responds to the hydraulic effects of acceleration by physiological reflexes which increase the heart rate and force of contraction and which constrict blood vessels. These reflexes help to circulate the blood that would otherwise pool peripherally or leak through vascular walls. Compensatory reflexes begin at about 5 seconds, are fully active in 15 to 20 seconds, and may persist for minutes or even hours if the g forces are not too great. With excessive g, the reflexes will fail, marking the person's time limit for tolerating that particular level of acceleration.

To withstand prolonged acceleration, both posture and type of restraint are major variables. Even such a

slight change as inclining the head and torso 20 degrees will increase tolerance to transverse g in the seated position from 5 to 6 g to 9 to 10 g (Ballinger and Dempsey, 1952). Figures 89 and 96 (Bondurant *et al.,* 1958) summarize several Air Force experiments with prolonged acceleration.

Weightlessness, a special case of prolonged acceleration at zero g, will not be discussed here, except to mention the metabolic alterations anticipated in prolonged space flight. Lack of exercise and of the normal muscular activity in bearing weight could lead to calcium loss from bones, with possible formation of kidney stones. Fractured bones during re-entry and muscular weakness after a flight could result. For a review, see Di Giovanni and Chambers (1964).

Two types of acceleration effect on the whole body which will not be discussed here are spinning and tumbling, and vibration. For spinning and tumbling, see the *Human Engineering Guide to Equipment Design,* pp. 466–473; for vibration effects, see the excellent summaries of Goldman and von Gierke (1960) and Magid and Coermann (1963).

Except for the analysis of free falls by DeHaven (1942) and Snyder (1963), knowledge of the effects of acceleration on body tissues has come chiefly from the experimental laboratory. Even free falls are but one kind of accident and not the most important from a design standpoint, beyond providing basic biomechani-

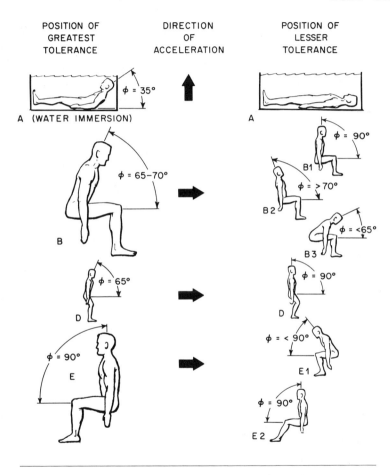

POSITION OF GREATEST TOLERANCE	DIRECTION OF ACCELERATION	POSITION OF LESSER TOLERANCE

A (WATER IMMERSION) $\phi = 35°$

B $\phi = 65-70°$

D $\phi = 65°$

E $\phi = 90°$

A

B1 $\phi = 90°$

B2 $\phi = >70°$

B3 $\phi = <65°$

D $\phi = 90°$

E1 $\phi = <90°$

E2 $\phi = 90°$

96. Effect of posture on tolerance for acceleration (Bondurant et al., 1958).

cal information on the human body. More relevant for the designer are aircraft and vehicular accidents. Data from USAF aircraft accidents involving 8,416 occupants (Moseley *et al.,* 1958), from automobile crashes involving 1,678 persons (Braunstein, 1957), and from light plane and motorcycle accidents agree in singling out the head as by far the most frequent site of injury, followed by the spine. Skull fracture—the usual though not invariable concomitant of serious brain damage—therefore receives the most attention in the following section. Concussion without fracture is of relatively minor importance (Haddon and McFarland, 1957–58).

SKULL FRACTURES

For good summaries, see the work of Gurdjian *et al.* (1950) and Evans (1957), sources for much of the following data.

Experimental Techniques

Stress-sensitive lacquer and strain gage studies on dry skulls can validly predict the tensile strain *pattern* of living bone subjected to similar experimental conditions. However, the *magnitude* of the strain is different because of excellent energy absorption by the scalp, a thin layer of soft tissue. In one set of experiments the scalp increased by ten-fold the force necessary to produce a linear fracture, from 40 inch-pounds in the dry skull to 400 inch-pounds in the fresh, intact cadaver skull.

Cadavers with the skull intact, studied within 24 hours after death, gave results similar to those obtained from blows on the living skull.

Linear fractures of the skull result from the *tensile,* not the compressive stress created in the bone by the blow. Frequently the most extensive deformation occurs at some distance from the point of impact. After a single linear fracture has begun, very little additional energy is required to produce multiple fractures and complete skull destruction.

The strongest resistance to fracture occurs in the buttressed portions of the skull, the zygomatico-frontal region, at the lateral inferior borders of the forehead, and the petrous ("stony") portion of the temporal bone, containing the inner ear. The weakest regions are near the foramen magnum (opening at skull base) and the parietotemporal area ("temples"), where fractures can occur without loss of consciousness.

Time Factors

The initial deformation from a blow occurs within 0.0002 to 0.0005 second. The total elapsed time for the deformation pattern to appear is 0.004 to 0.005 second. During the deformation, 3 to 6 cycles, with frequencies from about 9,000 to 12,000 cycles per second, occur in other portions of the skull.

A large amount of energy can be absorbed without fracture if it is absorbed slowly. In one head dropped experimentally on a steel block, deformation began in 0.0006 second, and the fracture occurred in 0.0006 second more, for a total time of 0.0012 second. On a deformable automobile panel, fractures occurred at 0.003, 0.005, and 0.009 second, but not at longer durations. In this respect, bone behaves differently from the whole body in free falls, where the shortest duration of force, below 0.0006 second, produces the least injury (Snyder, 1963).

Forces

Dropping fifty-five intact human cadaver heads on a steel block established the energy required to produce a single linear fracture as between 400 to 900 inch-pounds. Variations in the shape and thickness of the skull and scalp produced a 100 per cent variation in the amount of energy required for fracture—for example, 425 to 803 inch-pounds in the midforehead region. Such differences far outweigh the differences between the average energy required in various regions of the skull.

In another experiment, Evans *et al.* (1958) dropped from five to ten embalmed adult human heads onto an automobile instrument panel. Most of the kinetic energy was expended in denting the panel. It was concluded that the embalmed human head can tolerate, without fracture, peak impact accelerations as high as 686 foot-pounds and available kinetic energy as great as 577 foot-pounds, provided that a deformable surface is impacted over a long enough time. Skull fracture was produced by from 33 to 75 foot-pounds of energy.

Caution: Note the more than 100 per cent individual variation in threshold; the small number of skulls; the fact that they came chiefly from elderly hospital patients, who tolerate less force than healthy young persons; and the greater strength of living tissue than of dead tissue.

THE SPINE

The spine, or vertebral column, is the second most frequent site of injury in aircraft and vehicular crashes. As with the head, spinal injury is potentially more serious than injury to a limb or other less vital region. The spine consists of twenty-four individual bones, the vertebrae, separated by fibrocartilaginous discs. Either vertebrae or discs can be injured by mechanical force. Such injury can produce pain, limitation of movement to the extent of incapacity, or—by damaging the enclosed spinal cord—paralysis or death.

Spine as a Whole

USAF aircraft accidents producing vertebral fracture showed no relationship between fracture and the victim's age, height, or weight (Neely and Shannon, 1958). Most fractures occurred at or near the thoracolumbar junction and resulted from two forces, first headward and later spineward. Shoulder harness and lap belt restraints were deemed inadequate, since they produced counterforces compressing and flexing (bending forward) the spine, thereby enhancing the likelihood of vertebral fracture. Chest and inverted "V" leg straps, or other improved torso restraints, were recommended.

Positioning is a major determinant of spine fracture. With headward g in the seated position, as in aircraft crashes and upward ejection, maximum tolerance (about 35 G) is attained by preventing flexion (forward bending) of the spine (Stapp, 1957; Carter, 1958). One force causing flexion is the location of the center of gravity of the seated airman, close to the sternum (breastbone). The precise location of the *CG* depends on arm position, seat angle, and equipment worn on the back. (See p. 180.) This force is minor in amount and can be overcome by a face curtain or armrests. Secondly, the seat must align the axis of the spine with the thrust line of the catapult. And finally, resilient seat cushions, which magnify the force applied to the body, should be avoided.

In hogs with vertebral columns in optimal alignment, more than 75 g at higher than 5,000 g per second rate of onset were required to produce vertebral fractures. Chimpanzees have been similary exposed to 65 g without fractures.

With force applied transversely, from back to chest (transverse prone), a restrained hog disarticulated his vertebrae at 25 ± 2 G. The hog spine and pelvis are less flexible than those of the human, so that more force is presumably required to produce similar effects in man. One human subject wearing a lap belt felt spine pain at 23 to 27 G and 700 to 900 G per second onset, with transverse G (Stapp, 1955).

Caution: There is a high prevalence of spinal abnormalities, without symptoms, in healthy men who have passed physical examinations—as high as 45 per cent in one X-ray survey of 1,500 men (Colcher and Hursh, 1952)—deemed serious enough to disqualify them from railroad employment. This makes it hazardous to predict "safe limits" for individuals. Structures and supports should be designed to withstand at least 25 g, although men have tolerated much more.

It must also be emphasized that the foregoing discussion concerns only healthy young airmen. Women's bones, being smaller and lighter than men's, are weaker; the bones of persons over 60 years of age are also relatively weak. The universal—if quantitatively variable—osteoporosis (rarefaction or softness of bone) in postmenopausal women makes them especially subject to fractures from even slight trauma.

Breaking Load of Separate Vertebrae and Discs

In supporting body weight, both the vertebrae and the discs play a role, the latter probably the major one. In fact, experiments are said to show that the discs alone support vertical loads (Hirsch and Nachemson, 1954). Certainly the discs have been more thoroughly studied. In one of the few studies of static vertebral breaking load, Ruff (1950) estimated that the entire spine can absorb about 10 kg meters (72.3 foot-pounds) before the first vertebra, probably L-1, would collapse. Tables 128 and 129 contain Ruff's data on the breaking loads and percentages of total body weight supported by the individual vertebrae.

Caution: Ruff's figures were based on small samples and static loading, whereas the equipment designer is

Table 128. Breaking Load (in Kilograms) of Vertebrae by Age.
(Ruff, 1950)

Vertebra		19	21	21	23	33	36	38	43	44	46
Thoracic	8		640		540	600					
	9				616		720		700		
	10		800			660	700			730	
	11	750		720					860		755
	12		900	690		800		800			
Lumbar	1	720		840					900	800	800
	2		990			800		830			
	3	900							940		1100
	4				1100			900		950	
	5		1020						1000		1200

Table 129. Percentage of the Total Body Weight Supported
by the Vertebrae.
(Ruff, 1950)

Vertebra		Percentage of Total Body Weight
Thoracic	3	21
	6	25
	7	29
	8	33
	9	37
	10	40
	11	44
	12	47
Lumbar	1	50
	2	53
	3	56
	4	58
	5	60

more concerned with dynamic loading.

The intervertebral disc has a semifluid, plastic nucleus surrounded by a fibrous ring. The physical properties of the nucleus, which play an important role in normal vertebral function, can be altered by force, age, and disease, resulting in rupture. Defective or diseased discs are more variable and weaker than normal discs. Being elastic, discs react to force by deformation. Static loads averaging 773 to 935 pounds have been applied to discs without rupture (Virgin, 1951; Evans, 1957), while deformations were reversible up to loads of 287 pounds applied for 5 minutes or less (Hirsch and Nachemson, 1954).

The normal disc is elastic, with considerable power of recovery; strains or deformations produced in it dis-

appear soon after the load is removed. The discs are normally subject to pure compression, whereas the bones can bend and are thus subject to tensile and shearing stress as well.

Virgin (1951) loaded 1 to 5 discs from each of fifty-one cadavers, with the results shown in Figures 97 and 98. Note the wide variation in individual discs.

Residual disc deformation was greatest in the lower lumbar region, less in thoracic and upper lumbar regions. When a disc was loaded twice, the residual deformation was always less the second time. Loads of 50 pounds or less continuously applied for 40 hours resulted in small deformations, with rapid recovery after the load was removed.

Results of four tests on sections of whole spines were similar to those for single discs.

Table 130 presents the findings of Lissner and Evans (Evans, 1957) on average compressive stress and deformation in lumbar intervertebral discs, based on embalmed specimens from nineteen men averaging 57 years of age.

Static and dynamic loading were studied by Hirsch and Nachemson (1954) on ninety-four normal, unembalmed discs and vertebrae from persons mostly between 30 and 60 years of age. Disc deformations were reversible up to loads of 130 kg (287 lb), applied for no longer than 5 minutes. The 4th lumbar disc was somewhat more compressible than the 2nd. A static load of 40 kg (88 lb) compressed a disc about 1 mm, and a 100-kg load (220 lb) increased the compression to 1.4 mm. The authors estimated that in a person weighing 70 to 90 kg (154 to 198 lb) and standing erect, the spine bears approximately 40 kg (88 lb), a value quite close to Ruff's (1950) 60 per cent of body weight, cited in Table 129.

In 100 healthy discs, 40 kg (88 lb) expanded the disc anteriorly and posteriorly 0.5 mm each. With a 100-kg load (220 lb), disc expansion increased to 0.75 mm.

No factors other than the discs were found that supported vertical loads in the spine.

Since the spine, like other parts of the body, is seldom in static equilibrium, static measurements have only limited value. Dynamic loading, vertically applied, with loads lasting from 1 to 2 seconds, produced disc compression and lateral bulging, which disappeared immediately after the load was removed. Compression experiments could be repeated indefinitely without

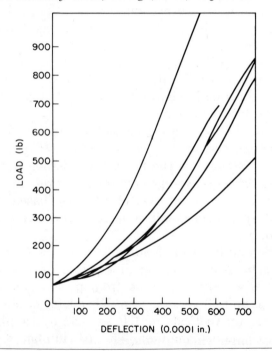

97. Elastic limit of six different intervertebral discs (after Virgin, 1951).

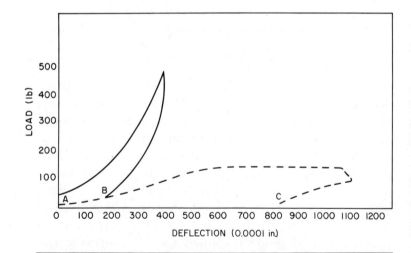

98. Effect of 500-pound loading and unloading on intervertebral disc. Solid line indicates normal disc, dotted line immature disc (after Virgin, 1951).

observable disturbance in disc elasticity.

Rapidly acting blows, regardless of size, always caused brief, small-intensity oscillations within 0.2 to 0.4 second. Very small, rapidly applied forces produced disc oscillations measurable in tenths of a millimeter. The frequency of disc oscillation paralleled that of the application of force. With loads of 40 kg (88 lb) or less, the oscillations were irregular. Great (but undeter-mined) lumbar stress was produced by loads applied rapidly (during tenths of seconds), whatever the force, during these experiments.

More recently Hirsch and Nachemson (1963) have made clinical and radiographic observations on the spine of fifty-five ejected pilots. Thirteen per cent suffered vertebral fractures, chiefly thoracic, none of which caused serious sequelae.

Table 130. Average Compressive Stress and Deformation
in Lumbar Intervertebral Discs.
(Evans, 1957)

Disc Between	Compressive Load (Pounds)		Compressive Stress (PSI)		Deformation (Inches)	
	Average	Range	Average	Range	Average	Range
T12-L1 (18 Specimens)	801	(230-1560)	430	(164-1100)	0.0796	(0.050-0.158)
L1-L2 (15 Specimens)	794	(200-1495)	445	(130-1234)	0.0782	(0.043-0.100)
L2-L3 (16 Specimens)	925	(240-1975)	400	(145-812)	0.095	(0.048-0.148)
L3-L4 (17 Specimens)	935	(185-1505)	373	(116-705)	0.104	(0.054-0.197)
L4-L5 (17 Specimens)	885	(185-1660)	377	(122-883)	0.121	(0.044-0.289)
L5-Sacrum (13 Specimens)	773	(375-1240)	292	(145-850)	0.205	(0.053-0.239)

From Evans, F. Gaynor, Stress and Strain in Bones, 1957. Courtesy of Charles C Thomas, Publisher, Springfield, Illinois.

STRESS AND STRAIN IN LONG BONES (Evans, 1957)

The femur will serve as an example, but it must be emphasized that no data are available for whole limbs, including not only the other bones, but also muscle, other soft tissues, and fluids, all of which increase tolerance to force. Studies with strain-sensitive lacquers show that under certain conditions long bones behave like elastic bodies, in that they can be deformed with relatively small amounts of energy or light loads, returning to their original shape and dimensions when the load is removed.

A small load suddenly applied can produce a deformation pattern as extensive as that arising from a much greater load gradually applied. Under both static and dynamic loading, the magnitude of tensile strain at any point on the femur is proportional to the perpendicular distance of that point from the load axis of the bone. With loading of the femoral head, the direction of both tensile and compressive strain in the shaft is parallel to its long axis.

Table 131 shows the effect of loading the femur with increasing weight. Local deformation is slight and directly proportional to load; the femur, despite its structural diversity, follows Hooke's law. Load deflection curves under increasing and decreasing load are not identical, indicating hysteresis. The modulus of elasticity is not constant throughout the femoral shaft.

The apparent tensile stress created in the femur by application of rather small amounts of energy is high, the greatest (2430 lb/in.²) arising from application of energy (11.8 in.-lb) to the point one-fourth of the way distally, down the lateral aspect of the shift. Applying energy to the midpoint of the posterior aspect of the shaft creates minimal tensile stress in the opposite side of the bone, but in one specimen this value reached 920 lb/in.² for an energy of 7.9 inch-pounds.

Relation of Stress and Strain to Fracture

Fractures result from tensile stress in long bones, since bone is weaker in tension than in compression. Typical breaking loads, in bending, for femurs varied from 1330 kg/cm² (18,917 lb/in.²) in an 82-year-old person to 1940 kg/cm² (27,593 lb/in.²) in two persons 57 and 58 years of age (Messerer, cited by Evans, 1957). Static loading of the femoral head produced a transverse fracture of the neck at 1280 pounds.

Table 131. Loading of Femoral Head, with Femur Vertical[a]
(from Küntscher, 1936)

Strain Site	Load (Kg)			
	100	200	400	800
Highest compressive site - shaft	0.0027 mm/mm	0.0055 mm/mm	0.011 mm/mm	0.22 mm/m
Highest tensile site - neck	.0003	.0065	.013	.026
Second highest tensile site - shaft	.002	.0041	.0081	.0162

[a] Number of samples not given - probably one.

Fractures of the femoral head and neck do not involve torsion, since the two move as a unit. However, spiral fractures of the shaft result from torsion loads of 166 to 498 inch-pounds of torque. These fractures arise from tensile, not shearing stresses, as commonly believed.

Table 132 indicates the static and dynamic forces required to fracture the femur.

Pelvis: The pelvis behaves like an elastic body. Its deformations can be demonstrated with relatively small loads, as little as 33 inch-pounds of energy applied upward to the defleshed ischial tuberosities ("sitting bones"). The presence of 0.5 inch of gluteus maximus

muscle and skin over the tuberosities resulted in only minimal deformation by as much as 450 inch-pounds.

Normally, the body weight tends to push the wedge-shaped sacrum (solid base of spine) between the ilia ("wings" of the pelvic brim), forcing the latter apart. This thrust is resisted by the sacroiliac and iliolumbar ligaments, which are thus under tensile strain. In sitting, the pubic symphysis (anterior juncture between the two pelvic halves) is compressed.

Pelvic fractures, like those of the skull and long bones, result from tensile stresses and strains and occur, both clinically and experimentally, in regions showing deformation patterns on stress-sensitive lacquer studies. A fracture has been experimentally produced by a dynamic vertical load of 240 inch-pounds applied to the embalmed lumbar spine and pelvis of a 79-year-old man. Static vertical loading of the isolated lumbar spine and pelvis produced pelvic fracture at 112 pounds in a man of unknown age, 738 pounds in an 85-year-old man, and 1,350 pounds in one of 60 years (Evans, 1957).

Caution: Note the wide variation, more than tenfold, and the fact that these few specimens from elderly persons lack the excellent energy absorption provided by soft tissues.

In more extensive studies, Fasola, Baker, and Hitchcock (1955) noted sacral fracture at an average of 1,925 pounds applied statically to the sacroiliac joint. Dy-

Table 132. Loads Necessary to Fracture Femur.
(Based on 115 Femurs)
(Smith, 1953)

Conditions	Site of Application	Average Force to Fracture	
		Static (Lb)	Dynamic (Ft-Lb)
Disarticulated femur	Femoral head, vertically	2092	327
Disarticulated femur	Greater trochanter, from side	1648	306
Articulated, with intact hip joint	Across femoral neck	900	36.4
Same . . .	Ilium, vertically	2149	250

namic loading disrupted the joint, by tearing the intact ligaments from their cartilaginous attachments to the bones, at 775 and 830 pounds in two specimens. After fracture of the pubic bones, the sacroiliac joint was disrupted by only 480 pounds dynamically applied. Following sacroiliac joint disruption, pubic bones fractured at 350 pounds and at 595 pounds in two specimens.

Pelvic injuries are relatively rare in light-plane and automobile accidents, occurring in 3.2 per cent of 800 survivors of light-plane crashes and 2.7 per cent of 1,678 persons injured in automobile accidents (Braunstein, Moore, and Wade, 1957).

Clavicle: Force applied statically, normal to a point at the junction of the medial ⅔ and lateral ⅓ of the clavicle, produced a fracture about 3 centimeters medial to the acromioclavicular (shoulder) joint, at about 230 pounds (Fasola *et al.,* 1955).

Fatigue test of foot bones: March, stress, or fatigue fractures have been noted in soldiers. Lease and Evans (1959), in the first experimental study of the small bones of the foot (metatarsals), tested fifty-one normal bones from eleven persons 54 to 100 years of age. Repetitions to fracture with a 15-pound load varied from 150,000 to 13,908,000 for moistened bones. There was no relationship between fatiguability and age, size, or shape of bone.

Mandible (lower jaw): Fractures result from tensile strains. Forces applied to the chin parallel to the long axis of the mandible produce tensile strain in both the long axis of the lower border of the jaw and the neck of the mandibular condyle (where the jaw joins the skull, just forward of the ear). No data are available on the forces necessary for fracture (Evans, 1957).

SOFT TISSUES

The data quoted above on automobile and aircraft crashes concern gross injuries. Autopsies reveal other, often unsuspected damage to soft tissues. Almost all severe decelerations involve bleeding into the lung, with fragments of fat or other tissue being forced into the blood vessels and found in the lung. Sometimes torn aortas, ruptured livers, and ruptured hearts are encountered (Mason, 1957; Silliphant and Stembridge, 1958). The frequency of significant injury to the heart, pericardium, or major blood vessels was 13.5 per cent of 3,400 fatal aircraft and parachuting accidents (Gable and Townsend, 1963). It should be kept in mind that fatalities can result from disturbance of cardiac conduction without visible anatomical damage (Robinson, Hamlin, Wolff, and Coermann, 1963).

The only experimental data on forces required to produce structural injury are for the lung. In chimpanzees, impact loads against harness of more than 75 lb/in.^{2} attained in less than 15 milliseconds produced fatal damage to the lungs, consisting of hemorrhagic areas of ruptured air sacs (Stapp, 1957).

Data on the strength of soft tissues are much scarcer than on bone. For general orientation, we reproduce Table 133, after Ruff (1950). Information on the experimental conditions and the number and type of subjects is unavailable, but the table is included because the data are unique.

The above values pertain to isolated tissues, whereas the equipment designer is more concerned with the living body as a whole or with intact segments. Research is needed on the breaking points of limbs or viscera in their normal relationships. In the absence of such data, only fragmentary values can be presented for soft tissues, as in Table 133 and the following paragraphs.

Table 133. Impact Tensile Strength of Tissues.
(Ruff, 1950)

	Lbs./in.^{2}
Striated ("voluntary") muscle (of limbs and trunk)	35.5
Elastic connective tissue (subcutaneous)	56.8
Hyaline cartilage (at ends of long bones)	14.2
Collagenous connective tissue (ligaments, tendons)	142.0
Skeleton	35.5

Tendons

Tendons are fibrous cords attaching muscles to bones and transmitting muscular forces; normally, the direction of the muscle and tendon fibers coincide, exerting tension; abnormally, shearing forces may result in a "sprain."

In one study, Marvin (1946) determined the elasticity and tensile strength of "several" animal tendons, site and species unstated. Under constant loading at 40 pounds per minute, with a hydraulic Schopper tester, yield strength ranged from 6,000 to 11,000 pounds per square inch, and elongation was about 10 per cent, with a modulus of elasticity 40,000 to 90,000 pounds per square inch (Fig. 99). These properties resemble those of nylon, which has great toughness in relation to rigidity. For further comparison, annealed electrolytic copper has a yield strength of 10,000 pounds per square inch, and pure aluminum 5,000 pounds per square inch. As the loading rate increased, the yield strength was greater but the percentage of elongation less.

Caution: Note the wide variation of about 100 per cent in tensile strength and elasticity among several samples. Test conditions also had a marked influence. Another investigator in the same laboratory found greater tensile strength and less elongation, possibly because of a higher rate of loading.

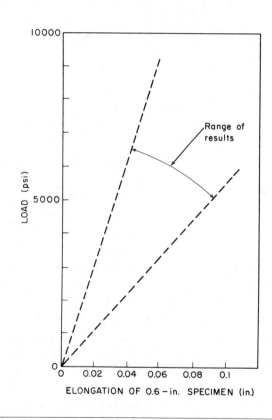

99. Summary of tensile tests of tendons (after Marvin, 1946).

Fascia (literally, "band")

This term refers to a sheet of fibrous tissue which surrounds and connects muscles and may help attach muscle to bone. The largest, toughest fascial sheet in the body is the "fascia lata" ("broad band") enclosing the lateral thigh muscles. Gratz (1931) obtained fresh material from "several" persons during surgery and applied tension in the direction of the fibers; fascia has very little strength in the transverse direction. Results are shown in Figure 100.

The strength of fascia lata is proportional to its area of cross-section. Thickness in the specimens studied varied from 0.014 to 0.03 inch; its specific gravity was about 1.31, and its average ultimate tensile strength about 7,000 pounds per square inch. Soft steel has a specific gravity of 7.83 and an ultimate strength of about 45,000 pounds per square inch. Weight for weight, therefore, fascia lata is nearly as strong as soft steel.

Individual variation is relatively slight, as seen in Table 134.

Compressibility of buttocks

Unpublished work by Silverman (1957) showed that with a 50- to 75-pound shoulder loading, a heavy subject's buttocks were about $\frac{1}{4}$ inch more compressible than a lean subject's; with a 100- to 250-pound loading, about $\frac{1}{2}$ inch more compressible.

Compressibility of Foot

The same author found that the heel bone (tuberosity of calcaneus) approaches the ground or floor by 0.042 inch per 100-pound load, with loads between 25 and 200 pounds, because of soft tissue compression under the

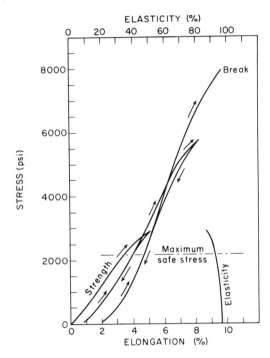

100. Tensile strength and elasticity test on living human fascia lata (after Gratz, 1931).

heel. The ankle bone (lateral malleolus) was depressed by 0.125 inch per 100 pounds because of soft tissue compression under the heel and spreading of the bony structure of the foot.

Distribution of Weight Between Heel and Ball of Foot

In a comfortable standing position, the proportion of body weight on heel and ball of foot ranged from 1 to 1 to 25 to 1, with 1.5 to 1 the best over-all figure (Silverman, 1957).

PROTECTING THE BODY AGAINST FORCE

Engineering to Prevent Accidental Injury

While the designer should strive to prevent or reduce

Table 134. Individual Variation in Thickness and Strength of Fascia Lata.

(Gratz, 1931)

Subject	Average Thickness (Inches)	Maximum Tensile Strength (Lb/In.^{2})
Adult female, age 74	0.017	6222
Athletic male, age 35	.014	7860
Paralytic female, age 15 (both legs involved for 10 years)	.030	6375

the occurrence of accidents by following the precepts of the *Human Engineering Guide to Equipment Design* (1963), accidents will continue to occur. Injury, however, is a function of design and not the inevitable result of accident. Proper design can minimize the number of deaths and injuries in those accidents which cannot be prevented.

The Automotive Crash Injury Research at Cornell University (Braunstein, 1957; Braunstein *et al.*, 1957) illustrates a successful application of human engineering. The procedure and results of this continuing pro-

gram can guide engineers in other fields. Their procedure, briefly, was as follows: (1) Gather the facts. What patterns of injuries are caused by accidents at varying speeds, angles of impact, type of vehicle, seat occupied, and the like? (2) Isolate the design features responsible, such as door latches, steering wheel, dashboard, and windshield. (3) Redesign to reduce injury potential: "package the passenger," streamline and pad the vehicle interior. (4) Collect more data to evaluate the effect of redesign. The initial problem is shown by Tables 135 and 136, which list by body area and sever-

Table 135. Percentage of Injuries to Each Body
Area, by Category of Severity.
(Braunstein, 1957)

		Area Injured				
	Head	Neck and Cervical Spine	Thorax and Thoracic Spine	Abdomen, Pelvis and Lumbar Spine	Upper Extremities	Lower Extremities
Minor	61.8	43.9	52.6	46.3	69.6	73.0
Nondangerous	17.9	14.9	22.2	22.6	20.0	17.5
Dangerous	4.2	6.1	6.0	13.2	0.4	0.4
Fatal	5.4	16.7	5.4	3.5	0	0.4
Not reported	10.8	18.4	13.8	14.4	9.9	8.7
Total	100.1	100.0	100.0	100.0	99.9	100.0

ity the injuries sustained by 1,673 persons in automobile crashes.

As the result of engineering redesign, seat belts reduced injury, of all degrees, up to 60 per cent. Safety locks halved the percentage of doors opening in crashes, from 50 to 25 per cent, and reduced the risk of dangerous and fatal accidents up to 30 per cent. Padding the dashboard panel with energy-absorbing plastic reduced head injuries by 30 per cent, and recessed-hub steering wheels reduced severe chest injuries by 50 per cent.

Recommendations for Equipment Design to Protect the Operator

Equipment involved: Dummies; crash protection items (helmets, harness, padding); pressure suits, ejection seats; all vehicles, including space vehicles.

Methods of reducing harmful effects of mechanical force:

(1) Absorb force on structures other than the human body, such as chassis, seat, or restraining harness; that is, prevent the man from colliding with solid objects;

Table 136. Percentage of Injured Persons Receiving Injury to
Each Body Area, by Category of Severity.
(Braunstein, 1957)

	Head	Neck and Cervical Spine	Thorax and Thoracic Spine	Abdomen, Pelvis and Lumbar Spine	Upper Extremities	Lower Extremities
Minor	44.7	3.0	19.2	7.1	20.5	34.3
Nondangerous	12.9	1.0	8.1	3.5	5.9	8.2
Dangerous	3.0	0.4	2.2	2.0	0.1	0.2
Fatal	3.9	1.1	2.0	0.5	0	0.2
Not reported	7.8	1.3	5.1	2.2	2.9	4.1
Total	72.3	6.8	36.6	15.3	29.4	47.0

"package the passenger." Relocate objects which can become dislodged and act as missiles.

Harnesses, seats, tie-downs, and similar items should be stressed to at least the G tolerance of the passenger, namely 50 G. Current operational harnesses are satisfactory only for brief exposures (under 1 second), as in aircraft or vehicle crashes. Thereafter, pressure points can damage the skin and underlying tissues. Experimental suits, seats, and body molds have been designed to meet the requirements of severe deceleration, but protection for ordinary travelers has lagged far behind.

(2) Dissipate forces on energy-absorbing materials, such as dashboard padding, honeycombed collapsible buffers, and immersion media. For attenuating forces before they reach the man, lightweight metal or honeycombed paper have been shown experimentally to absorb rather than to transmit impact forces.

The effect of force on the body can be reduced by padding the impact surface with energy-absorbing plastic. Several varieties have been developed, such as expanded polyvinylchloride, polystyrene, and the like. Since eggs can be dropped 120 feet on a $\frac{1}{2}$-inch pad of such material without breaking, its value for helmet liners or dashboard padding is clear.

The scalp, a layer of soft tissue $\frac{1}{8}$ to $\frac{1}{4}$ inch thick, absorbs about 90 per cent of the force impinging on the head (Evans, 1957). In general, the protection against head injury affected by energy-absorbing material is reported (Girling and Topliff, 1958) to be inversely pro-portional to the density of the material for a given thickness; while for any material the protection afforded varies directly as the thickness.

(3) Expose less vulnerable portions of the body.

(4) Maximize the body surface exposed.

Methods (3) and (4) are utilized in rearward seating, where forces are distributed over a large area of the back and a narrow region of the shanks. In experiments with a swing seat, Ruff (1950) found that backward seating increased G tolerance from 18 to 20 G to 28 to 30 G at a duration of 0.1 second (speed of onset not given but probably 1000 G/sec).

More recent work (Stapp, 1957; Bondurant *et al.*, 1958) has shown little difference between forward and backward seated tolerance to linear deceleration *assuming proper restraint in the forward-facing position.* This condition has been met to date only in special experiments. For ordinary operation, military or civilian, backward seating affords much greater crash protection.

SUMMARY—BODY TOLERANCE TO FORCE

Systematic study of human tolerance to physical and mechanical force has just begun. Data have been presented from two sources, laboratory experiments and the "natural experiments" provided by accidental falls and collisions. Recommendations have been offered for designing equipment to protect the operator against physical and mechanical force.

Design Recommendations ——————————— V

1. Introduction

The design recommendations contained in the following sections have been derived from the anthropometric and biomechanical data in Chapters II and III and from the literature on equipment design cited in the present chapter. It is sometimes difficult to evaluate published recommendations, especially if there is but one study in an area, but when several experimenters agree, their conclusions are likely to be correct. Moreover, so many variables affect human capabilities in operating controls that the results of one study cannot always be generalized, or even applied to an apparently similar situation. A few differences between one workspace layout and another may have a decisive effect on the operator's performance on identical controls.

In the current state of knowledge, the recommendations for control design should be considered more as general guides than as established rules. When more precise values are needed which cannot be derived from the data in Chapters II and III, the engineer may have to conduct his own experiment or seek help from the sources listed on page 43.

2. Some General Principles of Control Design

The following very general considerations apply to the design of all kinds of controls, whatever their purpose or mode of operation.

(1) The maximum force, speed, accuracy, or range of body movement required to operate a control must not exceed the limits of the least capable operator. Normal requirements for control operation should be considerably less than the maximum abilities of most operators.

(2) In any control system the most important items should be given priority in design and location. The less important ones should then be fitted into the remaining space.

(3) The design and location of the operator's seat and controls should not require him to expend unnecessary amounts of energy merely to maintain his body position. Control movements are easiest when the limbs are moderately flexed; extreme body positions should be avoided. The neutral position of a control should correspond to the operator's most comfortable resting posture.

(4) For any control task, determine and use the bio-

mechanically most efficient design.

(5) Keep the number of controls as few as possible, for simplicity of operation.

(6) Use the seated position where possible, though poorly designed seats may be worse than none at all.

(7) A backrest and footrest help the seated operator to maintain a stable body position and increase his force in pushing or pulling controls.

(8) Foot control is less precise but stronger than hand control. Use hand controls where accuracy is important, foot controls where accuracy is secondary to force.

(9) Control movements that are "natural" for the operator are more efficient and less fatiguing than those which seem awkward or difficult.

(10) Controls should be readily distinguishable by sight or touch, reducing the likelihood of inadvertent operation.

(11) For many kinds of repetitive or continuous control operation, muscular activity should be distributed about the body rather than concentrated in one part.

(12) Keep control movements as short as possible, consistent with the requirements of accuracy and "feel."

(13) Controls should be stressed enough to eliminate inadvertent activation by the weight of the controlling body member.

(14) The operator of some kinds of equipment, like high-speed aircraft or space vehicles, cannot apply enough force unaided to operate the controls, and power-boosted or fully powered control systems are necessary. In such cases, artificial resistance cues should be fed back.

3. Design Recommendations for Hand Controls

WHEN TO USE

Use hand controls in preference to foot controls if one or more of the following conditions pertain: (1) Accuracy of control positioning is important. The hands and arms are much more accurate than the feet (Grether, 1946). (2) Speed of control positioning is required. (3) Continuous or prolonged application of moderate to large forces (20 lb or more) is not necessary.

WHAT KIND TO USE

For start-stop or on-off controls, use either push buttons or toggle switches. When speed of operation is paramount, use push buttons, but if the controls must be spaced closely together (less than 1 inch between centers), use small toggles with relatively large resistances (Bradley and Wallis, 1959).

For fine adjustment tasks requiring low forces, like radio tuning, use rotary knobs, either finger- or hand-operated.

For discrete control settings operated by applying moderate to large forces, as with a gear shift or hand brake, use control sticks or levers.

For continuous adjustment or tracking tasks, sticks, levers, or wheels may be used. Where very precise control settings are necessary, or in tracking fast-moving targets, handwheels or hand cranks should be used with high wheel speeds and low gear ratios.

HOW TO OPERATE

Body Position

Controls should not be operated from awkward or uncomfortable positions like stooping, kneeling, or crouching, unless absolutely unavoidable.

The seated position is preferable to the standing position—especially if the controls are to be operated for any length of time. Standing is more fatiguing and usually results in less efficient, weaker, or less accurate control movements. For the seated operator, backrests and footrests greatly increase comfort and efficiency. Backrests permit large forces to be exerted in push movements, and footrests (as well as seat belts or restraining harnesses) favor pull movements.

The prone position can be used for operating controls, at some disadvantage in force. From the true prone position, only 71 per cent of the strength exertable when seated can be applied to controls (Hunsicker,

1955). A modified prone position, however (Fig. 87), permits upward and lateral movements stronger than those from the seated position (Brown *et al.,* 1949).

From any body position, forces greater than those exerted in a single effort can be attained if the operator can reposition his body and hands to better advantage during the course of the control movement (McFadden and Swearingen, 1958).

One or Two Hands

For precision and speed, one-handed controls are preferable to those operated with both hands, except that large-diameter control wheels with reciprocal rotary motion (like vehicle steering wheels) are best operated with two hands.

Controls requiring great force should be operated with two hands, which for most controls approximately doubles the amount of force which can be applied (Hertzberg, n.d.). Variations depend on control type and location as well as on the kind and direction of movement, as follows: (1) When two hands are used on wheel controls, rotary forces are effectively doubled in most cases (Hertzberg, n.d.; Provins, 1955). (2) When two hands are used on stick or lever controls located in the body midline, pull force almost doubles; push force doubles near the body but is only slightly stronger at distances farthest from the body; the force in moving right or left increases roughly 50 per cent (Hertzberg, n.d.). (3) When two hands are used on stick or lever controls away from the body midline, at or beyond the shoulder, pull force roughly doubles; push is not greatly increased except close to the body.

One-handed operation of an "L"-shaped door handle permits about 75 per cent as much force to be exerted as two-handed operation (McFadden and Swearingen, 1958).

Where two stick or lever controls are used, pulling one and pushing the other at the same time is more efficient than alternating (Lehmann, 1927).

Two-handed controls should be operable temporarily by one hand, in case of injury or when another control may occasionally have to be operated at the same time.

Parts of the Body Used

Where speed, accuracy, or strength of the control movement are important, use the preferred hand, if possible. (This option holds only for the operator of a midline control. The designer should plan for right-handed operation of the more important controls.)

In operating wheel, stick, or lever controls, use the entire arm, including the shoulder—not only the elbow and wrist, or wrist alone (Orlansky, 1948). This reduces fatigue, especially when large resistances must be overcome. Handwheels can be turned from the shoulder with 50 per cent greater force than from the elbow (Provins and Salter, 1955).

Grips, Knobs

A handgrip where thumb and forefinger overlap is much better than a wider grip separating the two (Fox, 1957). For whole-hand grasping or "trigger" grips, the fingers should be placed around the main shaft and the heel of the hand used to close the movable part (Dupuis et al., 1955). For small knob controls, the maximum force can be applied by the thumb and forefinger.

Tractionless Conditions (See also p. 263.)

Under tractionless conditions, as at zero g, more force can be applied to a control or tool with a minimum of resultant body movement if a handhold is available for one hand. The long axis of this handhold should parallel a line from its midpoint to the point of force application. The handhold should not be perpendicular to this line.

Without a handhold, maximum torques could be applied, in one set of experiments, with a minimum of body movement when the long axis of the body was at right angles to the rotational axis of the turning task. Large forces exerted for a short time resulted in less body movement than smaller forces exerted for a longer time (Dzendolet and Rievley, 1959).

Caution: The above tentative principles were derived from limited experiments with a "frictionless" device and may require modification under true weightlessness.

LOCATION OF HAND CONTROLS

The area available for the location of hand controls is much larger than for foot controls, since the arms and hands have a wider effective range of movement. Though hand controls could be placed anywhere within the operator's functional reach area, critical and frequently used controls take precedence and should be placed where they are most easily operated. For hand cranks, a specialized and thoroughly studied type of control, large work decrements have been demonstrated for obviously inconvenient or awkward locations, but Krendel (1960) believes that for this kind of control "common sense is about as good a guide as the data available." For other types of hand controls, some general guides to placement can be given, based on both experience and on "common sense."

Fore-and-Aft Location

Outer limits: The farthest distance at which hand controls can be placed varies with arm reach, body movement, vertical and lateral location of the control, and type of control. Ideally, all forward distances should be measured horizontally from the seat reference point (page 135). However, since the backrest angle and the amount of permitted body movement both influence reach distance, the recommendations below are measured from the plane of the backrest immediately behind

the shoulder. The following limits are intended as rough guides only. (See also page 138.)

In order to accommodate 5th percentile men, controls operated with the whole hand should be located no farther than 27 inches from the back of the shoulder. Finger-operated controls should be located no farther than $29\frac{1}{2}$ inches from the back of the shoulder. Controls located at or near these maximum distances should not be critical or frequently used, except that controls requiring maximum pull are best located far from the body.

Best Location: Hand controls should be placed not at the outer limits of the workspace, but where the worker can operate them when his upper arm and forearm make a 120-degree angle at the elbow. (A fuller discussion of elbow angle and hand control placement is found below).

Vertical Location

The best vertical zone for the location of hand controls is between shoulder and elbow (Caldwell, 1959b), where controls can be reached quickly and operated most accurately and efficiently. Some controls can be located above the shoulder or below the waist if they are not critical or frequently used, and if speed and accuracy are not important. The best single vertical location is probably at elbow height. For hand cranks, the region from 6 inches below the shoulder to waist level seems best (Reed, 1949; Provins, 1953).

Hand controls that should be located outside the shoulder-waist zone are those that require maximum force for pulling up or down. Pull-down controls are best placed 12 to 24 inches above the shoulder (Lehmann, 1927; Garry, 1930; Ross, 1956), whereas pull-up controls should be located below seat level (Hugh-Jones, 1947). Push is strongest at shoulder level; right-handed persons move controls to the left most strongly around waist level, but moving force to the right is little affected by vertical location (Caldwell, 1959a).

From the semiprone position, the best vertical location for hand controls is 5 inches below shoulder level (Brown *et al.,* 1949).

Lateral Location

The best area for manual operation is about 24 inches wide, 12 inches to either side of the midline (Ely, Thomson, and Orlansky, 1963b). When speed and accuracy of continuous or frequent control movements are critical, however, hand controls should be located in front of the operator between the parallel vertical planes of the two shoulders (Herbert, 1957), an area about 16 to 18 inches wide. Controls for the right hand should be within 8 to 9 inches to the right of the body midline; controls for the left hand, the same distance to the left of the midline. If both hands are used in positioning movements, the control should be located

directly in front of the operator's midline (Briggs, 1955).

Where strength is important, left and right movements are strongest when the hand is directly in front of the shoulder, down and push are strongest at about 30 degrees to the side, whereas for up and pull, lateral position makes little difference (Caldwell, 1959a).

It may sometimes be desirable to support the forearm on a horizontal armrest, especially to reduce the effects of acceleration. Under such conditions the control grip should be located 9 to 12 inches from the midline (Brissenden, 1957).

Location of Hand Controls for Exertion of Maximum Strength and Endurance

The position of a hand control at which maximum force can be exerted is generally also the position at which submaximal forces can be applied and maintained with the least fatigue (Krendel, 1960; Caldwell, 1961). Since maximum force is more reliably and more easily measured than duration, the former may be used to determine the location of controls for economy of muscular effort over periods of continuous or intermittent use.

Where strength and endurance are important, hand control location depends on the operator's elbow angle. The best elbow angle for exerting force in the seated position is about 120 degrees, with some variation for certain types of movement, as follows:

Pull forces are strongest when the controls are at or near full extension of the elbow (Hugh-Jones, 1947; Hunsicker, 1955; Caldwell, 1959b). In addition, for maximum pull at shoulder level, a footrest should be used, positioned so that the angle at the knee is 150 degrees and the long axis of the thigh is 20 degrees from the horizontal. In general, hand pull is strongest when the legs can exert the greatest force against the footrest (Caldwell, 1960).

Push forces are greatest at an elbow angle of from 150 to 160 degrees (Caldwell, 1959a).

For right-left movements, fore-and-aft location makes little difference, though forces are slightly stronger closer to the body (Hertzberg, n.d.; Hunsicker, 1955; Caldwell, 1959a).

Up-down forces are maximum at an elbow angle of about 120 degrees (Hunsicker, 1955), or when the force is exerted close to the body (Caldwell, 1959a).

For rotary movements, there is little difference in the forces exertable at most fore-and-aft locations, though the farthest points should be avoided (Hertzberg, n.d.).

From the prone position, the best fore-and-aft locations for applying force differ from those in the seated position, as follows: (1) Push-pull force is greatest at intermediate to farthest distances (elbow angles of 120 to 180 degrees), and is least close to the body (Hunsicker, 1955). (2) Up-down force is strongest close to the body, weakening as the distance increases (Hun-

sicker, 1955). (3) For right-left force, fore-and-aft control location makes little difference, though forces are somewhat stronger close to the body (Hunsicker, 1955).

Location of Hand Controls for Maximum Speed and Accuracy of Movement

The speed and accuracy of visually controlled positioning movements are greatest when controls are close to the operator (7 inches from the body), becoming progressively worse as the distance increases (Briggs, 1955). It should be noted that the positions from which maximum forces can be exerted on controls (see preceding section) are generally *not* the positions which favor the greatest precision (Krendel, 1960).

SPACING BETWEEN HAND CONTROLS

Controls located at arms' length and grasped by the hand without the aid of direct vision should be separated, in forward areas below shoulder level, by at least 6 inches. For areas behind the worker or above his shoulder to either side, the separation distance should be 12 inches. Both distances could be reduced for highly trained operators (Fitts, 1947). If direct vision is used, the separation distance between hand controls need only permit hand clearance; $2\frac{1}{2}$ to 3 inches will suffice. Minimum separation distance for rotary knob controls about $\frac{1}{2}$ inch in diameter, operated by the thumb and forefinger, should be $\frac{3}{4}$ to $1\frac{1}{4}$ inches between edges, preferably the latter, to avoid inadvertent operation (Bradley, 1957). Push-buttons about $\frac{1}{2}$ inch in diameter should have a minimum separation distance of $\frac{1}{2}$ to $\frac{3}{4}$ inch between edges (Bradley and Wallis, 1958). Where the separation distance between adjacent controls cannot exceed 1 inch, toggle switches with small dimensions and relatively large resistances should be used rather than push buttons, in order to minimize inadvertent operation of adjacent controls (Bradley and Wallis, 1959).

AMOUNT OF MOVEMENT FOR HAND CONTROLS

Finger-operated knobs used for fine adjustment should have 1 to 2 inches of pointer movement for one complete turn of the knob. If less pointer movement is desired, provide lower ratios (less pointer movement per turn of knob); for more pointer movement, higher ratios. In general, accuracy increases as the ratio decreases (Jenkins and Connor, 1949).

Push buttons should travel about $\frac{1}{8}$ to $\frac{1}{4}$ inch, with a minimum of $\frac{1}{16}$ inch and a maximum between $\frac{1}{2}$ and $\frac{3}{4}$ inch.

Toggle switches should move between 20 and 135 degrees, preferably about 40 degrees (in part from Bradley and Wallis, 1959).

Rotary selector switches, whether pointers or knobs, should have a minimum movement between stops of 15 degrees if positioned by sight, and 30 degrees if positioned by touch. Maximum movements can be as much as 90 degrees. The best movement for all purposes is 40 degrees (Ely, Thomson, and Orlansky, 1963a).

Hand levers should move no less than 2 inches, to insure precise adjustment (Hick, 1944). Their maximum excursion is limited only by comfortable reach distance. Handgrips in front of the operator's midline can be comfortably rotated to the left or right a total of 80 degrees. When the neutral or resting position of the handgrip is 19 degrees to the left of the midline, as is usually the case, a right-handed worker can comfortably rotate the control 46 degrees to the left of this neutral position and 34 degrees to the right. Extreme values for 66 Air Force men averaged 110 degrees to the left and 54 degrees to the right of the midline (Daniels and Hertzberg, 1952).

Handgrips beside the operator, as on armrests, can be rotated to both left and right in a vertical plane about 90 degrees, though half of this distance, or 45 degrees, should be considered the operational limit for exerting torque. Such a handgrip can be rotated 60 degrees forward of vertical and 35 degrees aft, though again operational limits are about half this figure (Brissenden, 1957).

DIRECTION OF MOVEMENT FOR HAND CONTROLS

Controls should move in the "expected" direction, producing a machine or display movement in a similar direction. For example, a control movement to the right or clockwise should cause the machine or display to which it is linked to move to the right (Mitchell and Vince, 1951).

Where great strength must be exerted upon a control, push-pull movements should be used. Rotating a wheel produces the next highest force, followed in descending order by up-down and right-left movements. In the weakest direction, right-left, control movements are only about $\frac{1}{3}$ as strong as those exerted in the strongest direction, push-pull.

Right-left movements are, in general, slightly faster than left-right for right-handers (Brown and Slater-Hammel, 1949), and arm flexion is faster than extension (Glanville and Kreezer, 1937).

For applying force to a handle the best direction of movement is up, with the subject standing and lifting. The next best direction is down, utilizing body weight (McFadden and Swearingen, 1958). For fine adjustments, push is slower than movement in any other direction (Herbert, 1957). For precision, a single control moving in two or three dimensions is better than separate ones, each moving in one dimension (Orlansky, 1948). Right-handed workers move indicator knobs most precisely between 9 and 12 o'clock with the right hand and between 12 and 3 o'clock with the left hand (Chapanis, 1951a).

Handwheels and hand cranks should be turned forward or clockwise for maximum speed and efficiency (Baines and King, 1950; Provins, 1953). Vertical cranking, either parallel or perpendicular to the transverse body plane, appears slightly more efficient than horizontal cranking (Reed, 1949). For small cranks of less than 4-inch radius, the direction of movement is unimportant (Krendel, 1960).

SPEED OF MOVEMENT FOR HAND CONTROLS

In hand control movement, speed and accuracy are inversely related. Increasing precision means increasing the operating time. Where speed is crucial, control resistance should be minimal, though a pound or two makes little difference; control speed decreases as the load increases. Pilots can operate control sticks at rates up to 75 inches per second with resistances less than 35 pounds, 50 inches per second with loads of 35 pounds, and 10 inches per second with a load of 100 pounds (Orlansky, 1948). The advantage of low resistance must be balanced against the possibility of inadvertent activation (see the following section, Resistance of Hand Controls).

For maximum speed, the distance moved should be as short as possible, since longer movements take more time. However, the increased time is not a linear function of distance. A 4-inch movement may take 0.8 second, whereas a 16-inch movement will take only 1.0 second, because reaction time and starting time are constant regardless of distance moved, and longer movements permit increased rates of movement.

The movement should not change in direction. Continuous, curved movements should be used whenever possible. For speed the preferred directions of movement for hand controls are horizontal rather than vertical, and fore-and-aft rather than lateral. One can push against resistance on a joystick faster than he can pull by about 25 per cent (Orlansky, 1948). For short linear movements about 2 to 3 inches long, requiring precision, vertical movements are fastest, followed by lateral and fore-and-aft movements (Herbert, 1957).

Two-handed simultaneous positioning movements, as in reaching for objects, are fastest within the area 30 degrees to left and right of the center line (Barnes and Mundell, 1939). Knobs about $\frac{1}{4}$ inch in diameter can be turned fully as fast as those of $\frac{3}{4}$- to 2-inch or greater diameter. The smaller knobs can be rolled between the thumb and index finger, whereas the larger knobs are manipulated by wrist action (Stump, 1953).

Where possible, control movement should be terminated by a fixed, mechanical stop rather than by muscular control guided by sight or touch (Barnes, 1936).

Steering wheels are turned most rapidly when the wheel is nearly vertical (Lehmann, 1958). For hand cranks, speed of movement varies with control resistance and radius (slower speed with larger resistance and radius), but 180 rpm is a good average figure (Helson, 1949; Reed, 1949).

RESISTANCE OF HAND CONTROLS

Hand controls must offer some resistance to movement in order to eliminate inadvertent operation by the weight of hand and arm, hand tremor, body sway, vibration, and the like. The resistance should not be so great as to prevent some users' satisfactory operation. In general, operators tend initially to apply too much force to overcome small resistances and too little to large resistances.

Normal

For controls requiring single applications, or short periods of applied force, a reasonable loading would be less than half the operator's greatest strength. For controls operated continuously or for long periods resistances should be much lower. Only about 15 per cent of maximum strength can be exerted throughout the day without muscle fatigue (Hettinger, 1961).

Minimum

Resistance for hand controls (excluding finger-operated controls such as knobs, push buttons, toggles, and rotary selector switches) should never be less than 2 pounds. Even below 5 pounds the pressure sensitivity of the hands is poor (Orlansky, 1948). If the full weight of the arm and hand rest on a control such as a lever, minimum resistance should be 10 to 12 pounds; if only the forearm and hand, 5 pounds; and if only the hand, 2 pounds (after Dempster, 1955a). Cranks used for steady, rapid turning should have minimum resistances of 2 pounds if their radius is less than $3\frac{1}{2}$ inches, and 5 pounds if their radius is from 5 to 8 inches or more. For handwheels, minimum resistances should be 5 pounds. Finger-operated push buttons and toggle switches should have resistances of at least 10 ounces, and rotary selector switches, 12 ounces (Ely, Thomson, and Orlansky, 1963a).

Maximum

Limits for the maximum resistance of hand controls are often difficult to determine, because of wide variation in operating populations, in the type and location of controls, and in the frequency, duration, direction, and amount of control movement. For example, there is a greater than fourfold difference in the push exertable on a control stick depending on whether it is located in the midline away from the operator, where his pushing force is greatest, or to the left near the (right-handed) operator, where his pushing force is least (Hertzberg, n.d.).

Despite these reservations, some general recommendations for maximum control resistance can be given. Individual design problems can best be solved by referring to the data on arm and hand strength (p. 210) and using the 5th percentile of arm strength as a guide. That is, control resistance should not exceed the strength of the 5th percentile operator.

Push buttons and finger-operated toggle switches should not require pressures greater than $2\frac{1}{2}$ pounds, and rotary selector switches should have resistances not exceeding 3 pounds (Ely, Thomson, and Orlansky, 1963a). Resistances close to these figures will help prevent inadvertent operation if controls are close together, but when there is no such danger a load of a few ounces may be acceptable (Bradley and Wallis, 1959).

Knobs gripped and turned with the thumb and forefinger can have a maximum resistance of 10 pounds (Lobron and Hedberg, 1954). In practice lower resistances are preferable. For fine adjustment tasks, knobs 1 inch in diameter should have maximum resistances of $4\frac{1}{2}$ ounces, larger knobs 6 ounces. Cranks used in steady, rapid turning should have maximum resistances of 5 pounds if their radii are $3\frac{1}{2}$ inches or less, 10 pounds if their radii are 5 to 8 inches or more. Handwheels should have a maximum resistance of 30 pounds if operated by one hand, 50 pounds if operated by both hands (Ely, Thomson, and Orlansky, 1963a).

For levers grasped by the hand, maximum resistances are given in Table 137. This table provides a rough guide to the maximum control forces that seated men of the 5th percentile in strength can normally be expected to exert within the forward working area. For controls located in the best position for exerting force, considerably greater resistances than these would be feasible. For inefficiently located controls, resistance loads should be less.

The maximum values in Table 137 are for single applications of force; for frequent or continuous appli-

cation, much lower resistances are indicated.

Emergency hand controls, such as door-release handles (in transport aircraft, a rubber-covered steel ring 52½ inches above the floor), can be operated by almost all healthy young to middle-aged women—and therefore by even more men—if the horizontal or vertical pull required for 5 seconds does not exceed 30 pounds from the seated position or 35 pounds from the standing position. If a single jerk will suffice, resistance may be 65 pounds (McFadden *et al.,* 1959).

ANGULATION OF HAND CONTROLS

The handgrip of a control should be placed so that its neutral position coincides with the "neutral" or comfortable position of the hand at rest. For an aircraft stick-control with a shaped handgrip located 13½ inches above and 19 inches in front of the SRP (seat reference point), the fore-and-aft plane of the handle in the resting position should be 19 degrees to the left of the midline (Daniels and Hertzberg, 1952). For a laterally placed control, in the fore-and-aft plane of the armrest, the handgrip should be angled 15 degrees forward of the vertical and 8 degrees to the right—for the right hand (Brissenden, 1957). For steering wheels, the maximum force can be exerted when the wheel is almost horizontal (Lehman, 1958).

Table 137. Maximum Resistance for Hand Levers.

(Derived from Hertzberg, n.d.; Orlansky, 1948; and Hunsicker, 1955.)

Movement	One Hand (Lb.)	Both Hands (Lb.)
Push	35	55
Pull	40	85
Up	20	--
Down	20	--
Out	15	35
In	20	35

SIZE OF HAND CONTROLS

Handgrips

The best diameter for a handgrip is between $\frac{3}{4}$ inch and $1\frac{1}{2}$ inches (Müller, 1934; Hertzberg, 1956). Minimum diameters depend on the forces to be exerted: for 10 to 15 pounds the diameter should be no less than $\frac{1}{4}$ inch, preferably larger; for 15 to 25 pounds, a minimum of $\frac{1}{2}$ inch; and for 25 or more pounds, a minimum of $\frac{3}{4}$ inch (Hertzberg, 1956). Maximum diameters should not greatly exceed $1\frac{1}{2}$ inches.

The length of a handgrip should be at least $3\frac{3}{4}$ inches to accommodate the full hand breadth.

For grasping or "trigger" handles in which two elements are squeezed together, the most force can be applied when $2\frac{1}{2}$ inches separate the "trigger" and the heel in the open position (Hertzberg, 1955). In the closed position this distance should be $1\frac{1}{2}$ to 2 inches (Dupuis *et al.*, 1955).

Hand Levers

For handles or levers upon which moderate to large forces are exerted, force increases directly with the length from the fulcrum, so that the proper length will depend on the mechanical advantage required. In one study of the lever type of door handle, $2\frac{1}{2}$ times as much force could be exerted upon an 11-inch handle as on a 5-inch handle (McFadden and Swearingen, 1958).

Cranks and Handwheels

For general-purpose hand-cranking, Craik and Vince (1963) found that crank radii of 3 to 5 inches were best at fairly light loads—between $\frac{3}{8}$ and 8 pounds per inch —and fairly high speeds of about 150 to 200 rpm. Helson (1949) and Davis (1949) found that for light loads, fast speeds, or both, radii of about $1\frac{1}{2}$ to 4 inches were preferable; for heavy loads, radii of 4 to 7 inches or more. For very heavy loads, the radius should not exceed 20 inches; for wheels operated by two hands, like steering wheels, the minimum radius should be 7 inches and the maximum 21 inches (Ely, Thomson, and Orlansky, 1963a). With unstressed hand cranks, the cranking rate of large cranks decreases as the radius increases, but with torque loads of 10, 30, 50, or 70 inch-pounds, the cranking rate increases with increasing radius (Katchmar, 1957).

Knobs

Knob diameters have little effect on speed and accuracy. Diameters of $\frac{1}{2}$ to 2 inches are generally acceptable (Craik and Vince, 1945; Jenkins and Connor, 1949). Two-inch diameters provide smooth operation at any resistance, though smaller diameters can be used with moderate resistances (1.75 to 3.5 inch-ounces or less). If resistances are above 5.25 to 7 inch-ounces, knob diameters should be at least $1\frac{1}{2}$ inches (Bradley and Arginteanu, 1956). In general, knobs should be increased in size as greater torques are required. For knobs located on a panel directly in front of a seated operator, these diameters will permit the following *maximum* torques during a single application: 0.5 inch, 42.6 inch-ounces; 1.0 inch, 115.6 inch-ounces; 2.0 inches, 244.5 inch-ounces; 3.0 inches, 444.3 inch-ounces; 4.0 inches, 694.8 inch-ounces; and 5.0 inches, 898.5 inch-ounces (Sharp, 1962).

Where many knobs are required and space is at a premium, the best arrangement is obtained with knobs about $\frac{1}{2}$ inch in diameter, with $\frac{3}{4}$ to $1\frac{1}{4}$ inches between the edges (Bradley, 1957). Where three knobs are concentric, the best diameters are $\frac{1}{2}$ to 1 inch for the front knob, 2 inches for the middle one, and $3\frac{1}{4}$ inches for the back knob (Bradley and Stump, 1955).

Minimum knob depth should be about $\frac{1}{2}$ inch; the maximum can be as large as desired. The best depth is about $\frac{3}{4}$ inch.

Push Buttons

Push buttons should have diameters of $\frac{1}{2}$ to 1 inch. If space between adjacent buttons is less than $1\frac{1}{2}$ inches, select the smaller (half-inch) diameter (Bradley and Wallis, 1958).

Toggle Switches

Toggle switches should have lever-tip diameters between $\frac{1}{8}$ and 1 inch. The length of the lever arm should be at least $\frac{1}{2}$ inch and at most 2 inches (Ely, Thomson, and Orlansky, 1963a).

Rotary-Selector Switches

When the rotary-selector switch is a moving pointer on a fixed scale, the width of the switch should not exceed 1 inch, the length should be at least 1 inch, and the depth, or height, can range between $\frac{1}{2}$ and 3 inches. If a knob is used for this type of control, diameters may range between 1 and 4 inches, and depths, or heights, between $\frac{1}{2}$ and 3 inches (Ely, Thomson, and Orlansky, 1963a).

SHAPE OF HAND CONTROLS

Handgrips

These may be either round or oval in cross-section, or contour-molded to the shape of the hand. Handles on which large forces are exerted (as in opening emergency doors) should be "T"-shaped (Table 102). The double "L" is the next most efficient and the single "L" the poorest (McFadden and Swearingen, 1958). With only light resistance, there is little difference among these handles.

Push Buttons

These should be round, with concave surfaces.

Knobs

These should be round, with knurled or serrated edges. They may also be shape-coded for tactile discrimination where controls may be confused.

The dimensions of the space envelope occupied by the hand (from fingertip to wrist) of 95th percentile men while performing various tasks are given in Table 138 (from Baker, McKendry, and Grant, 1960). See also Fig. 101.

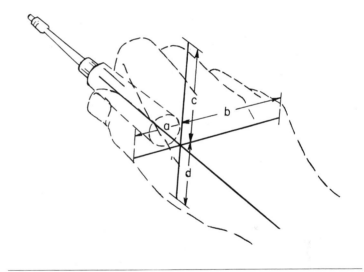

Table 138. Clearances Needed for the Right Hand in Various Actions by a 95 Percentile Man. (Baker, McKendry, and Grant, 1960)

| Hand Action | Minimum Dimensions for Hand Clearance (in Inches, as Measured from Reference Point. See Fig. 101.) | | | |
	(a) To Left	(b) To Right	(c) Up	(d) Down
Manipulating plug-in units (tubes 2-1/16" diameter)	1.59	2.24	2.56	1.70
Turning screwdriver (8" length) or spintite wrench (6" length)	1.49	2.42	2.44	1.67
Grasping, turning and cutting with needle-nosed pliers (5-1/2" length) or wire-cutters (5" length)	2.08	2.89	1.83	2.66
Turning socket wrench (3/8" base, 3-1/4" shaft)	2.12	3.25	3.25	2.88
Turning Allen wrench (2" length)	1.42	3.38	3.72	2.54

101. A 3-dimensional representation of the space envelope needed for hand tool use. Required clearance is measured, from the intersection of the long axis of arm and tool and the two planes perpendicular to it, in four directions: a) to left; b) to right; c) up; d) down (Baker, McKendry, and Grant, 1960).

4. Design Recommendations for Foot Controls

WHEN TO USE

Use foot controls rather than hand controls: (1) when a continuous control task is required—if precision of control positioning is not of primary importance; (2) when the application of moderate to large forces, greater than about 20 to 30 pounds, is necessary—whether intermittently or continuously; and (3) when the hands are in danger of becoming overburdened with control tasks.

WHAT KIND TO USE

Pedals on which pressure is applied from the whole leg, like a brake pedal, should be used when forces above 10 to 20 pounds are required.

Pedals on which pressure is applied mainly from the ankle, like an accelerator pedal, should be used when small forces, of about 10 pounds or less, and continuous operation are required.

Foot switches or buttons, like the headlight dimmer switch, are advantageous where small forces and intermittent operation are required.

HOW TO OPERATE

Seated or Standing Position

Whenever possible foot controls should be operated from the seated position. More force can thereby be exerted on the pedal, with less effort required to keep the body upright and balanced (Koch, 1941).

For the seated position, a backrest should always be provided and—if both feet are not used fully for control operation—a footrest as well.

In the standing position, the most force can be exerted when the body is directly over the pedal (Koch, 1941).

Right or Left Side

There is little difference between the legs with regard to strength, speed, or accuracy, but most people prefer the right leg, especially for critical tasks.

For tiring, noncritical tasks, use the feet interchangeably whenever possible (Koch, 1941).

Applying the Foot to the Pedal

For all but light pedal pressures (under 10 to 20 pounds), the foot should be applied to the pedal so that the long axis of the tibia (lower leg) is immediately over, and in line with, the axis or pivot of the pedal. The long axes of the foot and lower leg should form a 90-degree angle, which requires the least muscular effort to hold the foot in position. The next-best position is with the arch of the foot over the pedal axis (Lauru and Brouha, 1957); the toe and heel are least effective for heavy pressures (Müller, 1936; Le Gros Clark and Weddell, 1944).

For small pressures, the toe may be satisfactory. Where rapid, continued pedal movements are required, the pedal should be toe-operated, with the fulcrum at the base of the heel (Barnes *et al.,* 1942).

Knee Angle

For small pedal forces, the knee angle should be at least 90 degrees and preferably 135 degrees or more. The leg exerts maximum force with the knee angle about 130 to 150 degrees (Müller, 1936) or 160 to 165 degrees (Hugh-Jones, 1947).

The most comfortable knee angle for operating brake-rudder pedals in U.S. military and naval aircraft was found to be 111 $\pm$5 degrees (Elbel, 1949; Swearingen, 1949).

LOCATION

Every operator has an optimum pedal location, or range of locations, requiring the least energy expenditure and providing the most comfort. This location is specified with reference to the seat reference point (SRP), where the back of the seat surface intersects the backrest, in the midline (see p. 135).

Fore-and-Aft Location

Fore-and-aft SRP-pedal distance is an important determinant of the amount of pressure exertable on a foot control. The shorter the distance, the greater the force exertable (Martin and Johnson, 1952).*

When the pedal is at seat level, the fore-and-aft SRP-pedal distance must be less than the length of the outstretched leg by at least 2 to 4 inches. As seat height increases above the pedal, the SRP-pedal distance should be correspondingly decreased (Müller, 1936).

Since fore-and-aft SRP-pedal distance bears a definite relationship to leg length and consequently to stature, it can be estimated from stature. Where maximum pedal pressures are sought, this distance should be about 47.5 per cent of stature when the pedal is 2.5 inches above the SRP. *Example:* For an operator 67 inches tall, fore-and-aft SRP-pedal distance should be about 32 inches (Martin and Johnson, 1952).

---

*It is impossible to reconcile the conflicting results of Hugh-Jones (1947) on 6 men, Müller (1936) on 3 men, and Martin and Johnson (1952) on 166 men. A definitive study is greatly needed.

Where great force is not needed, the distance should be increased for comfort. Under such conditions a distance of 55 per cent of stature is satisfactory. *Example:* For an operator 67 inches tall, fore-and-aft SRP-pedal distance should be about 37 inches, when the center of the pedal is about 5.5 inches below the SRP (Dupuis *et al.,* 1955). A range of 37 to 44 inches has been suggested by Hugh-Jones (1947) but the upper limits of this range would be excessive for any but the tallest persons. Ely, Thomson, and Orlansky (1963b) recommend optimum SRP-pedal distances, when the thigh is horizontal, of 27 inches for a pedal 18 inches below the seat. This distance should increase to a maximum of 40 inches as the pedal is raised to 5 inches below the seat.

Vertical Location

For maximum pressures, the center of the pedal surface should be at or *above* the SRP level, never more than 4 inches above or below seat level (Müller, 1936). The best vertical SRP-pedal distance is about 3.5 per cent of the operator's stature; for example, 2.3 inches above the SRP for a 67-inch man (Martin and Johnson, 1952).

Where comfort matters more than the force exertable, pedals can be placed *below* the SRP by vertical distances varying with the type of task and the subjects' evaluation of comfort. This distance should not exceed 16 inches. For tractors, Dupuis *et al.* (1955) recommend that the center of the pedal be about 5.5 inches below the SRP.

Pedals operated from a standing position should never be more than 10 inches from the floor, preferably 8 inches or less (Koch, 1941).

Lateral Location

As pedals are moved laterally from the midline, the force exertable decreases, and discomfort increases. The maximum force exertable falls to 90 per cent when the pedal is placed 3 inches to either side; to 73 per cent, with a 6.7-inch shift; and to 63 per cent with a 10.2-inch shift. If midline location is not feasible, pedals should be displaced no farther than 3 to 5 inches from the midline (Dupuis *et al.,* 1955).

RESISTANCE OF FOOT CONTROLS

Maximum

Pedal resistance must never exceed the pressure exertable by the weakest operator (Tables 87–91).

For male military groups and for most male workers, the resistance of leg-operated pedals should not exceed 200 pounds for single brief applications (Hugh-Jones, 1947; Elbel, 1949). The United States National Advisory Committee for Aeronautics, predecessor of the National Aeronautics and Space Administration (NASA), established for aircraft rudder controls a maximum resistance of 180 pounds (Orlansky, 1948).

For frequently but not continuously used leg-operated pedals, a pressure of about 30 per cent of the maximum exertable is reasonable; that is, 60 to 120 or more pounds depending on pedal location (Dupuis *et al.*, 1955).

For toe-operated pedals in continuous use, like an automobile accelerator, resistance should not exceed 10 pounds (Dupuis, 1958).

For pedals used continuously over long periods, resistances should not require more than 15 per cent of the operator's maximum exertable strength.

Minimum

The minimum resistance for leg-operated pedals should be about 8 to 10 pounds, in order to exceed the 7-pound average pressure exerted on the pedal by the weight of the leg (Orlansky, 1948). Where the leg does not rest on the control, the minimum can be 4 pounds (Ely, Thomson, and Orlansky, 1963a).

Minimum resistance for toe-operated controls may be less; to wit, about 3 to 4 pounds.

Optimum

The best range of resistance for leg-operated pedals is from 8 to 60 pounds (Orlansky, 1948).

For toe-operated pedals, the best resistance is from 6.5 to 9 pounds. The lower figure "corresponds well to the pressure that the foot exerts merely by its own weight when all the muscles are totally relaxed" (Lehmann, 1958).

DISPLACEMENT OF FOOT CONTROLS

Pedals operated by the entire leg should generally have a 2- to 4-inch displacement, with an additional 2 or 3 inches permissible for pedals of the brake type (Dupuis *et al.*, 1955). Displacement of 3 to 4 inches or more should be coupled with pedal resistance greater than 10 pounds.

Pedals operated by ankle action should have a maximum travel of about 2 inches, corresponding to an angle of about 10 to 12 degrees. The angle should not exceed 30 degrees (Dupuis *et al.*, 1955; Dupuis, 1958), which is about half the total range of ankle movement.

Heavy footgear makes it difficult to gauge pedal travel, resulting in excessive movement and pressure. If heavy footgear is anticipated, pedal travel should be increased above the values just given (Le Gros Clark and Weddell, 1944).

DIRECTION OF TRAVEL

For all foot controls the direction of travel should be down, or away from the body, in line with the long axis of the lower legs, and roughly parallel to the midsagittal plane of the body.

ANGLE WITH FLOOR

The angulation of pedals operated by leg action at the hip and knee should permit the foot to be placed on the pedal surface with the ankle at a 90-degree angle. Pedal angulation will thus vary considerably with vertical and fore-and-aft pedal location.

Example: Hertzberg (1954) found that the forces applied to aircraft brake pedals were sharply reduced as the pedal angle (with the vertical) decreased below 20 degrees or increased above 40 degrees. Maximum forces were at pedal angles of about 30 degrees.

The angulation of most pedals operated by ankle action also varies with vertical and fore-and-aft location. In general the angle of the pedal surface should permit the foot and lower leg to form an angle of at least 90 degrees but never more than 130 degrees. Foot-leg angles less than 90 degrees should be avoided except when greater pressures are needed briefly, but high pressures should normally not be required with an ankle-operated pedal.

Exception: In the prone position, foot-leg angles of 75 to 90 degrees are acceptable and they may be used to achieve greater pedal pressures (Brown *et al.,* 1950b), though 90 degrees and above remain best for continuous operation.

SIZE OF FOOT CONTROLS

The following dimensions will insure a reasonably large area of contact between shoe and pedal. (1) *Width:* Pedals should be as wide, or almost as wide, as the sole of the shoe, *i.e.* at least 3.5 inches. Maximum width matters little if there is enough clearance between adjacent pedals. (2) *Length:* Pedals used intermittently should be at least 3 inches long. Pedals used for control or rest purposes continuously or for long periods should be about 11 to 12 inches long, to provide support.

SHAPE OF FOOT CONTROLS

Pedal shape is not very important as long as the pedal surface is flat and affords a large enough area of contact with the shoe (see preceding paragraph).

For continuously used pedals, the shape should be rectangular and conform roughly to footgear dimensions.

Intermittently used pedals may be of any shape—square, rectangular, circular, or oval—provided that they are at least 3 inches long and 3.5 inches wide.

For pedals on which large forces must be exerted, that is 200 pounds or more, a pedal bar (or recessed heel section) will prevent the foot from slipping off the pedal and will assist the operator to locate the pedal by feel. This is particularly advantageous in cold weather or when large, heavy boots are worn (Le Gros Clark and Weddell, 1944).

5. The Design of Seats and the Seated Workspace

INTRODUCTION

It is impossible to give design specifications for *a* seat or *a* workspace, since both will vary in dimensions according to their intended purpose and the physical characteristics of their users. "Executive" chairs, typists' chairs, school children's chairs, aircraft pilots' seats, factory workbenches, and their surrounding workspaces all differ markedly in size, shape, construction, and components. General-purpose design, however, in home, office, or passenger transport vehicles, as contrasted to specialized work areas, should accommodate as many of the general population as possible.

Table 139 contains recommendations for nine seat dimensions or other characteristics from twenty studies, grouped by kind of seat.

The following text contains a brief discussion of each seat and workspace dimension and other characteristics, with recommended values for the general population, based on the anthropometric data of chapter II.

THE SEAT

Dimensions

Height: Seat height above floor, measured vertically to the front of the sitting surface, is a major determinant of comfort. Many seats are too high. As a rule, tall people can accommodate to a low seat more easily than short people to a high seat. If forced to choose between too low and too high a seat, choose the lower.

In a Swiss study of office seats with a constant seat-to-desk-top distance of 11 inches, the average seat height preferred by men was 17.4 inches, with "acceptable" limits averaging 15.4 and 19.3 inches, and the average seat height preferred by women was 17.1 inches, with "acceptable" limits averaging 15.5 and 18.9 inches (Burandt and Grandjean, 1963).

On the basis of the popliteal heights of the American civilian population presented in chapter II, it would appear that the maximum seat height to accommodate persons larger than the 1st percentile, wearing shoes, is 15 inches; to accommodate the 5th percentile and larger persons, 16 inches. A footrest would permit these seat heights to be increased, while vertical adjustment above these heights would insure comfort for those who might find such a seat too low.

Where possible, the space between the floor and the front of the seat should be open, to permit placing the feet underneath the seat. This permits change in position and, by moving the center of gravity of the body farther to the rear, facilitates rising from the seat (Keegan, 1962).

Length: Seat length (or depth) is measured from the front edge of the seat to the intersection of the rear edge with the backrest or backrest plane (SRP). Too long a seat presses into the leg behind the knee, causing discomfort and possibly danger (from thrombophlebitis, or blood clotting), unless the sitter shifts his buttocks forward. Short people are particularly affected. Too short a seat, by failing to support the lower thighs, affects chiefly tall persons.

Current seats vary considerably in length, from 13 inches in a typist's chair to the clearly excessive 24 inches or more in some upholstered furniture. The best general-purpose seat length, based on the buttock-popliteal length of the shorter members of the civilian population, is about 16 to 17 inches.

Width: Seat width matters only as a minimum, not a maximum dimension. Current seat widths range from 14 through 19 inches, though anything below 17 inches is too small. A reasonable figure is 18 inches.

Multiple seating requires additional space for shoulders and elbows, plus a few inches for associated clothing. At least 24 inches per person should be allowed for comfort. Several persons side by side can overlap by turning their trunks, occupying much less space than if they sat "foursquare." Precise data on such seating, as in automobiles or troop transports, are greatly needed. For auditorium seats divided by a single armrest, 26 inches should be allowed for each seat. If each seat has two armrests, 28 inches should be allowed (Thomson, Covner, Jacobs, and Orlansky, 1963).

BACKREST

Dimensions

Height: High backrests are necessary only in resting or semireclining seats. From 18 to 20 inches will provide support up to the shoulder level, 25 inches will support the shoulders, and about 35 inches will provide support for the head. In work seats or in chairs in-

---

NOTES TO TABLE 139

[a]A maximum for seats of fixed height, without footrests.

[b]For shoulder support; for head support, 34 inches; for lumbar support only, 5 to 6 inches, with bottom edge 5 inches above seat.

[c]For shoulder support; for lumbar support only, 12 to 13 inches will suffice.

[d]Ideally should be adjustable between 15 and 19 inches.

[e]19 inches between armrests.

[f]Lower edge 8 inches above seat.

[g]Minimum for lumbar support; bottom edge should be 4.5 inches above seat, leaving a free space for rearward projection of sacrum, and should be vertically curved to provide support for 4th and 5th lumbar vertebrae.

[h]105 degrees for erect sitting, but at least 115 degrees for sustained relaxing comfort.

[i]"Approximate" acceptable range given for each dimension.

[j]16 inches without armrests, 19 inches with armrests.

[k]For desk heights between 28.3 and 30.7 inches.

[l]Lower edge should be adjustable 5.5 to 9.4 inches above seat.

[m]Adjustable through this range.

[n]Excluding 6-inch headrest.

Table 139. Recommendations for Seat Design According to Various Studies (Dimensions in Inches, Angles in Degrees)

Kind of Seat	Source of Data	Seat Height	Seat Length	Seat Breadth (Minimum Only)	Backrest Length	Backrest Breadth (Minimum Only)	Vertical Adjustment (Minimum Only)	Fore-and-Aft Adjustment (Minimum Only)	Seat Angle	Backrest Angle (From Horizontal)
General Seating	Present authors	15-16[a]	16-17	18	18-20[b]	20[c]	4[d]		6-7	115
General Seating	Floyd and Roberts, 1958	17	15	16[e]	5[f]		4[d]			
General Seating	Keegan, 1962,1964	16	16		9[g]				5	105-115[h]
Wooden Chair	Kocker and Frey, 1932	18	16	17						
Wooden Chair	Åkerblom, 1948, 1954	15-16	15 3/4						3-7	115-120
Upright Chair[i]	General Services Adm., 1950	17.25-18	15-16.5	16-18	15.25-17.25	12.25-15				
Office Chair (British)	Meade, n.d.	17	15	16-19[j]					3-5	95-105
Office Chair (Swiss)	Burandt and Grandjean, 1963	15.7-20.9[k]	13.8-15.7		7.9[l]				3	100
Secretarial Chair[i]	General Services Adm., 1950	16-22	13-16	16-18	7.5-11.5	11.5-16	3			
Secretarial Chair	Woodson, 1954	15-18[m]	15	15		12	3			105
Executive Chair	Woodson, 1954	18	18	19		19				105
Upholstered Chair[i]	General Services Adm., 1950	17-18.5	17-20	18.5-24.5	15.25-20	16-21				
Upholstered Chair, Swivel[i]	General Services Adm., 1950	17-22	17-20	18 1/2-24 1/2	15.25-20	16-21	3			
Stadium Seat	Washburn, 1932	18		17-18						
Railway (Passenger)	Hooton, 1945	16.9	20		28					
Aircraft (Passenger)	McArthur, 1945	15-15.5								
Aircraft (Passenger)	Lippert, 1950	16.8	18							
Aircraft (Pilot-Commercial)	Nat. Air Stand., n.d.	10-16[m]	16-17.5	17-18	23-25[n]	17-18	6	10	2-12	103-123
Aircraft (Pilot-Military)	Randall et al., 1946		15	16	25	16	7	3	4-10	103
Automobile (Driver)	Lay and Fisher, 1940	16.5-17.75	16.5-17.75						6-7	110-114
Automobile (Driver)	McFarland and Stoudt, 1961	10-14	18	18	18-21	20	4.5	8	7	112
Truck (Driver)	Dunlap and Kephart, 1954	14.7-17.4	16.1						9.2-11.4	98.6-114[l]
Truck (Driver)	McFarland, Damon and Stoudt, 1958	15.5	17	19	18-20	21	4	6		
Tractor	Dupuis, 1958	15.8	13.8-15.8	17.7				5.9		

tended for erect sitting it is better if the backrest does not extend behind the shoulder. Lumbar support, the most important function of any backrest, can be provided by a backrest 5 to 6 inches high, with the bottom edge 7 to 8 inches above the seat surface. In any backrest, there should be an open space or recess no less than 4 inches high between the seat surface and the back of the lumbar support to provide space for protrusion of the buttocks.

Width: Where the sitter can rest or relax, at least 20 inches will provide full support across the shoulders. In seats where only lumbar support is needed, as for typists and in certain factory jobs, a minimum of 12 to 13 inches will suffice.

Inclination: In most seated positions the backrest is angled backward from 103 to 115 degrees. Any angle in this range is probably satisfactory, though the larger angles are more appropriate in chairs designed primarily for resting rather than working.

Curvature: Backrests which afford only lumbar support, as in typists' chairs or some factory seating, should not have a lateral curvature deeper than that of a circle 7.3 inches in radius (Darcus and Weddell, 1947). A shallower curvature is preferable. In fact, since the human lumbar curve is convex forward, there is good reason for designing seats with a convex rather than a concave lumbar support. Åkerblom's (1948) chair has such a convex support.

For backrests at shoulder or head height, as in general-purpose seating, curvatures should have radii of from 16 to 18 inches, never less than 12 inches (Floyd and Roberts, 1958). Such backrests need not be curved at all.

Adjustability

Fore-and-Aft: Six inches of adjustment, in 1-inch increments, will bring most persons into adequate contact with hand and foot controls and with the workspace surface. Eight inches will include the extremes of the general population, both male and female.

Vertical: This will not be required in seats used for rest, travel, or relaxation. To bring the eyes to a specific level, as in operating a sighting or tracking station with a fixed eyepiece, 9.5 inches of vertical adjustment will fit the entire civilian population. For a male group, 7 inches will suffice. Where it is acceptable to locate the eyes within a larger vertical area, as in operating a vehicle, these figures can be reduced to about 6 inches for the general civilian population and 4 inches for males only.

SEAT SURFACE

The area of contact of the buttocks and thighs on a flat, hard seat varied, in one study of 104 men, from 101 to 268 square inches, with a mean of 179 square inches. The average sitting pressure on the occupied seat ranged from 0.69 to 1.23 pounds per square inch, with a mean of 0.92 pound per square inch (Swearingen, Wheelright, and Garner, 1962). However, nearly half of the body weight is concentrated on the 8 per cent of the sitting area located under or adjacent to the ischial tuberosities, the bony protuberances underlying the gluteal muscles and fat upon which one sits. Here pressures may be much higher, perhaps up to 60 pounds per square inch.

The total body weight supported by the seat surface can be reduced 39.4 per cent by means of a footrest, armrests, and a slightly sloping backrest. A footrest alone will reduce body weight on the sitting area by 18.4 per cent, armrests alone by 12.4 per cent and a backrest alone by 4.4 per cent. It should be noted that a footrest may reduce the sitting area proportionately more than it reduces pressure on the seat, thereby increasing the pressure per unit area of contact averaged over the whole seat (Swearingen et al., 1962).

Because the skin immediately over the ischial tuberosities has a blood supply better adapted to support weight than the rest of the buttocks (Edwards and Duntley, 1939), some designers with medical training have prescribed firm, unshaped seat surfaces on which the ischial tuberosities carry most of the body weight (Darcus and Weddell, 1947; Åkerblom, 1948, 1954).

Such designs make a virtue of necessity in that the specialized blood supply—specifically, the absence of vessels immediately under the tuberosities—represents an adaptation to the "hard" facts of life before man could modify his environment. Common experience and actual tests (Randall et al., 1946; Slechta et al., 1957) show that people prefer and perform more efficiently for longer periods in soft seats—not "supersoft," but certainly cushioned—than on hard surfaces. From 1 to 2 inches of compression will suffice.

Shape: Most seats should be flat rather than shaped, because of the varied conformation of the human buttocks and perineal region as well as the difficulty of changing position in a shaped seat (Darcus and Weddell, 1947). However, preliminary studies of seats in which the occupant is virtually immobile for many hours, as in fighter aircraft, indicate that seats with cut-outs or depressions under the ischial tuberosities are more comfortable and efficient than flat seats. Cut-outs remove part of the load from the tuberosities and spread it over the remaining area of the buttocks (Hertzberg, 1955).

Pulsating cushions: Flat cushions or those with ischial tuberosity cut-outs can be supplemented by an air bladder which is alternately inflated and deflated. When the cut-out cushion is inflated, pressure is applied to all parts of the buttocks except the ischial tuberosities. With the cushion deflated, the ischial tuberosities bear most of the body weight. Alternating cycles of 20 seconds each for inflation and deflation have been found most comfortable (Burns and Stockman, 1958). A pulsating cushion can, however, have any rate and duration of the cycle. Faster and shallower cycles, around twelve or more per minute, can stimulate circulation in the buttocks. In extensive use tests in the Air Force (Hertzberg, 1949, 1955; Burns and Stockman, 1958) and Navy (Hanna and Libber, 1958), pulsating cushions have significantly prolonged comfort and delayed fatigue for flyers sitting immobilized for many hours.

Nylon net: Subjective evaluation tests have shown that the nylon net support originally developed by Hertzberg and Colgan (1948) for prone position beds can also serve as comfortable seats (Hertzberg, 1961; Forrest *et al.,* 1958). One subject occupied such a seat for 32 consecutive hours with but minor discomfort (Duddy and Dempsey, 1958).

Inclination: Angling back the whole seat or its surface prevents the buttocks from sliding forward on the seat and permits the backrest to support part of the body weight. The current range varies from 0 (horizontal) to 12 degrees; the best angle is around 5 to 7 degrees.

ARMREST

When armrests do not interfere with necessary body movements, they may increase the sitter's comfort.

Armrest height, the most critical dimension, should be from 8 to 10 inches above the seat surface.

THE SEATED WORKSPACE

Clearances Around the Seat

In order to provide proper "fit" or spatial accommodation for a seated operator, *minimum values* are needed for a series of workspace dimensions. These values, derived from the anthropometric data in chapter II and presented here, should be adequate for the static body size of about 99 per cent of most populations. With body movement, additional space will be required, the precise amount depending on the kind and extent of the movement. See "Dynamic Fit," page 134.

Vertical workspace clearances: A vertical seat-to-roof distance of 39 inches will provide minimum head clearance for 99th percentile persons. However, to allow for headgear and to provide a small safety factor, a better minimum figure would be 41 inches. With cushioned seats this measurement should be made from the occupied or compressed cushion.

For clearance above the shoulder region of the tallest men when sitting erect and clothed, 28.5 inches will suffice. For clearance above the knees, as in the distance from the underside of desk or table to floor, or from foot pedal to steering wheel, 26 inches will accommodate the largest men wearing shoes. Vertical clearance for the thighs, as between the seat and the bottom of the table or desk, should be at least 8 inches.

Lateral workspace clearances: At least 24 inches should be allowed laterally at elbow level, the maximum body breadth. For heavy or bulky clothing allow at least 27 inches, preferably more. It should be noted that this dimension is applicable only when the elbows are at the sides of the body. For lateral arm movement, additional clearances of up to 41 inches will be needed at the shoulder level if the upper arms are maximally abducted, with bent elbows. For clearance across the hips at least 19.5 inches should be allowed; 21.0 inches if heavy bulky clothing is worn. For clearance across the shoe allow 5 inches for each foot.

Fore-and-aft workspace clearances: At chest level 12.0 inches should be provided, or 13.5 inches for heavy clothing. At waist level allow 15.5 or 16.5 inches, respectively; between the seat back and the front of the knees at knee level, at least 26.5 inches plus an additional 0.5 inch for heavy clothing. Shoe length requires a minimum of 13 inches.

Shape of the Workspace

The workspace can be considered as a three-dimensional region surrounding the worker, defined by the outermost points touched by the various parts of the body and by the tools, controls, or other equipment used. Each such workspace, or "kinetosphere" (Demp-ster, 1955a), can be measured and its shape described. The boundaries of different kinetospheres can be compared, and their variability treated statistically. However, since kinetospheres differ widely from person to person and task to task according to the amounts of body movement necessary, it is not possible to define any universally applicable sizes or shapes. For applications of the kinetosphere technique to specific situations the reader is referred to Dempster (1955a, 1955b), and Dempster, Gabel, and Felts (1959). Simpler techniques for determining workspace limits and for locating controls and other items used by the worker are described in sections 3 (on hand controls) and 4 (on foot controls) of this chapter.

When the normal workspace—that is, an area not requiring the outer limits of reach—is confined to a horizontal surface such as a desk, tabletop, or workbench, the shape of the most desirable area can be more easily determined than if it were three-dimensional. It was formerly believed that the working area on such surfaces could be defined by two intersecting semicircles, one for each arm, but Squires (1956) has shown that because of biomechanical restrictions on the range of arm movements, the area is more correctly described as a prolate epicycloid (Fig. 102). The dimensions of this normal work area will vary with the worker's arm length (see p. 95, forearm-hand length, and p. 138, functional reach measurement) and the location of his elbow on the horizontal surface.

Dimensions Within the Workspace

The best height for tables, desks, or workbenches used by the general population from the seated position is about 29 inches, though this dimension is closely related to the height of the accompanying seat. For typewriting, a desk height of 25 to 25.5 inches has been recommended by Meade (n.d.), and 25.6 inches by Burandt and Grandjean (1963). A vertical distance of 10 to 12 inches is desirable between seat and table surfaces. These dimensions should result in an eye-to-table-top distance of between 12 to 16 inches when a person is sitting in a comfortable reading or writing position; this distance affords satisfactory vision. The same seat-to-table-top vertical distance of 10 to 12 inches should enable most people to rest their elbows and forearms comfortably on the table (Karvonen, Koselka, and Noro, 1962). At least a 25-inch distance is desirable between the floor and the underside of the working surface.

Where groups of persons are seated at a table, at least 30 inches should be allowed from side to side per person (Thomson, Covner, Jacobs, and Orlansky, 1963). The best work-surface height for the standing position is about 3 inches below the elbow. This height is reported to combine the most speed and efficiency of manipulation with the least muscle strain. If the height of work surfaces for standing operation must remain fixed, about 41 inches will accommodate the largest number of persons (Ellis, 1951).

The depth of horizontal work surfaces is of less consequence than height and area. Depth may vary considerably with the type of use, though many such surfaces, especially laboratory benches, are commonly too deep to serve efficiently as part of the normal working area. Excessive depth may require the worker to rise from his chair, if seated, or to use one hand for stability, if standing, in order to reach the back of the work surface.

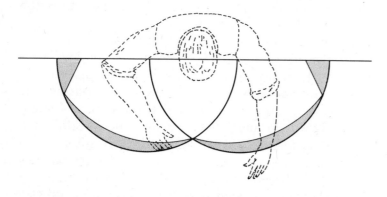

102. The shape of the normal working area is defined, not by semi-circles, as formerly believed, but by prolate epicycloids which eliminate the stippled area in the figure above. The left arm indicates maximum, as opposed to normal, reach (after Barnes, 1949, and Squires, 1956).

6. Design Recommendations for Passageways, Doorways, and Escape Hatches

PASSAGEWAYS

The dimensions recommended here for various kinds of passageways, including corridors, aisles, and tunnels, will accommodate 99 per cent of the population wearing light to medium clothing—shoes, trousers, jacket, and hat. These dimensions could be decreased if fewer than 99 per cent of the population were to be provided for, or if the users would stoop, bend, or otherwise adapt to inadequate openings. On the other hand, the dimensions would have to be increased to allow for bulky or specialized clothing or equipment, for which Table 6 gives the appropriate increments. For most purposes, the passageway dimensions are minimal, with additional room to be supplied if possible. Unless otherwise indicated, the following recommendations are derived from data in Hertzberg, Emanuel, and Alexander (1956), and in chapter II of this book. Procedures like those used by King, Ostrich, and Richardson (1954) and by Roebuck (1957, 1961) in studying emergency escape from aircraft can be applied to passageways under normal as well as emergency conditions.

For Walking
(including corridors, aisles, catwalks, tunnels)

The following dimensions permit passage of single persons: (1) Height: at least 77 inches; 63 inches or less would suffice if stooping is permitted, but this is generally undesirable. (2) Width: at least 25 inches. If space is at a premium, the passageway can be trapezoidal in shape, 25 inches wide at shoulder level and 12 inches wide at the feet. A 15-inch width will permit only sideways movement, which is undesirable.

The following dimensions permit passage of two or more persons: (1) Height: as for a single person. (2) Width: at least 44 inches, preferably 54 inches; if one of the two persons stands against a wall, at least 30 inches, preferably 36. For three persons abreast, at least 60 inches but preferably 72 (Thomson *et al.,* 1963).

For One Person Crawling on Hands and Knees

(1) Height: Allow at least 32 inches, less if head is held down and elbows are flexed. (2) Width: Allow at least 25 inches.

For Prone Crawling

(1) Height: Allow at least 17 inches, preferably more. (2) Width: Allow at least 25 inches. To provide for lateral elbow movement in prone crawling, an additional 15 inches are highly desirable.

DOORWAYS OR OTHER ENTRANCES AND EXITS

The preceding recommendations for passageways hold as well for doorways and similar openings. Since traversing doorways is virtually instantaneous, doorways can be smaller than passageways. A man 75 inches tall—77 inches when fully clothed—could stoop to pass through a much lower doorway, or in emergencies could even squeeze through an opening 16 inches high. But this is undesirable. Doorways should have the minimum dimensions of other passageways.

STAIRS

(1) One-way: at least 20 inches wide, preferably 22 inches. (2) Two-way: at least 48 inches wide, preferably 51 inches (Thomson *et al.,* 1963).

Handrails or Railings

The top of the rail should be 36 inches above the floor level. On stairs the top of the railing should be at least 34 inches above the top of the front edge of each step (Thomson *et al.,* 1963). The best handrail diameter is 2 inches, 1.75 inches at a minimum (Hall and Bennett, 1956).

ESCAPE HATCHES

The following dimensions are adequate for emergency escape hatches used by U.S. Air Force personnel as tall as 73 inches and as heavy as 215 pounds nude. During tests, these subjects wore standard Air Force interme-diate clothing, plus experimental outfits of heavy arctic flying clothing, plus a survival kit. For the general population, therefore, these escape hatch dimensions should more than suffice. Difficulties for this group would result not from escape hatch dimensions but from lack of agility due to age or poor physical condition. The following dimensions are from Randall *et al.* (1946) and White *et al.* (1952).

Vertical: minimum dimensions, 31 by 20 inches.

Horizontal: on floor, minimum dimensions, 20 by 29 inches. Overhead, minimum dimensions, 22 by 22 inches. A step or ledge should be provided not more than 45 inches below the hatch.

7. Recommendations for Lifting and Carrying

LIFTING

Location of Object

The most important determinant of lifting force is the distance of the feet from the grasping axis. Lifting force is greatest when the weight lifted is in the midsagittal plane of the body, and decreases sharply as the weight moves away from the body (Whitney, 1958).

The best height for lifting (vertical pull) is at or slightly above the level of the middle fingertip, measured with the subject wearing shoes and standing erect with arms hanging by his sides (Bedford and Warner, 1937). Middle fingertip height averages about 28 inches and varies from 25 inches in men 65 inches tall to 31 inches in men 75 inches tall. Above midfingertip height, lifting power decreases very rapidly; below it, more slowly. With a load near the floor, lifting force may be only 75 to 80 per cent of that at the best height (Reijs, 1921; Vernon, 1924; Dempster, 1958).

Body Position

In general, "leg lift," with back vertical and legs bent, affords a stronger vertical pull than "back lift," with legs straight and back bent (Bedford and Warner, 1937; Clarke, 1945), although at lifting heights from 5 to 20 inches Whitney (1958) found no appreciable difference between the two methods. Leg lift entails so much less risk of back injury that it is certainly preferable.

Maximum Weights That Can Be Lifted to Various Heights

Table 140 lists the heaviest weights that can be lifted to various heights by men of the 5th percentile in lifting strength, as calculated from tests on nineteen healthy young men (Emanuel, Chaffee, and Wing, 1956). The weight, an aircraft ammunition case 25 inches long, 11 inches high, and 6 inches wide, with handles at either end, was lifted mainly with the legs (Fig. 80). The tests were not carried to the point of fatigue.

The maximum weight of compact equipment (boxes 12 × 6 × 12 inches, equipped with handles) that can be lifted from the floor six times, one minute apart, without strain by at least 99 per cent of a young, healthy male population is as follows (Switzer, 1962): to knee level, 62 pounds; to waist level, 38 pounds; and to shoulder or eye level, 27 pounds.

Lifting Pattern for Repetitive Work

For some tasks which involve continuous lifting, rest pauses can decrease the amount of human energy needed. In lifting weights ranging from 4.4 to 22.0 pounds from one table to another through vertical distances varying between 2.0 and 17.3 inches, the most efficient rate—in terms of mechanical efficiency—was

Table 140. Lifting Ability of Fifth Percentile Men, Calculated from Tests on 19 Healthy Young Men. (Emanuel, Chaffee, and Wing, 1956)

Height Lifted from Floor (Ft)	Maximum Weight Lifted by 5th Percentile Men (Lb)
1	142
2	139
3	77
4	55
5	36

about twenty to thirty lifts per minute. An 11-pound weight lifted 4.7 inches and back ten times in 17 seconds, followed by a 13-second rest, required only half the energy of the same work carried out in ten lifts evenly spaced over the same 30-second period. However, for smaller loads of 4.4 pounds or heavier loads of 22 pounds, or when the rate was increased to forty lifts per minute, no difference was found between the two techniques (Ronnholm, Karvonen, and Lapinleimu, 1962). Thus, optimum rates and patterns need to be established for each distinct type of repetitive lifting task.

CARRYING

The literature on portability, summarized by Carlock, Weasner, and Strauss (1963), as well as their own experiments, indicates a lack of correspondence between performance and physiological cost. It is performance, of course, which concerns the human factors specialist. Research on portability is greatly needed.

Handles

In general, objects weighing more than 10 pounds should have convenient carrying handles, located above the center of gravity and at least 4.5 inches long, with the inner surface of the handle at least 2 inches from the surface of the object (Folley *et al.,* 1963).

Size of Containers

The more compact the container, the more easily can it be carried. For bulky equipment, the center of gravity should not be more than 20 inches from the carrier's body (Hertzberg, 1956).

Maximum Weights

Solely by muscle power, a normal man can transport about 200 tons a day through a horizontal distance of 1 meter, or 50 tons through a vertical distance of 1 meter (Müller *et al.,* 1958). The maximum individual loads, if carried in either hand by means of handgrips, should be about 60 pounds for short distances and 35 pounds for longer distances.

The weight of bulky articles, around 30 inches to a side, should not exceed 20 pounds (Hertzberg, 1956). In general, a weight is "heavy" when it reaches 35 per cent or more of body weight (Bedale, 1924). Carrying one heavy weight a given distance costs less physiologically than carrying two weights, each weighing half as much, the same distance in two trips (Müller *et al.,* 1958).

Conclusion ——————— VI

If equipment were designed in a logical way, the first things to establish would be the basic, relatively simple requirements of size, shape, and biomechanical suitability for the intended consumers. The designer could then consider the technically more complex refinements of psychophysiological integration and larger systems coordination. In reality, equipment has been designed in the reverse order. The sophistication of displays and controls, of cybernetics and feedback mechanisms, and of systems engineering has far outstripped the application of physical anthropology or even its acceptance in principle. Without attempting to account for this discrepancy on philosophic, historical, or economic grounds, we have tried in this book to provide the required anthropometric data and instructions for their use. From the mass of detail some general conclusions can be drawn as to the present status of knowledge and practice of applied physical anthropology.

Static descriptions of physique are fairly well mapped out for the population of the United States, with which this book has been chiefly concerned. Populations elsewhere in the world remain to be investigated on a similar scale. In the United States additional "human engi-

neering" measurements, beyond the basic ten of the National Health Examination Survey, are needed for the general population, while both the basic and more extensive measurements should be extended to children and to various specific groups—regional, ethnic, racial, and occupational. Anthropometric surveys will always be needed for such selected groups as soldiers, airmen, and truck drivers as criteria of selection change, and as equipment becomes more specialized. Long-term changes in physique require remeasurement of a population. The anthropometrist can look forward to employment for many years to come.

On the whole, one may anticipate only minor additions and modifications of our present knowledge of the static dimensions of the human body. Research is more urgently required on dynamic body measurements and biomechanical abilities as they affect human performance and equipment design. Under ideal conditions, design recommendations like those in Chapters II and V should not be based on static body dimensions alone, but sometimes they have had to be. Even when derived from such measurements as arm and leg reach, strength, endurance, or speed of response, design recommendations must be regarded as tentative, as we have pointed out, in the absence of firm, experimental data obtained in the appropriate workspace and environment. The empirical mock-up approach outlined in Chapter I, based on a few men at the extremes of a distribution and a few others scattered throughout the range, is useful as a first approximation, but only as such. Definitive recommendations should rest not on assumption and extrapolation but on experiment. Spatial and bio-mechanical accommodation should be systematically reduced, and the effects on performance noted.

There will probably never be enough anthropologists to conduct such research unaided. Other biomedical scientists—anatomists, physicians, physiologists, and psychologists—as well as engineers and designers will have to enter the field. Working together, the human biologist and the designer can materially improve man's comfort, efficiency, health, and safety in a world increasingly man-made.

References

Sources of Anthropometric Data in Tables 7–45

1. Bayer, L. M., and H. Gray, "Anthropometric standards for working women," *Human Biology,* 6:472–488, 1934.
2. Bowles, G. T., *New Types of Old Americans at Harvard and at Eastern Women's Colleges,* Harvard University Press, Cambridge, Mass., 1932.
3. Carter, I. G., "Physical measurements of 'Old American' college women," *American Journal of Physical Anthropology,* 16:497–514, 1932.
4. Damon, A., "Physique and success in military flying," *American Journal of Physical Anthropology,* 13:217–252, 1955.
5. Damon, A., "Constitutional factors in acne vulgaris: prevalence in white soldiers," *A.M.A. Archives of Dermatology,* 76:172–78, 1957.
6. Damon, A., *Heights and Weights of White and Negro Women in New York City,* unpub. data.
7. Damon, A., H. K. Bleibtreu, O. Elliot, and E. Giles, "Predicting somatotype from body measurements," *American Journal of Physical Anthropology,* 20:461–473, 1962.
8. Damon, A. and H. W. Stoudt, "The functional anthropometry of old men," *Human Factors,* 5:485–491, 1963.
9. Daniels, G. S., H. C. Meyers, and E. Churchill, *Anthropometry of Male Basic Trainees,* WADC Technical Report 53-49, Aero Medical Laboratory, Wright-Patterson Air Force Base, Ohio, 1953.
10. Daniels, G. S., H. C. Meyers, and S. H. Worrall, *Anthropometry of WAF Basic Trainees,* WADC Technical Report 53-12, Aero Medical Laboratory, Wright-Patterson Air Force Base, Ohio, 1953.
11. Diehl, H. S., "Height and weights of American college men," *Human Biology,* 5:445–479, 1933.
12. Diehl, H. S., "The heights and weights of American college women," *Human Biology,* 5:600–628, 1933.
13. Donelson, E. G., M. A. Ohlson, B. Kunerth, M. B. Patton, and G. M. Kinsman, "Anthropometric data on college women of the middle states," *American Journal of Physical Anthropology,* 27:319–332, 1940.
14. Elbel, E. R., "Body measurements of male students entering the University of Kansas," *Kansas Studies in Education,* 4:1–24, 1954.
15. Elbel, E. R., and J. K. Barland, "Height and weight of female students entering the University of Kansas," *University of Kansas Bulletin of Education,* 14:19–26, 1959.
16. Gibbons, T. B., I. A. Phillips, R. K. Budensiek, and J. R. Gilbertson, *Age, Height, and Weight of 2173 Men Entering Recruit Training During 1952 at the U.S. Naval Training Center, Great Lakes, Ill.,* Naval Medical Research Unit No. 4, Great Lakes Naval Training Center, Ill., 1953.
17. Gifford, E. C., *Compilation of Anthropometric Measures on U.S. Navy Pilots,* NAMC-ACEL-437, Air Crew Equipment Laboratory, Philadelphia, Pa., 1960.
18. Heath, C. W., *What People Are,* Harvard University Press, Cambridge, Mass., 1945.

19. Hertzberg, H. T. E., G. S. Daniels, and E. Churchill, *Anthropometry of Flying Personnel—1950,* WADC Technical Report 52-321, Aero Medical Laboratory, Wright-Patterson Air Force Base, Ohio, 1954.

20. Hooton, E. A., *A Survey in Seating,* Heywood-Wakefield Co., Gardner, Mass., 1945.

21. Jones, W. L., and E. C. Gifford, *Anthropometry of U.S. Navy Pilots,* Air Crew Equipment Laboratory, Naval Air Materiel Center, Philadelphia, Pa., n.d.

22. Karpinos, B. D., "Height and weight of Selective Service registrants processed for military service during World War II," *Human Biology,* 30:292–321, 1958.

23. Karpinos, B. D., "Current height and weight of youths of military age," *Human Biology,* 33:335–354, 1961.

24. King, B. G., D. G. Morrow, and E. P. Vollmer, *Cockpit Studies—The Boundaries of the Maximum Area for the Operation of Manual Controls,* Project X-651, Report 3, Naval Medical Research Institute, National Naval Medical Center, Bethesda, Md., 1947.

25. McCormick, N. Y., *Physical Characteristics of the 1946 Airline Transport Pilot Population,* Medical Service Publication, Civil Aeronautics Authority, U.S. Department of Commerce, Washington, D.C., 1947.

26. McFarland, R. A., A. Damon, and H. W. Stoudt, "Anthropometry in the design of the driver's workspace," *American Journal of Physical Anthropology,* n.s., 16: 1–23, 1958.

27. National Health Survey, *Weight, Height and Selected Body Dimensions of Adults: United States, 1960–62* (by H. W. Stoudt, A. Damon, R. A. McFarland, and J. Roberts), U.S. Public Health Service, Washington, D.C., 1965.

28. Newman, R. W., and R. M. White, *Reference Anthropometry of Army Men,* Report No. 180, Environmental Protection Section, Quartermaster Climatic Research Laboratory, Lawrence, Mass., 1951.

29. O'Brien, R., and W. C. Shelton, *Women's Measurements for Garment and Pattern Construction,* Miscellaneous Publication No. 454, U.S. Department of Agriculture, Washington, D.C., 1941.

30. Pett, L. B., and G. F. Ogilvie, "The Canadian weight-height survey," *Human Biology,* 28:177–188, 1956. See also, "The report on Canadian average weights, heights and skinfolds," *Canadian Bulletin on Nutrition,* 5:1–81, 1957.

31. Randall, F. E., *Survey of Body Size of Army Personnel, Male and Female: (4) Body Dimensions of Army Females —Methodology and General Considerations,* Report 123, U.S. Quartermaster Climatic Research Laboratory, Lawrence, Mass., 1947.

32. Randall, F. E., A. Damon, R. S. Benton, and D. I. Patt, *Human Body Size in Military Aircraft and Personal Equipment,* Army Air Forces Technical Report 5501, Air Materiel Command, Wright Field, Ohio, 1946.

33. Randall, F. E., and E. H. Munro, *Reference Anthropometry of Army Women,* Report No. 159 Environmental Protection Section, Quartermaster Climatic Research Laboratory, Lawrence, Mass., 1949.

34. Randall, F. E., E. H. Munro, and R. M. White, *Anthropometry of the Foot,* Environmental Protection Section Report No. 172, Quartermaster Climatic Research Laboratory, Lawrence, Mass., 1951.

35. Stoudt, H. W., A. Damon, and R. A. McFarland, "Heights and weights of white Americans," *Human Biology,* 32:331–341, 1960.

36. Tyroler, H. A., *Preliminary Cross-Sectional Documenta-*

tion of Physical Measurements of Male Populations at Risk by Geographic Location and Age, Health Research Foundation, Asheville, N.C., 1958.

37. U.S. Army Quartermaster Research and Engineering Command, Natick, Mass., unpub. data.
38. U.S. Marine Corps Research Laboratory, Marine Clothing Depot, Philadelphia, Pa., unpub. data.
39. U.S. Navy, "Height distribution of recruits," *Statistics of Navy Medicine,* 5(7):2–3, 1949.
40. U.S. Navy, "Weight distribution of recruits," *Statistics of Navy Medicine,* 5(6):2–3, 1949.
41. U.S. Navy, Naval Air Station, Pensacola, Fla., unpub. data.
42. U.S. Navy, Navy Medical Research Laboratory, Naval Submarine Base, New London, Conn., unpub. data.
43. White, R. M., *Anthropometry of Army Aviators,* Technical Report EP-150, Quartermaster Research and Engineering Center, Natick, Mass., 1961.

Bibliography

Advisory Committee for Aeronautics
1916. *Experiments on the Possible Rate at Which a Pilot Can Pull Back the Control Column in an Aeroplane,* R. and M. No. 282, Great Britain Technical Department of the Advisory Committee for Aeronautics.

Åkerblom, B.
1948. *Standing and Sitting Posture* (A. B. Nordiska Bokhandeln, Stockholm).
1954. "Chairs and sitting," in *Symposium on Human Factors in Equipment Design,* W. F. Floyd and A. T. Welford, eds. (H. K. Lewis and Co., London).

Amar, J.
1920. *The Human Motor* (E. P. Dutton Co., New York).

Arkin, A. M.
1941. "Absolute muscle power," *Archives of Surgery,* 42:395–410.

Ashe, W. F., P. Bodenman, and L. B. Roberts
1943. *Anthropometric Measurements,* Project No. 9, File No. 741-3, Armored Force Medical Research Laboratory, Fort Knox, Ky.

Asmussen, E., and K. Heebøll-Nielsen
1962. "Isometric muscle strength in relation to age in men and women," *Ergonomics,* 5:167–169.

Backman, G.
1924. "Körperlange und Tageszeit," *Upsala Läkar. Förhandl.,* 28:255–282.

Bailey, G. B., and R. Presgrave
1958. *Basic Motion Timestudy* (McGraw-Hill Book Company, New York).

Baines, R., and E. S. King
1950. "A study in the relationship between maximum cranking speed and cranking radius," *Motor Skills Research Exchange,* 2:24–28.

Baker, P. T.
1955. *Relationship of Desert Heat Stress to Gross Morphology,* Environmental Protection Division Technical Report EP-7, Quartermaster Research and Development Center, U.S. Army, Natick, Mass.

Baker, P. T., and F. Daniels, Jr.
1956. "Relationship between skinfold thickness and body cooling for two hours at 15 degrees C," *Journal of Applied Physiology,* 8:409–416.

Baker, P. T., E. E. Hunt, and T. Sen
1958. "The growth and interrelations of skinfolds and brachial tissues in man," *American Journal of Physical Anthropology,* 16:39–58.

Baker, P. T., J. M. McKendry, and G. Grant
1960. "Volumetric requirements for hand tool usage," *Human Factors,* 2:156–162.

Baker, P. T., and R. W. Newman
1957. "The use of bone weight for human identification," *American Journal of Physical Anthropology,* 15:601–618.

Ballinger, E. R., and C. A. Dempsey
1952. *The Effects of Prolonged Acceleration on the Hu-*

man Body in the Prone and Supine Positions, WADC Technical Report 52-250, Wright Air Development Center, Wright-Patterson Air Force Base, Ohio.

Barnes, R. M.
1936. "An investigation of some hand motions used in factory work," *University of Iowa Studies in Engineering,* Bulletin No. 6.
1949. *Motion and Time Study* (John Wiley and Sons, New York).

Barnes, R. M., H. Hardaway, and O. Podolsky
1942. "Which pedal is best?" *Factory Management and Maintenance,* 100:98–99.

Barnes, R. M., and M. E. Mundell
1939. "A study of simultaneous symmetrical hand motions," *University of Iowa Studies in Engineering,* Bulletin No. 17.

Barter, J. T.
1957. *Estimation of the Mass of Body Segments,* WADC Technical Report 57-260, Aero Medical Laboratory, Wright Air Development Center, Wright-Patterson Air Force Base, Ohio.

Barter, J. T., I. Emanuel, and B. Truett
1957. *A Statistical Evaluation of Joint Range Data,* WADC Technical Note 57-311, Aero Medical Laboratory, Wright Air Development Center, Wright-Patterson Air Force Base, Ohio.

Barter, J. T., E. I. Fry, and B. Truett
1956. "Anthropometry and Biokinematics of the Hand," unpub. ms., Aero Medical Laboratory, Wright Air Development Center, Wright-Patterson Air Force Base, Ohio.

Bartholomew, S. H.
1952. *Determination of Knee Moments During the Swing Phase of Walking, and Physical Constants of the Human Shank* (University of California, Berkeley).

Bass, D. E., P. F. Iampietro, and E. R. Buskirk
1959. "Comparison of basal plasma and blood volumes of Negro and white males," *Journal of Applied Physiology,* 14:801–803.

Baxter, B.
1942. "A study of reaction time using factorial design," *Journal of Experimental Psychology,* 31:430–437.

Bayer, L. M., and H. Gray
1934. "Anthropometric standards for working women," *Human Biology,* 6:472–488.

Bayly-Pike, D. F., and W. L. Kesteven
1950. *Investigation of Brake Lights and Drivers' Reactions,* Report No. 7/50, Army Operational Research Group (Great Britain).

Bedford, T., and C. G. Warner
1937. "Observations on the effects of posture or pull," *Lancet,* (2):1328–1329.

Beeler, F. de
1944. *Maximum Rates of Control Motion Obtained from Ground Tests,* National Advisory Committee for Aeronautics Wartime Report L-100, Washington, D.C.

Bennett, E., J. Degan, and J. Spiegel (eds.)
1963. *Human Factors in Technology* (McGraw-Hill Book Co., New York).

Birmingham, H. P., and F. V. Taylor
1954. *A Human Engineering Approach to the Design of Man-Operated Continuous Control Systems,* U.S. Naval Research Laboratory, Report 4333, Washington, D.C.

Bondurant, S., N. P. Clarke, W. G. Blanchard, H. Miller, R. R. Hessberg, and E. P. Hiatt
1958. *Human Tolerance to Some of the Accelerations Anticipated in Space Flight,* WADC Technical

Report 58-156, Aero Medical Laboratory, Wright Air Development Center, Wright-Patterson Air Force Base, Ohio.

Bowles, G. T.
1932. *New Types of Old Americans at Harvard and at Eastern Women's Colleges* (Harvard University Press, Cambridge, Mass.).

Bradley, J. V.
1957a. "Control knob arrangement can save aircraft instrument panel space," *Journal of Aviation Medicine,* 28:322–327.
1957b. *Glove Characteristics Influencing Control Manipulability,* WADC Technical Report 57-389, Aero Medical Laboratory, Wright Air Development Center, Wright-Patterson Air Force Base, Ohio.

Bradley, J. V., and J. Arginteanu
1956. *Optimum Knob Diameter,* WADC Technical Report 56-96, Aero Medical Laboratory, Wright Air Development Center, Wright-Patterson Air Force Base, Ohio.

Bradley, J. V., and N. E. Stump
1955. *Minimum Allowable Dimensions for Controls Mounted on Concentric Shafts,* WADC Technical Report 55-355, Aero Medical Laboratory, Wright Air Development Center, Wright-Patterson Air Force Base, Ohio.

Bradley, J. V., and R. A. Wallis
1958. *Spacing of On-Off Controls. I: Push Buttons,* WADC Technical Report 58-2, Aero Medical Laboratory, Wright Air Development Center, Wright-Patterson Air Force Base, Ohio.
1959. *Spacing of On-Off Controls. II: Toggle Switches,* WADC Technical Report 58-475, Aero Medical Laboratory, Wright Air Development Center, Wright-Patterson Air Force Base, Ohio.

Braune, W., and O. Fischer
1889. "The center of gravity of the human body as related to the equipment of the German infantry," *Treatises of the Mathematical-Physical Class of the Royal Academy of Sciences of Saxony,* No. 7, Leipzig. (U.S. Army Air Forces, Air Materiel Command Translation No. 379, Wright Field, Dayton, Ohio).
1892. "Bestimmung der Trägheitsmomente des menschlichen Körpers und seiner Glieder," *Abh. d. Math. Phys. Cl. d.K. Sachs. Gesell. d. Wiss. Leipzig,* S. Hirzel.

Braunstein, P.
1957. "Medical aspects of automotive crash injury research," *Journal of the American Medical Association,* 163:249–255.

Braunstein, P. W., J. O. Moore, and P. A. Wade
1957. "Preliminary findings of the effect of automotive safety design on injury patterns," *Surgery, Gynecology and Obstetrics,* 105:257–263.

Bresler, B., and J. P. Frankel
1950. "The force and movements in the leg during level walking," *Transactions of the American Society of Mechanical Engineers,* 72:27–36.

Briggs, S. J.
1955. "A Study in the Design of Work Areas," unpub. diss., Purdue University, Lafayette, Ind. [cited in McCormick, 1957].

Brissenden, R. F.
1957. *Some Ground Measurements of the Forces Applied by Pilots to a Side-Located Aircraft Controller,* National Advisory Committee for Aeronautics, Technical Note 4171, Washington, D.C.

British Standards Institution
1955. *School Dining Tables and Chairs,* British Standard

2639, British Standards House, London.

Brown, C. W., E. E. Ghiselli, R. F. Jarrett, E. W. Minium, and R. M. U'Ren
1949. *Magnitude of Forces Which May Be Applied by the Prone Pilot to Aircraft Control Devices. 1. Three-Dimensional Hand Controls,* MCREXD-694-4J, Engineering Division, Aero Medical Laboratory, Wright-Patterson Air Force Base, Ohio.
1950a. *Magnitude of Forces Which May Be Applied by the Prone Pilot to Aircraft Control Devices. 2. Two-Dimensional Hand Controls,* Air Force Technical Report No. 5954, Aero Medical Laboratory, Air Materiel Command, Wright-Patterson Air Force Base, Ohio.
1950b. *Magnitude of Forces Which May Be Applied by the Prone Pilot to Aircraft Control Devices. 3. Foot Controls,* U.S. Air Force Technical Report No. 5955, Aero Medical Laboratory, Air Materiel Command, Wright-Patterson Air Force Base, Ohio.

Brown, J. L., and R. E. Burke
1958. "The effect of positive acceleration on visual reaction time," *Journal of Aviation Medicine,* 29:48–58.

Brown, J. L., and M. Lechner
1956. "Acceleration and human performances," *Journal of Aviation Medicine,* 27:32–49.

Brown, J. S., and A. T. Slater-Hammel
1949. "Discrete movements in the horizontal plane as a function of their length and direction," *Journal of Experimental Psychology,* 39:84–95.

Brozek, J.
1952. "Changes in body composition in man during maturity and their nutritional implications," *Federation Proceedings,* 11:784–793.
1955. "Role of anthropometry in the study of body com-position: toward a synthesis of methods," *Annals of the New York Academy of Sciences,* 63:491–504.
1963. "Quantitative description of body composition: physical anthropology's 'fourth' dimension," *Current Anthropology,* 4:3–39.

Brozek, J., K. P. Chen, W. Carlson, and F. Bronczyk
1953. "Age and sex differences in man's fat content during maturity," *Federation Proceedings,* 12:21–22.

Brozek, J., E. Simonson, A. Keys, and A. Snowden
1952. "A test of speed of leg and arm movements," *Journal of Applied Physiology,* 4:753–760.

Brues, A. M.
1946a. "Regional differences in the physical characteristics of an American population," *American Journal of Physical Anthropology,* n.s., 4:463–482.
1946b. "Movements of the head and eye in sighting," pp. 211–219 in Randall, F. E., A. Damon, R. S. Benton, and D. I. Patt, *Human Body Size in Military Aircraft and Personal Equipment,* Technical Report No. 5501, U.S. Army Air Forces, Wright Field, Dayton, Ohio.

Bryan, W. L.
1892. "On the development of voluntary motor ability," *American Journal of Psychology,* 5:125–204.

Buck, C. A., F. B. Dameron, M. J. Dow, and H. V. Skowlund
1959. "Study of normal range of motion in the neck utilizing a bubble goniometer," *Archives of Physical Medicine and Rehabilitation,* 40:390–392.

Burandt, U., and E. Grandjean
1963. "Sitting habits of office employees," *Ergonomics,* 6:217–228.

Burns, H. L., and R. E. Stockman
1958. *Design and Development of a Pressure and Cycle Control for Dynamic Seat Cushions,* WADC Tech-

nical Report 58-616, Aero Medical Laboratory, Wright Air Development Center, Wright-Patterson Air Force Base, Ohio.

Caldwell, L. S.
1959a. *The Effect of Elbow Angle and Back Support Height on the Strength of Horizontal Push by the Hand,* Report No. 378, Psychology Division, U.S. Army Medical Research Laboratory, Fort Knox, Ky.
1959b. *The Effect of the Spatial Position of a Control on the Strength of Six Linear Hand Movements,* Report No. 411, U.S. Army Medical Research Laboratory, Fort Knox, Ky.
1960. *The Effect of Foot-rest Position on the Strength of Horizontal Pull by the Hand,* Report No. 423, U.S. Army Medical Research Laboratory, Fort Knox, Ky.
1961. *The Relationship between the Maximum Force Exertable by the Hand in a Horizontal Pull and the Endurance of a Sub-Maximal Holding Response,* Report No. 470, U.S. Army Medical Research Laboratory, Fort Knox, Ky.
1962. "Body stabilization and the strength of arm extension," *Human Factors,* 4:125–130.
1963. "Relative muscle loading and endurance," *Journal of Engineering Psychology,* 2:155–161.

Canfield, A. A., A. L. Comrey, and R. C. Wilson
1948. *An Investigation of the Maximum Forces Which Can Be Exerted on Aircraft Elevator and Aileron Controls,* Contract No. N6 OKI 77, Office of Naval Research, Washington, D.C.

Carlock, J., M. H. Weasner, and P. S. Strauss
1963. "Portability: A new look at an old problem," *Human Factors,* 5:577–581.

Carter, I. G.
1932. "Physical measurements of 'Old American' college women," *American Journal of Physical Anthropology,* 16:497–514.

Carter, R. L.
1958. "Vertebral injuries from ejection forces," Paper presented at 29th Annual Meeting, Aero Medical Assoc., Washington, D.C., March 24, 1958. Abstract, *Journal of Aviation Medicine,* 29:233.

Cathcart, E. P., E. M. Bedale, C. Blair, K. Macleod, and E. Weatherhead
1927. *The Physique of Women in Industry,* Report No. 44, Industrial Fatigue Research Board, Medical Research Council, London.

Cathcart, E. P., D. E. R. Hughes, and J. G. Chalmers
1935. *The Physique of Man in Industry,* Report No. 71, Industrial Health Research Board, Medical Research Council, London.

Chapanis, A.
1951a. "Studies of manual rotary positioning movements: I. The precision of setting an indicator knob to various angular positions," *Journal of Psychology,* 31:51–64.
1951b. "Theory and methods for analyzing errors in man-machine systems," *Annals of the New York Academy of Sciences,* 51:1179–1203.

Chen, K. P., A. Damon, and O. Elliot
1963. "Body form, composition, and some physiological functions of Chinese on Taiwan," *Annals of the New York Academy of Sciences,* 110:760–777.

Clarke, H.
1945. "Analysis of physical fitness index test scores of air crew students at the close of a physical conditioning program," *Research Quarterly,* 16:192–195.

Cochran, W. G.
1953. *Sampling Techniques* (John Wiley & Sons, Inc., New York).

Colcher, A. E., and A. M. W. Hursh
  1952. "Pre-employment low back x-ray survey," *Industrial Medicine and Surgery,* 21:319–321.
Comstock, G. W., and V. T. Livesay
  1963. "Subcutaneous fat determinations from a community-wide chest x-ray survey in Muscogee County, Georgia," *Annals of the New York Academy of Sciences,* 110:475–491.
Conferences on Body Composition
  1956. "The role of body measurements in the evaluation of human nutrition," *Human Biology,* 28:111–273, 1956; also Brozek, J. (ed.) *Body measurements and human nutrition* (Wayne University Press, Detroit, Mich.).
  1961. *Techniques for Measuring Body Composition,* National Academy of Sciences-National Research Council, Washington, D.C.
  1963. "Body composition," *Annals of the New York Academy of Sciences,* 110:1–1018.
Cotton, F. S.
  1932. "Studies in centre of gravity changes," *Australian Journal of Experimental Biology,* 10:16–34; and 225–247.
Craik, K. J. W., and M. A. Vince
  1945. *A Note on the Design and Manipulation of Instrument-Knobs,* MRC 46/272, A.P.U. 14.14.1. 1945. Applied Psychology Research Unit, Cambridge University.
  1963. "Psychological and physiological aspects of control mechanisms with special reference to tank gunnery," *Ergonomics,* 6:1–33.
Cullumbine, H., S. W. Bibile, T. W. Wikramanayake, and R. S. Watson
  1950. "Influence of age, sex, physique and muscular development on physical fitness," *Journal of Applied Physiology,* 2:488–511.
Damon, A.
  1943. *Effect of Flying Clothing on Body Measurements of Army Air Forces Flyers,* Memorandum Rpt. ENG-49-695-32, Engineering Division, Aero Medical Laboratory, ATSC, Wright Field, Ohio.
  1955. "Physique and success in military flying," *American Journal of Physical Anthropology,* n.s., 13:217–252.
  1957. "Constitutional factors in acne vulgaris: prevalence in white soldiers," *A.M.A. Archives of Dermatology,* 76:172–178.
  1960. "Host factors in cancer of the breast and uterine cervix and corpus," *Journal of the National Cancer Institute,* 24:483–516.
  1964. "Notes on anthropometric technique: I. Stature against a wall and standing free," *American Journal of Physical Anthropology,* 22:73–78.
Damon, A., H. K. Bleibtreu, O. Elliot, and E. Giles
  1962. "Predicting somatotype from body measurements," *American Journal of Physical Anthropology,* 20:461–473.
Damon, A., and R. F. Goldman
  1964. "Predicting fat from body measurements: densitometric validation of ten anthropometric equations," *Human Biology,* 36:32–44.
Damon, A., and R. A. McFarland
  1955. "The physique of bus and truck drivers, with a review of occupational anthropology," *American Journal of Physical Anthropology,* n.s., 13:711–742.
Damon, A., and F. E. Randall
  1944. "Physical anthropology in the Army Air Forces," *American Journal of Physical Anthropology,* n.s., 2:293–316.

Damon, A., and H. W. Stoudt
1963. "The functional anthropometry of old men," *Human Factors,* 5:485–491.

Daniels, G. S., and E. Churchill
1952. *The "Average Man?"* Technical Note WCRD 53-7, Wright Air Development Center, Air Research and Development Command, U.S. Air Force, Wright-Patterson Air Force Base, Ohio.

Daniels, G. S., and H. T. E. Hertzberg
1952. "Applied anthropometry of the hand," *American Journal of Physical Anthropology,* n.s., 10:209–215.

Daniels, G. S., H. C. Meyers, and E. Churchill
1953. *Anthropometry of Male Basic Trainees,* WADC Technical Report 53-49, Aero Medical Laboratory, Wright Air Development Center, Wright-Patterson Air Force Base, Ohio.

Daniels, G. S., H. C. Meyers, and S. H. Worrall
1953. *Anthropometry of WAF Basic Trainees,* WADC Technical Report 53-12, Aero Medical Laboratory, Wright Air Development Center, Wright-Patterson Air Force Base, Ohio.

Darcus, H. D.
1951. "The maximum torques developed in pronation and supination of the right hand," *Journal of Anatomy,* 85:55–67.
1953. "A strain-gauge dynamometer for measuring the strength of muscle contraction and for re-educating muscles," *Annals of Physical Medicine,* 1:163–176.
1954. "The range and strength of joint movement," in *Proceedings of the Ergonomics Society,* II, *Symposium on Human Factors in Equipment Design,* W. F. Floyd and A. T. Welford, eds., London.

Darcus, H. D., and A. G. M. Weddell
1947. "Some anatomical and physiological principles concerned in the design of seats for naval war weapons," *British Journal of Industrial Medicine,* 4:77–83.

Davenport, C. B., and A. G. Love
1921. *Army Anthropology,* U.S. Government Printing Office, Washington, D.C.

Davis, L. F.
1949. "Human factors in design of manual machine controls," *Mechanical Engineering,* 71:811–816, 837.

De Haven, H.
1942. "Mechanical analysis of survival in falls from heights of 50 to 150 feet," *War Medicine,* 2:586–596.

Dempsey, C. A., and I. Emanuel
n.d. "Arm reach capability," unpub. report, Anthropology Section, Aero Medical Laboratory, Wright Air Development Center, Wright-Patterson Air Force Base, Ohio.

Dempster, W. T.
1955a. *Space Requirements of the Seated Operator,* WADC Technical Report 55-159, Aero Medical Laboratory, Wright Air Development Center, Wright-Patterson Air Force Base, Ohio.
1955b. "The anthropometry of body action," *Annals of the New York Academy of Sciences,* 63:559–585.
1958. "Analysis of two-handed pulls using free-body diagrams," *Journal of Applied Physiology,* 13:469–480.
1961. "Free-body diagrams as an approach to the mechanics of human posture and motion," in *Biomechanical Studies of the Musculo-Skeletal System,* F. G. Evans, ed. (Charles C Thomas, Springfield, Ill.).

Dempster, W. T., W. C. Gabel, and W. J. L. Felts
1959. "The anthropometry of the manual work space for the seated subject," *American Journal of Physical Anthropology,* 17:289–317.

Deupree, R. H., and J. R. Simon
  1963. "Reaction time and movement time as a function of age, stimulus duration and task difficulty," *Ergonomics,* 6:403–411.

Diehl, H. S.
  1933a. "Height and weight of American college men," *Human Biology,* 5:445–479.
  1933b. "The heights and weights of American college women," *Human Biology,* 5:600–628.

Di Giovanni, C., Jr., and R. M. Chambers
  1964. "Physiologic and psychologic aspects of the gravity spectrum," *New England Journal of Medicine,* 270:35–41, 88–94, and 134–139.

Dill, D. B.
  1938. *Life, Heat and Altitude* (Harvard University Press, Cambridge, Mass.).

Donelson, E. G., M. A. Ohlson, B. Kunerth, M. B. Patton, and G. M. Kinsman
  1940. "Anthropometric data on college women of the middle states," *American Journal of Physical Anthropology,* 27:319–332.

Dreyfuss, H.
  1960. *The Measure of Man* (Whitney Library of Design, New York).

Drillis, R. J.
  1959. "The use of gliding cyclograms in the biomechanical analysis of movements," *Human Factors,* 1:1–11.

Duddy, J. H., and C. A. Dempsey
  1958. *Light-Weight Seating: Design Research and Development of a Net Seat for Project Manhigh,* WADC Technical Report 58-307, Aero Medical Laboratory, Wright Air Development Center, Wright-Patterson Air Force Base, Ohio.

Duggar, B. C.
  1963. "Continuous Control Performance of the Human Upper Extremity," unpub. diss., Harvard School of Public Health, Boston, Mass.

Dunlap, J. W., and N. C. Kephart
  1954. *Human Factors in the Design of Vehicle Cab Areas,* Final Report, Occupational Research Center, Purdue University, Lafayette, Indiana, to the U.S. Armed Forces Epidemiological Board, Washington, D.C.

Dupuis, H.
  1958. "Some standards for the design of the tractor driver's work place," paper presented at the 1958 Annual Meeting of the American Society of Agricultural Engineers, June 22–25.

Dupuis, H., R. Preuschen, and B. Schulte
  1955. "Zweckmässige Gestaltung des Schlepperführerstandes," *Schriftenreihe Landarbeit und Technik,* Heft 20.

Dusek, E. R.
  1958. *Encumbrance of Arctic Clothing,* Technical Report EP-85, Environmental Protection Research Division, U.S. Army Quartermaster Research and Engineering Center, Natick, Mass.

Dvorak, A., N. I. Merrick, W. C. Dealey, and G. C. Ford
  1936. *Typewriting Behavior* (American Book Co., New York) [cited in *Handbook of Human Engineering Data for Design Engineers,* Tufts College, Medford, Mass., 1952].

Dzendolet, E., and J. F. Rievley
  1959. *Man's Ability to Apply Certain Torques While Weightless,* WADC Technical Report 59-94, Aero Medical Laboratory, Wright Air Development Cen-

ter, Wright-Patterson Air Force Base, Ohio.

Eberhart, H. D., and V. T. Inman
1951. "An evaluation of experimental procedures used in a fundamental study of human locomotion," *Annals of the New York Academy of Sciences,* 51:1213–1228.

Edelman, I. S., and J. Leibman
1959. "Anatomy of body water and electrolytes," *American Journal of Medicine,* 27:256–277.

Edwards, E. A., and S. Q. Duntley
1939. "The pigments and color of living human skin," *American Journal of Anatomy,* 65:1–34.

Eiband, A. M.
1959. *Human Tolerance to Rapidly Applied Accelerations: a Summary of the Literature,* National Aeronautics and Space Administration Memo 5-19-59E, Washington, D.C.

Elbel, E. R.
1949. "Relationship between leg strength, leg endurance and other body measurements," *Journal of Applied Physiology,* 2:197–207.
1954. "Body measurements of male students entering the University of Kansas," *Kansas Studies in Education,* 4:1–24.

Elbel, E. R., and J. K. Barland
1959. "Height and weight of female students entering the University of Kansas," *University of Kansas Bulletin of Education,* 14:19–26.

Elftman, H.
1941. "The action of muscles in the body," *Biological Symposia,* 3:191–209.

Ellis, D. S.
1951. "Speed of manipulating performance as a function of work-surface height," *Journal of Applied Psychology,* 35:289–296.

Ely, J. H., R. M. Thomson, and J. Orlansky
1963a. "Design of controls," in *Human Engineering Guide to Equipment Design* (McGraw-Hill Book Company, New York), chap. 6.
1963b. "Layout of workplaces," in *Human Engineering Guide to Equipment Design* (McGraw-Hill Book Company, New York), chap. 7.

Emanuel, I.
n.d. "U.S. Air Force clothing increment," unpub. report, Anthropology Section, Aero Medical Laboratory, Wright-Patterson Air Force Base, Ohio.

Emanuel, I., M. Alexander, E. Churchill, and B. Truett
1959. *A Height-Weight Sizing System for Flight Clothing,* WADC Technical Report 56-365, Aero Medical Laboratory, Wright Air Development Center, Wright-Patterson Air Force Base, Ohio.

Emanuel, I., J. W. Chaffee, and J. Wing
1956. *A Study of Human Weight Lifting Capabilities for Loading Ammunition into the F-86H Aircraft,* WADC Technical Report 56-367, Aero Medical Laboratory, Wright Air Development Center, Wright-Patterson Air Force Base, Ohio.

Evans, F. G.
1957. *Stress and Strain in Bones* (Charles C Thomas, Springfield, Ill.).

Evans, F. G., H. R. Lissner, and M. Lebow
1958. "The relation of energy, velocity, and acceleration to skull deformation and fracture," *Surgery, Gynecology and Obstetrics,* 107:593–601.

Fasola, A. F., R. C. Baker, and F. A. Hitchcock
1955. *Anatomical and Physiological Effects of Rapid*

Deceleration, WADC Technical Report 54-218, Aero Medical Laboratory, Wright Air Development Center, Wright-Patterson Air Force Base, Ohio.

Fenn, W. O.
1938. "The mechanics of muscular contraction in man," *Journal of Applied Physiology,* 9:165–177.

Fidanza, F., A. Keys, and J. T. Anderson
1953. "Density of body fat in man and other mammals," *Journal of Applied Physiology,* 6:252–256.

Fischer, O.
1906. *Theoretische Grundlagen für eine Mechanik der lebenden Körper* (B. G. Teubner, Berlin).

Fisher, M. B., and J. E. Birren
1946. "Standardization of a test of hand strength," *Journal of Applied Psychology,* 30:380–387.
1947. "Age and hand strength," *Journal of Applied Psychology,* 31:490–497.

Fitts, P. M.
1947. "A study of location discrimination ability," in *Psychological Research on Equipment Design,* Research Report No. 19, Army Air Forces Aviation Psychology Program, Wright Field, Dayton, Ohio.

Floyd, W. F., and D. F. Roberts
1958. "Anatomical and physiological principles in chair and table design," *Ergonomics,* 2:1–16.

Fogel, L. J.
1963. *Biotechnology: Concepts and Applications* (Prentice-Hall, Inc., Englewood Cliffs, N.J.).

Folley, J. D., J. W. Altman, A. Chapanis, and J. S. Cook
1963. "Design for ease of maintenance," *Human Engineering Guide to Equipment Design* (McGraw-Hill Book Company, New York), chap. 9.

Forbes, G.
1945. "The effect of certain variables on visual and auditory reaction times," *Journal of Experimental Psychology,* 35:153–162.

Forbes, R. M., A. R. Cooper, and H. H. Mitchell
1953. "The composition of the adult human body as determined by chemical analysis," *Journal of Biological Chemistry,* 203:359–366.

Forrest, J., E. A. Wade, W. K. Carter, and R. F. Slechta
1958. *Light-Weight Seating: Design Research on a Nylon Net Seat,* WADC Technical Report 58-309, Aero Medical Laboratory, Wright Air Development Center, Wright-Patterson Air Force Base, Ohio.

Fox, K.
1957. *The Effect of Clothing on Certain Measures of Strength of Upper Extremities,* Technical Report EP-47, Environmental Protection Research Division, U.S. Army Quartermaster Research & Engineering Command, Natick, Mass.

Foxboro Company
1942. *A Study of Factors Determining Accuracy of Tracking by Means of Handwheel Control,* Office of Scientific Research and Development Report No. 3451, Washington, D.C.
1943a. *Handwheel Speed and Accuracy of Tracking,* Office of Scientific Research and Development Report No. 3453, Washington, D.C.
1943b. *Inertia, Friction, and Diameter in Handwheel Tracking,* Office of Scientific Research and Development Report No. 3454, Washington, D.C.

Frankford Arsenal
1954. *Engineering Manual—Cartridge Actuated Devices*

for Aircraft Use, Philadelphia, Pa. (cited in Eiband, 1959).

Gamble, J. L., and R. S. Shaw
1947. *Pathology in Dogs Exposed to Negative Acceleration,* Memorandum Report TSEAA-695-74B, Wright Field, Dayton, Ohio (cited in Eiband, 1959).
1948. *Animal Studies on Impact Negative Acceleration,* Report MCREXD-695-74G, Wright Field, Dayton, Ohio (cited in Eiband, 1959).

Gamble, W. D., and F. M. Townsend
1963. "An analysis of cardiovascular injuries resulting from accelerative force," *Aerospace Medicine,* 34:929–934.

Garn, S. M., and M. M. Gertler
1950. "An association between type of work and physique in an industrial group," *American Journal of Physical Anthropology,* 8:387–397.

Garner, J.
1936. "Proper seating," *Industrial Medicine,* 5:324–327.

Garry, R. C.
1930. "The factors determining the most effective push or pull which can be exerted by a human being on a straight lever moving in a vertical plane," *Arbeitsphysiologie,* 3:330–346.

Gauer, O., and S. Ruff
1939. "Die Erträglichkeitsgrenzen für Fliehkräfte in Richtung Rücken-Brust," *Luftfahrtmedizin* 3:225–230.

Gauer, O. H., and G. D. Zuidema
1961. *Gravitational Stress in Aerospace Medicine* (Little Brown and Co., Boston).

Gavan, J. A., S. L. Washburn, and P. H. Lewis
1952. "Photography: an anthropometric tool," *American Journal of Physical Anthropology,* 10:331–352.

Geertz, A.
1946. *Limits and Special Problems in the Use of Seat Catapults,* Aero Medical Center Translation, U.S. Army Air Forces (cited in Eiband, 1959).

General Services Administration
1950. *Specification, Chairs, Office, Steel,* No. 791b, Washington, D.C.

Geoghegan, B.
1953. "The determination of body measurements, surface area, and body volume by photography," *American Journal of Physical Anthropology,* 11:97–120.

Gibbons, T. B., I. A. Phillips, R. K. Budensiek, and J. R. Gilbertson
1953. *Age, Height, and Weight of 2173 Men Entering Recruit Training During 1952 at the U.S. Naval Training Center, Great Lakes, Ill.,* Naval Medical Research Unit No. 4, Great Lakes Naval Training Center, Ill.

Gifford, E. C.
1960. *Compilation of Anthropometric Measures on U.S. Navy Pilots,* NAMC-ACEL-437, Air Crew Equipment Laboratory, Philadelphia, Pa.

Gilliland, A. R.
1921. "Norms for amplitude of voluntary joint movement," *Journal of the American Medical Association,* 77:1357.

Girling, F., and E. D. L. Topliff
1958. *Dynamic Testing of Energy Absorbing Materials,* DRB Project No. D50-93-20-02, Defence Research Medical Laboratories, Toronto, Canada.

Glanville, A. D., and G. Kreezer
  1937. "The maximum amplitude and velocity of joint movements in normal male human adults," *Human Biology,* 9:197–211.
Goldman, D. E., and H. E. Von Gierke
  1960. *The Effects of Shock and Vibration on Man,* Lecture and Review Series No. 60-3, U.S. Naval Medical Research Institute, Bethesda, Md.
Gratz, C. M.
  1931. "Tensile strength and elasticity tests on human fascia lata," *Journal of Bone and Joint Surgery,* 13:334–340.
Green, M. R., and F. A. Muckler
  1959. *Speed of Reaching to Critical Control Areas in a Fighter-Type Cockpit,* WADC Technical Report 58-687, Aero Medical Laboratory, Wright Air Development Center, Wright-Patterson Air Force Base, Ohio.
Grether, W. F.
  1946. *Study of Several Design Factors Influencing Pilot Efficiency in the Operation of Controls,* Engineering Division Memorandum Report TSEAA-694-9, Aero Medical Laboratory, Wright Air Development Center, Wright Field, Dayton, Ohio.
Guibert, A., and C. L. Taylor
  1951. *The Radiation Area of the Human Body,* Air Force Technical Report 6706, Wright Air Development Center, Dayton, Ohio.
Gurdjian, E. S., J. E. Webster, and H. R. Lissner
  1950. "Biomechanics: fractures, skull," in *Medical Physics,* 2:98–104. O. Glasser, ed. (Chicago, Year Book Publishers).

Haddon, W. A., Jr., and R. A. McFarland
  1957- *A Survey of Present Knowledge of the Physical*
  1958   *Thresholds of Human Head Injury from an Engineering Standpoint,* Annual Report of Commission on Accidental Trauma, U.S. Armed Forces Epidemiological Board, Dept. of Defense.
Haggard, H. W.
  1946. "Mechanics of human muscles," *Mechanical Engineering,* 68:321–324.
Hall, A. L., J. A. Greenwood, and H. L. Dodson
  1950. *The Various Profile Areas of a 'Light,' 'Medium,' and 'Heavy' Airman Clothed in Various Types of Naval Flight Equipment,* Project No. PTR MED-7144 (NM 001 061), U.S. Naval Air Test Center, Patuxent River, Md.
Hall, N. B., and E. M. Bennett
  1956. "Empirical assessment of handrail diameter," *Journal of Applied Psychology,* 40:381–382.
Handbook of Biological Data
  1956. W. S. Spector, ed. (W. B. Saunders Co., Philadelphia, Pa.).
Handbook of Growth
  1962. P. L. Altman and D. S. Dittmer, eds., Federation of American Societies for Experimental Biology, Washington, D.C.
Hanna, T. D., and L. N. Libber
  1958. *The Present Status of the Navy Fatigue-Relieving Pneumatic Seat Cushion,* Air Crew Equipment Laboratory, Naval Air Materiel Center, Philadelphia, Pa.
Hansen, R., D. Y. Cornog, and H. T. E. Hertzberg, eds.
  1958. *Annotated Bibliography of Applied Physical An-*

thropology in Human Engineering, WADC Technical Report 56-30, Aero Medical Laboratory, Wright Air Development Center, Wright-Patterson Air Force Base, Ohio.

Hansen, S.
1912. "On the increase in stature in certain European populations," pp. 23–27, *Problems in Eugenics,* Proceedings of the First International Eugenics Congress, Eugenics Education Society, London.

Heath, C. W.
1945. *What People Are* (Harvard University Press, Cambridge, Mass.).

Hellebrandt, A., G. Genevieve, and R. H. Tepper
1937. "The relation of the center of gravity to the base of support in stance," *American Journal of Physiology,* 119: 331–332.

Hellebrandt, F., A. M. Parrish, and S. J. Houtz
1947. "Cross education, the influence of unilateral exercise on the contralateral limb," *Archives of Physical Medicine,* 28:76–85.

Helson, H.
1949. "Design of equipment and optimal human performance," *American Journal of Psychology,* 62:473–497.

Herbert, M. J.
1957. *Speed and Accuracy with Which Six Linear Arm Movements Can Be Visually Positioned from Two Different Control Locations,* Report No. 260, U.S. Army Medical Research Laboratory, Fort Knox, Ky.

Hertel, H.
1930. *Determination of the Maximum Control Forces and Attainable Quickness in the Operation of Airplane Controls,* Technical Memorandum No. 583, National Advisory Committee for Aeronautics, Washington, D.C.

Hertzberg, H. T. E.
n.d. "Arm strength in operation of aircraft controls," unpub. report, Aero Medical Laboratory, Wright-Patterson Air Force Base, Ohio.

1949. *Comfort Tests of the Pulsating Seat Cushion and Lumbar Pad,* Memorandum Report MCREXD-695-82, Engineering Division, Aero Medical Laboratory, Wright-Patterson Air Force Base, Ohio.

1954. "Maximum brake pedal pressures at various angles," unpub. report, Aero Medical Laboratory, Wright-Patterson Air Force Base, Ohio.

1955. "Some contributions of applied physical anthropology to human engineering," *Annals of the New York Academy of Sciences,* 63:616–629.

1956. Revision of *Handbook of Instructions for Aircraft Ground Support Equipment Design,* Paragraph c. 1-2.3.1, 23 October 1956. Anthropology Unit, Aero Medical Laboratory, Wright-Patterson Air Force Base, Ohio.

1960. "Dynamic anthropometry of working positions," *Human Factors,* 2:147–155.

1961. *Nylon Net Seat for a Modified RB-57 Aircraft,* ASD Technical Report 61-206, Aerospace Medical Research Laboratories, Wright Air Development Center, Wright-Patterson Air Force Base, Ohio.

Hertzberg, H. T. E., E. Churchill, C. W. Dupertuis, R. M. White, and A. Damon
1963. *An Anthropometric Survey of Turkey, Greece, and Italy* (Pergamon Press, London).

Hertzberg, H. T. E., and J. W. Colgan
   1948. *A Prone Position Bed for Pilots,* Memorandum Report MCREXD-695-71D, Aero Medical Laboratory, Wright-Patterson Air Force Base, Ohio.

Hertzberg, H. T. E., and G. S. Daniels
   1950. *The Center of Gravity of a Fully Loaded F-86 Ejection Seat in the Ejection Position.* Memorandum Report No. MCREXD-45341-4-5, USAF Air Materiel Command, Wright-Patterson Air Force Base, Ohio.

Hertzberg, H. T. E., G. S. Daniels, and E. Churchill
   1954. *Anthropometry of Flying Personnel—1950,* WADC Technical Report 52-321, Wright Air Development Center, Wright-Patterson Air Force Base, Ohio.

Hertzberg, H. T. E., C. W. Dupertuis, and I. Emanuel
   1957. "Stereophotogrammetry as an anthropometric tool," *Photogrammetric Engineering,* 24:942–947.

Hertzberg, H. T. E., I. Emanuel, and M. Alexander
   1956. *The Anthropometry of Working Positions. 1. A Preliminary Study,* WADC Technical Report 54-520, Wright-Patterson Air Force Base, Ohio.

Hettinger, T.
   1961. *Physiology of Strength* (Charles C Thomas, Springfield, Ill.).

Hettinger, T., and E. A. Müller
   1953. "Muskelleistung und Muskeltraining," *Arbeitsphysiologie,* 15:111–126 (cited in Hunsicker, 1955).

Hewitt, D.
   1928. "Range of active motion at the wrist of women," *Journal of Bone and Joint Surgery,* 10:775–787.

Hick, W. E.
   1944. "Psychological aspects of flying controls," *Aeronautics,* 10:34–40.

   1952. "On the rate of gain of information," *Quarterly Journal of Experimental Psychology,* 4:11–26.

Hirsch, C., and A. Nachemson
   1954. "New observations on the mechanical behavior of lumbar discs," *Acta Orthopedica Scandinavica,* 23: 254–283.

   1963. "Clinical observations on the spine in ejected pilots," *Aerospace Medicine,* 34:629–632.

Homans, J.
   1954. "Thrombosis of deep leg veins due to prolonged sitting," *New England Journal of Medicine,* 250: 148–149.

Hooton, E. A.
   1945. *A Survey in Seating* (Heywood-Wakefield Co., Gardner, Mass.).

Hugh-Jones, P.
   1947. "The effect of limb position in seated subjects on their ability to utilize the maximum contractile force of the limb muscles," *Journal of Physiology,* 105:332–344.

Hulse, F. S.
   1957. "Exogamie et hétérosis," *Archives Suisses d'Anthropologie Générale,* 22:103–125.

Human Engineering Guide to Equipment Design
   1963. U.S. Joint Armed Services, C. L. Morgan, A. Chapanis, J. L. Cook, and M. Lund, eds. (McGraw-Hill Book Company, New York).

Hunsicker, P. A.
   1955. *Arm Strength at Selected Degrees of Elbow Flexion,* WADC Technical Report 54-548, Aero Medical Laboratory, Wright-Patterson Air Force Base, Ohio.

1957. *A Study of Muscle Forces and Fatigue,* WADC Technical Report 57-586, Aero Medical Laboratory, Wright-Patterson Air Force Base, Ohio.

Hunsicker, P., and G. Greey
1957. "Studies in human strength," *Research Quarterly,* 28:109-122.

Hutchinson, F. W.
1950. "Experimental evaluation of human shape factors with respect to floor areas," *Transactions of the American Society of Mechanical Engineers,* 72: 889-891.

Hutchinson, F. W., and M. Baker
1952. "Shape factors for the average man" (abstract), *Heating and Ventilating,* 49:825.

Ikai, M., and A. H. Steinhaus
1961. "Some factors modifying the expression of human strength," *Journal of Applied Physiology,* 16:157-163.

Ishii, Y.
1957. "Studies on the physique and physical strength of workers (Report No. 7)," *Journal of Science of Labour* (Japan), 33:259-269.

Janoff, I. Z., L. H. Beck, and I. L. Child
1950. "The relation of somatotype to reaction time, resistance to pain, and expressive movement," *Journal of Personality,* 18:454-460.

Jeanneret, P. R., and W. B. Webb
1963. "Strength of grip on arousal from full night's sleep," *Perceptual and Motor Skills,* 17:759-761.

Jenkins, L. J., and M. B. Connor
1949. "Some design factors in making settings on a linear scale," *Journal of Applied Psychology,* 33:395-409.

Jones, W. L., and E. C. Gifford
n.d. *Anthropometry of U.S. Navy Pilots,* Air Crew Equipment Laboratory, Naval Air Materiel Center, Philadelphia, Pa.

Karpinos, B. D.
1958. "Height and weight of Selective Service registrants processed for military service during World War II," *Human Biology,* 30:292-321.
1961. "Current height and weight of youths of military age," *Human Biology,* 33:335-354.

Karpovich, P. V.
1953. *Physiology of Muscular Activity* (W. B. Saunders Co., Philadelphia, Pa.).

Karpovich, P. V., E. L. Herden, and M. M. Asa
1960. "Electrogoniometric study of joints," *U.S. Armed Forces Medical Journal,* 11:424-450.

Karvonen, M. J., A. Koselka, and L. Noro
1962. "Preliminary report on the sitting postures of school children," *Ergonomics,* 5:471-477.

Katchmar, L. T.
1957. *Physical Force Problems: I. Hand Crank Performance for Various Crank Radii and Torque Load Combinations,* Technical Memorandum No. 3-57, Human Engineering Laboratory, U.S. Army Ordnance, Aberdeen Proving Ground, Md.

Keegan, J. J.
1962. "Evaluation and improvement of seats," *Industrial Medicine and Surgery,* 31:137-148.
1964. *The Medical Problem of Lumbar Spine Flattening in Automobile Seats,* Society of Automotive Engineers, New York.

Kelly, H. J., H. J. Souders, A. T. Johnston, L. E. Bound,

H. A. Hunscher, and I. G. Macy
  1943. "Daily decreases in the body total and stem lengths of normal children," *Human Biology,* 15:65–72.
Kennedy, K. W.
  1964. *Reach Capability of the USAF Population. Phase I. The Outer Boundaries of Grasping-Reach Envelopes for the Shirt-Sleeved, Seated Operator.* AMRL-TDR-64-59, Aerospace Medical Research Laboratories, Wright-Patterson Air Force Base, Ohio.
Keys, A., and J. Brozek
  1953. "Body fat in adult man," *Physiological Reviews,* 33:245–325.
Keys, A., J. Brozek, A. Henschel, O. Mickelsen, and H. L. Taylor
  1950. *The Biology of Human Starvation* (University of Minnesota Press, Minneapolis).
King, B. G., D. G. Morrow, and E. P. Vollmer
  1947. *Cockpit Studies—The Boundaries of the Maximum Area for the Operation of Manual Controls,* Project X-651, Report 3, Naval Medical Research Institute, National Naval Medical Center, Bethesda, Md.
King, B. G., R. Ostrich, and M. C. Richardson
  1954. *Emergency Escape Procedures,* Medical Division, Civil Aeronautics Administration, Department of Commerce, Washington, D.C.
Kobrick, J. L.
  1956. *Spatial Dimensions of the 95th Percentile Arctic Soldier,* Technical Report EP-39, Environmental Protection Research Division, U.S. Quartermaster Research and Development Center, Natick, Mass.
  1957. *Dimensions of the Lower Limit of Body Size of the Arctic Soldier,* Technical Report EP-51, Environmental Protection Research Division, U.S. Quartermaster Research and Development Center, Natick, Mass.
Koch, H.
  1941. "Fussbediente Arbeitsmaschinen und Ermüdung," *Arbeitsschutz,* 3:52–59.
Kocker, A. L., and A. Frey
  1932. "Seating heights and spacing," *Architectural Record,* 71:261–269.
Koepke, C. A., and L. S. Whitson
  1940. "Power and velocity developed in manual work," *Mechanical Engineering,* 62:383–389.
Krendel, E. S.
  1960. "Man-generated power," *Mechanical Engineering,* 82:36–39.
Küntscher, G.
  1936. "Die Spannungsverteilung am Schenkelhals," *Archiv für Klinische Chirurgie,* 185:308–321.
Lauru, L., and L. Brouha
  1957. "Physiological study of motions," *Advanced Management,* 22:17–24.
Lawrence, J. H.
  1956. "Some studies of function and structure in medicine," *Quarterly Bulletin,* Northwestern University Medical School, Chicago, 30:215–225.
Lay, W. E., and L. C. Fisher
  1940. "Riding comfort and cushions," *Society of Automotive Engineers Journal,* 47:482–496.
Lease, G. O'D., and F. G. Evans
  1959. "Strength of human metatarsal bones under repetitive loading," *Journal of Applied Physiology,* 14:49–51.

Le Gros Clark, W. E.
  1946. "Contribution of anatomy to the War," *British Medical Journal,* 1:39–43.
Le Gros Clark, W. E., and G. Weddell
  1944. *The Pressure Which Can Be Exerted by the Foot of a Seated Operator with the Leg in Various Positions,* R.N.P. 44/153, Royal Naval Personnel Research Committee Report, Great Britain.
Lehmann, G.
  1927. "Arbeitsphysiologische Studien:IV," *Pflüger's Archiv für Gesamte Physiologie,* 215:329–364.
  1958. "Physiological basis of tractor design," *Ergonomics,* 1:197–206.
Lippert, S.
  1950. "Designing for comfort in aircraft seats," *Aeronautical Engineering Review,* 9:39–41.
Lobron, C. M., and R. C. Hedberg
  1954. *The Maximum Torque a Man Can Apply to a $1\frac{1}{8}$ Inch Knob,* Human Engineering Report No. 5, Frankford Arsenal Laboratory, Philadelphia, Pa.
Loftus, J. P., and L. R. Hammer
  1961. *Weightlessness and Performance, A Review of the Literature,* ASD Technical Report 61-166, Aerospace Medical Laboratory, Wright-Patterson Air Force Base, Ohio.
Lombard, C. F., *et al.*
  1948. *The Effects of Negative Radial Acceleration on Large Experimental Animals (Goats),* Project NR 161-014, Office of Naval Research, University of Southern California, Los Angeles (cited in Eiband, 1959).
Lombard, C. F., A. A. Canfield, M. D. Warren, and D. R. Drury

  1948. *The Influence of Positive (Head to Foot) Centrifugal Force Upon a Subject's (Pilot's) Ability to Exert Maximum Pull on an Aircraft Control Stick,* Research Report Con. N6 ori-77, Tasks 1 and 3, Office of Naval Research, Washington, D.C.
Mackworth, N. H.
  1950. *Researches on the Measurement of Human Performance,* Special Report Series No. 268, Medical Research Council, London.
Magid, E. B., and R. R. Coermann
  1963. "Human response to vibration," pp. 86–119 in *Human Factors in Technology* (E. Bennett, J. Degan, J. Spiegel, eds.), (McGraw-Hill Book Company, New York).
Magnus-Levy, A.
  1910. "Über den Gehalt normaler menschlicher Organer an Chlor, Calcium, Magnesium und Eisen sowie an Wasser, Eiweiss, und Fett," *Biochemische Zeitschrift,* 24:363–380.
Mainland, D.
  1963. *Elementary Medical Statistics* (W. B. Saunders Co., Philadelphia, Pa.).
Malling-Hansen, R. (1886)
  1929. Cited in G. Nylin, "Periodical variations in growth, standard metabolism, and oxygen capacity of the blood in children," *Acta Medica Scandinavica,* Supplementum 31.
Martin, E. G.
  1921. "Tests of muscular efficiency," *Physiological Reviews,* 1:454–475.
Martin, W. B., and E. E. Johnson
  1952. *An Optimum Range of Seat Positions as Determined by Exertion of Pressure Upon a Foot Con-*

trol, Report No. 86, U.S. Army Medical Research Laboratory, Fort Knox, Ky.

Marvin, P. R.
1946. "Biomechanics," *Mechanical Engineering,* 68: 569–570.

Mason, J. K.
1957. *Failed Escape from Aircraft,* Memorandum No. 3, Joint Commission on Aviation Pathology, U.S. Armed Forces Institute of Pathology.

McArthur, W.
1945. "Aircraft seating in transport airplanes," unpub. ms.

McCormick, E. J.
1957. *Human Engineering* (McGraw-Hill Book Company, New York).

McCormick, N. Y.
1947. *Physical Characteristics of the 1946 Airline Transport Pilot Population,* Medical Service Publication, Civil Aeronautics Authority, U.S. Department of Commerce, Washington, D.C.

McFadden, E. B., and J. J. Swearingen
1958. "Forces that may be exerted by man in the operation of aircraft door handles," *Human Factors,* 1(1):16–22.

McFadden, E. B., J. J. Swearingen, and C. D. Wheelright
1959. "The magnitude and direction of forces that man can exert in operating aircraft emergency exits," *Human Factors,* 1(4):16–27.

McFarland, R. A.
1937. "Psycho-physiological studies at high altitude in the Andes. II. Sensory and motor responses during acclimatization," *Journal of Comparative Psychology,* 23:227–258.

McFarland, R. A., A. Damon, and H. W. Stoudt
1958. "Anthropometry in the design of the driver's workspace," *American Journal of Physical Anthropology,* 16:1–23.

McFarland, R. A., and H. W. Stoudt
1961. *Human Body Size and Passenger Vehicle Design,* SP-142 A, Society of Automotive Engineers, New York, N.Y.

McKee, M. E.
1957. *The Effect of Clothing on the Speed of Movement in the Upper Extremity,* Technical Report EP-48, Environmental Protection Research Division, U.S. Army Quartermaster Research and Engineering Center, Natick, Mass.

Meade, D.
n.d. "Furniture and fittings," *Better Offices,* Institute of Directors, London.

Merkel, J.
1885. "Die zeitlichen Verhältnisse der Willensthätigkeit," *Philosophische Studien,* 2:73–127 (cited in Woodworth and Schlosberg, 1955).

Miles, D. W.
1937. "Preferred rates in rhythmic responses," *Journal of General Psychology,* 16:427–469.

Mitchell, M. J. H., and M. A. Vince
1951. *The Direction of Movement of Machine Controls,* M.R.C.—A.P.U. 137/50, Applied Psychology Research Unit, Cambridge University.

Morant, G. M.
1947. "Anthropometric problems in the Royal Air Force," *British Medical Bulletin,* 5:25–31. (Also, *Yearbook of Physical Anthropology,* 3:197–203).

Moseley, H. G.
1957. *Design Factors that Induce Pilot Error,* Publication M-8-57, Directorate of Flight Safety Research, Norton Air Force Base, Calif.

Moseley, H. G., F. M. Townsend, and V. A. Stembridge
1958. "Prevention of death and injury in aircraft accidents," *Archives of Industrial Health,* 17:111–117.

Mosso, A.
1884. "Application de la balance à l'étude de la circulation du sang chez l'homme," *Archives Italiennes de Biologie,* 5:130–143.

Müller, E. A.
1934. "Der beste Handgriff und Stiel," *Arbeitsphysiologie,* 8:28–42.
1936. "Die günstigste Anordnung im sitzen betätigter Fusshabel," *Arbeitsphysiologie,* 9:125–137 (translated as "Optimum arrangement of pedals to be operated from the sitting position," Air Technical Intelligence Translation 182580, F-TS-8336/V, Wright-Patterson Air Force Base, Ohio, n.d.).
1959. "Training muscle strength," *Ergonomics,* 2:216–222.

Müller, E. A., K. Vetter, and E. Blümel
1958. "Transport by muscle power over short distances," *Ergonomics,* 1:222–225.

Nareff, M. J.
1959. " 'Passenger phlebitis': a complication of long distance aerial travel," *Journal of Aviation Medicine,* 30:197 (abstract).

National Aircraft Standards Committee
n.d. *Specification—Aircraft Seats and Berths,* Aircraft Industries Association of America, Washington, D.C.

Neely, S. E., and R. H. Shannon
1958. "Vertebral fractures in survivors of military aircraft accidents," *Journal of Aviation Medicine,* 29:750–757.

Nemethi, C. E.
1952. "An evaluation of hand grip in industry," *Industrial Medicine and Surgery,* 21:65–66.

Newman, R. W.
1956. "Skinfold measurements in young American males," *Human Biology,* 28:154–164.
1963. "The body sizes of tomorrow's young men," chap. 8 in *Human Factors in Technology,* E. Bennett, J. Degan, and J. Spiegel, eds. (McGraw-Hill Book Company, New York).

Newman, R. W., and E. Munro
1955. "The relation of climate and body size in U.S. males," *American Journal of Physical Anthropology,* n.s., 13:1–18.

Newman, R. W., and R. M. White
1951. *Reference Anthropometry of Army Men,* Report No. 180, Environmental Protection Section, Quartermaster Climatic Research Laboratory, Lawrence, Mass.

Nicoloff, C.
1957. *Effects of Clothing on Range of Motion in the Arm and Shoulder Girdle,* Technical Report EP-49, U.S. Army Quartermaster Research and Engineering Center, Natick, Mass.

Nissley, H. R.
1952. "Is there an ideal factory chair?" *Management Review,* 41:175–176.

Noskoff, S.
1942. *Moments of Inertia of Man,* Douglas Aircraft Co. Report SM-3922.

O'Brien, R., and W. C. Shelton
1941. *Women's Measurements for Garment and Pattern Construction,* Miscellaneous Publications No. 454, U.S. Dept. of Agriculture, Washington, D.C.

Orlansky, J.
1948. *The Human Factor in the Design of Stick and Rudder Controls for Aircraft,* Contract N ori:-151, Project No. 20-M-lc, Special Devices Center, Office of Naval Research, Washington, D.C.

Pascale, L. R., M. I. Grossman, H. S. Sloane, and T. Frankel
1956. "Correlations between thickness of skinfolds and body density in 88 soldiers," *Human Biology* 28:165–176.

Paton, C. R., E. C. Pickard, and V. H. Hoehn
1940. "Seat cushions and the ride problem," *Society of Automotive Engineers Journal* (*Transactions*) 47:273–283.

Peters, W., and A. A. Wenborne
1936. "The pattern time of voluntary movements," *British Journal of Psychology,* 26:388–406.

Pett, L. B., and G. F. Ogilvie
1956. "The Canadian weight-height survey," *Human Biology,* 28:177–188.

Pett, L. B., and G. F. Ogilvie
1957. "The report on Canadian average weights, heights and skinfolds," *Canadian Bulletin on Nutrition* 5:1–81.

Pierson, W. R.
1962a. "The estimation of body surface area by mono-photogrammetry," *American Journal of Physical Anthropology,* 20:399–402.

1962b. "The relationship of body mass and composition to the rapidity of voluntary movement," *Journal of Sports Medicine and Physical Fitness,* 2:205–206.

Pierson, W. R., and H. J. Montoye
1958. "Movement time, reaction time and age," *Journal of Gerontology,* 13:418–421.

Provins, K. A.
1953. *A Study of Some Factors Affecting Speed of Cranking,* R.N.P. 53/755, O.E.S. 237, Royal Naval Personnel Research Committee, Medical Research Council (Great Britain).

1955. "Maximum forces exerted about the elbow and shoulder joints on each side separately and simultaneously," *Journal of Applied Physiology,* 7:390–392.

Provins, K. A., and N. Salter
1955. "Maximum torque exerted about the elbow joint," *Journal of Applied Physiology,* 7:393–398.

Pugh, G., and M. Ward
1953. "Physiology and medicine," appendix VII in *The Ascent of Everest,* by J. Hunt (Hodder and Stoughton, London).

Randall, F. E.
1947. *Survey of Body Size of Army Personnel, Male and Female: (4) Body Dimensions of Army Females— Methodology and General Considerations,* Report 123, U.S. Quartermaster Climatic Research Laboratory, Lawrence, Mass.

Randall, F. E., and M. J. Baer
1947. *Survey of Body Size of Army Personnel, Male and Female: (1) Methodology,* Report No. 122, U.S.

Quartermaster Climatic Research Laboratory, Lawrence, Mass.

Randall, F. E., A. Damon, R. S. Benton, and D. I. Patt
1946. *Human Body Size in Military Aircraft and Personal Equipment,* Army Air Forces Technical Report 5501, Air Materiel Command, Wright Field, Dayton, Ohio.

Randall, F. E., and E. H. Munro
1949. *Reference Anthropometry of Army Women,* Report No. 159, Environmental Protection Section, Quartermaster Climatic Research Laboratory, Lawrence, Mass.

Randall, F. E., E. H. Munro, and R. M. White
1951. *Anthropometry of the Foot,* Environmental Protection Section Report No. 172, Quartermaster Climatic Research Laboratory, Lawrence, Mass.

Rasch, P. J., and W. R. Pierson
1963. "Some relationships of isometric strength, isotonic strength, and anthropometric measures," *Ergonomics,* 6:211–236.

Reed, J. D.
1949. "Factors influencing rotary performance," *Journal of Psychology,* 28:65–92.

Reedy, J. D.
1953. *The Relation of Power and Endurance Training to Physical Efficiency,* Report No. 118, U.S. Army Medical Research Laboratory, Fort Knox, Ky.

Rees, J. E., and N. E. Graham
1952. "The effect of backrest position on the push which can be exerted on an isometric foot-pedal," *Journal of Anatomy,* 86:310–319.

Reijs, J. H. O.
1921. "Über die Veränderung der Kraft während der Bewegung," *Pflüger's Archiv für die gesamte Physiologie der Menschen und der Tiere,* 191:234–257.

Reynolds, E., and R. W. Lovett
1909. "A method of determining the position of the center of gravity in its relation to certain bony landmarks in the erect position," *American Journal of Physiology,* 24:289–293.

Roberts, D. F.
1960. "Functional anthropometry of elderly women," *Ergonomics,* 3:321–327.

Roberts, D. F., K. A. Provins, and R. J. Morton
1959. "Arm strength and body dimensions," *Human Biology,* 31:334–343.

Roberts, L. B.
1945. *Size Increase of Men Wearing Various Clothing Combinations,* Project No. 9 SPMEA 741–3, Armored Medical Research Laboratory, Fort Knox, Ky.

Robinson, F. R., R. L. Hamlin, W. M. Wolff, and R. R. Coermann
1963. Response of the rhesus monkey to lateral impact, *Aerospace Medicine,* 34:56–62.

Roebuck, J. A.
1957a. "Anthropometry in aircraft engineering design," *Journal of Aviation Medicine,* 28:41–56.
1957b. *Effects of Opening-Size Variables on Overwing Emergency-Exit Time,* Paper presented at the Air Safety Seminar sponsored by the Flight Safety Foundation, Inc., Palo Alto, Calif., November 12.
1961. "Aircraft ground emergency-exit design considerations," *Human Factors,* 3:174–209.

Ronnholm, N., M. J. Karvonen, and V. O. Lapinleimu
1962. "Mechanical efficiency of rhythmic and paced work

of lifting," *Journal of Applied Physiology,* 17:768–770.

Ross, D. M.
1956. *The Use of Mechanical Principles in the Evaluation of a Human Functional Dimension,* unpub. diss., University of Pittsburgh School of Public Health, Pittsburgh, Pa.

Rossen, R., H. Kabat, and J. P. Anderson
1943. "Acute arrest of cerebral circulation in man," *Archives of Neurology and Psychiatry,* 50:510–528.

Ruff, S.
1950. "Brief acceleration: less than one second," *German Aviation Medicine in World War II,* vol. I., pp. 584–597, U.S. Government Printing Office, Washington, D.C.

Ruff, S., and H. Strughold
1942. *Compendium of Aviation Medicine* (translated from the German), National Research Council, Washington, D.C.

Salter, N.
1955. "Methods of measurement of muscle and joint function," *Journal of Bone and Joint Surgery* 37-B:474–491.

Salter, N., and H. D. Darcus
1952. "The effect of the degree of elbow flexion on the maximum torques developed in pronation and supination of the right hand," *Journal of Anatomy,* 80:197–202.
1953. "The amplitude of forearm and of humeral rotation," *Journal of Anatomy,* 87:407–418.

Santschi, W. R., J. DuBois, and C. Omoto
1963. *Moments of Inertia and Centers of Gravity of the Living Human Body,* Technical Documentary Report No. AMRL-TDR-63-36, Wright-Patterson Air Force Base, Ohio.

Saul, E. V., and J. Jaffe
1955. *Effects of Clothing on Gross Motor Performance,* Technical Report EP-12, U.S. Army Quartermaster Research and Engineering Command, Natick, Mass.

Schneider, E. C.
1939. *Physiology of Muscular Activity,* 2d ed. (W. B. Saunders Co., Philadelphia, Pa.).

Schochrin, W. A.
1934. "Die Muskelkraft der Beuger und Strecker des Unterschenkels," *Arbeitsphysiologie,* 8:251–260.

Seale, R. U.
1959. "The weight of the dry fat-free skeleton of American whites and Negroes," *American Journal of Physical Anthropology,* 17:37–48.

Seashore, S. H., and R. H. Seashore
1941. "Individual differences in simple auditory reaction times of hands, feet and jaws," *Journal of Experimental Psychology,* 29:342–345.

Seltzer, C. C.
1946. "Chest circumference changes as a result of severe physical training," *American Journal of Physical Anthropology,* 4:389–393.

Sendroy, J., Jr.
1961. "Surface area techniques and their relationship to body composition," pp. 59–68 in *Techniques for Measuring Body Composition* (J. Brozek and A. Henschel, eds.) National Academy of Sciences—National Research Council, Washington, D.C.

Sendroy, J., Jr., and L. P. Cecchini

1954. "Determination of human body surface area from height and weight," *Journal of Applied Physiology,* 7:1–12.

Sendroy, J., Jr., and H. A. Collison
1960. "Nomogram for determination of human body surface area from height and weight," *Journal of Applied Physiology,* 15:958–959.

Sharp, E. D.
1962. *Maximum Torque Exertable on Knobs of Various Sizes and Rim Surfaces,* Technical Documentary Report No. MRL-TDR-G2-17, Aerospace Medical Research Laboratories, Wright-Patterson Air Force Base, Ohio.

Sheldon, W. H., S. S. Stevens, and W. B. Tucker
1940. *The Varieties of Human Physique* (Harper and Brothers, New York).

Silliphant, W. M., and V. A. Stembridge
1958. "Aviation pathology," *U.S. Armed Forces Medical Journal,* 9:207–223.

Sills, F. D.
1950. "A factor analysis of somatotypes and of their relationship to motor skills," *Research Quarterly,* 21: 424–437.

Silverman, E. N.
1957. *Study of Foot and Buttocks,* Personal communication from Alderson Research Laboratories, Inc., New York, N.Y.

Simon, J. R.
1960. "Changes with age in the speed of performance on a dial setting task," *Ergonomics,* 3:169–174.

Simonson, E.
1947. "Physical fitness and work capacity of older men,"

Geriatrics, 2:110–119.

Sinelnikoff, E., and M. Grigorowitsch
1931. "Die Beweglaigkeit der Gelenke als sekundäres geschlechtliches und konstitutionelles Merkmal," *Zeitschrift für Konstitutionslehre,* 15:679–693.

Slater-Hammel, A. T.
1952. "Reaction time and speed of movement," *Perceptual Motor Skills Research Exchange,* 4:110–113.

Slechta, R. F., E. A. Wade, W. K. Carter, and J. Forrest
1957. *Comparative Evaluation of Aircraft Seating Accommodation,* WADC Technical Report 57-136, Aero Medical Laboratory, Wright-Patterson Air Force Base, Ohio.

Smith, G. M., and H. K. Beecher
1959. "Amphetamine sulphate and athletic performance. I. Objective effects," *Journal of the American Medical Association,* 170:542–557.

Smith, L. D.
1953. "Hip fractures," *Journal of Bone and Joint Surgery,* 35-A:367–383.

Smyth, C. J. (ed.)
1959. "Rheumatism and arthritis," *Annals of Internal Medicine,* 50:366–801.

Snyder, R. G.
1963. "Human tolerances to extreme impacts in free-fall," *Aerospace Medicine,* 34:695–709.

Snyder, R. G., J. Ice, J. C. Duncan, A. S. Hyde, and S. Leverett, Jr.
1963. *Biomedical Research Studies in Acceleration, Impact, Weightlessness, Vibration, and Emergency Escape and Restraint Systems: A Comprehensive*

Bibliography, Civil Aeromedical Research Institute, Federal Aviation Agency, Oklahoma City, Okla.; Aerospace Medical Research Laboratories, USAF School of Aerospace Medicine, Brooks Air Force Base, Texas.

Society of Actuaries
1959. *Build and Blood Pressure Study,* Society of Actuaries, Chicago, Ill.

Society of Automotive Engineers
n.d. *Proposed S.A.E. Seating Manual,* New York, N.Y.

Squires, P. C.
1956. *The Shape of the Normal Work Area,* Report No. 275, U.S. Naval Medical Research Laboratory, New London, Conn.

Stapp, J. P.
1949. *Human Exposures to Linear Deceleration: I. Preliminary Survey of Aft-Facing Seated Position,* AF-TR-5915, Aero Medical Laboratory, Wright-Patterson Air Force Base, Ohio (cited in Eiband, 1959).

1951. *Human Exposures to Linear Deceleration: II. The Forward-Facing Position and the Development of a Crash Harness,* AF-TR-5915, Aero Medical Laboratory, Wright-Patterson Air Force Base, Ohio (cited in Eiband, 1959).

1955a. "Effects of mechanical force on living tissues," *Journal of Aviation Medicine,* 26:268–288; 1956, 27:407–413; 1957, 28:281–290.

1955b. "Tolerance to abrupt deceleration," in *Collected Papers on Aviation Medicine,* AGARDograph No. 6, Butterworth's Scientific Publications, London (cited in Eiband, 1959).

1957. "Human tolerance to deceleration," *American Journal of Surgery,* 93:734–740.

Stauffer, F. R.
1951. "Further evidence of fluid translocation during varied acceleration stresses: Gross pathological findings and weight changes in specific tissues," unpub. paper, Department of Aviation Medicine, University of Southern California, Los Angeles (cited in Eiband, 1959).

1953. "Acceleration problems of Naval air training. I. Normal variations in tolerance to positive radial accelerations," *Journal of Aviation Medicine,* 24:167–186.

Steindler, A.
1935. *Mechanics of Normal and Pathological Locomotion in Man* (Charles C Thomas, Springfield, Ill.).

1955. *Kinesiology of the Human Body* (Charles C Thomas, Springfield, Ill.).

Steinhaus, A.
1933. "Chronic effects of exercise," *Physiological Reviews,* 13:103–148.

Stevenson, P. H.
1928. "Calculation of the body-surface area of Chinese," *Chinese Journal of Physiology,* Report Series 1:13–24.

Stoudt, H. W., A. Damon, and R. A. McFarland
1960. "Heights and weights of white Americans," *Human Biology,* 32:331–341.

Stoudt, H. W., A. Damon, R. A. McFarland, and J. Roberts
1965. *Weight, Height and Selected Body Dimensions of Adults, United States 1960–1962,* Public Health Service Publication No. 1000, Series 11, No. 8, U.S.

Government Printing Office, Washington, D.C.

Stump, N. E.
1953. *Manipulability of Rotary Controls as a Function of Knob Diameter and Control Orientation,* Technical Note WCRC 53-12, Aero Medical Laboratory, Wright-Patterson Air Force Base, Ohio.

Sutro, P. J., and G. H. Kydd
1955. *Human Visual Capacities as a Basis for the Safer Design of Vehicles,* Annual Report to the Commission on Accidental Trauma of the Armed Forces Epidemiological Board, CAA Project No. 53-209, Civil Aeronautics Administration, U.S. Department of Commerce.

Swearingen, J. J.
1949. *Determination of the Most Comfortable Knee Angle for Pilots,* Report No. 1, Biotechnology Project No. 3-48, Civil Aeronautics Medical Research Laboratory, Oklahoma City, Okla.
1953. *Determination of Centers of Gravity of Man,* Civil Aeronautics Medical Research Laboratory, Oklahoma City, Okla.

Swearingen, J. J., C. D. Wheelwright, and J. D. Garner
1962. *An Analysis of Sitting Areas and Pressures of Man,* Civil Aeromedical Research Institute, Oklahoma City, Okla.

Switzer, S. A.
1962. *Weight-lifting Capabilities of a Selected Sample of Human Males,* Technical Documentary No. MRL-TDR-62-57, Aerospace Medical Research Laboratories, Wright-Patterson Air Force Base, Ohio.

Tanner, J. M.
1952. "The effect of weight-training on physique," *American Journal of Physical Anthropology,* n.s., 10:427–462.

Taylor, C. L.
1954. "The biomechanics of the normal and of the amputated upper extremity," in *Human Limbs and Their Substitutes,* P. E. Klopsteg and P. D. Wilson, eds. (McGraw-Hill Book Company, New York).

Taylor, C. L., and A. C. Blaschke
1951. "A method for kinematic analysis of motions of the shoulder, arm, and hand complex," *Annals of the New York Academy of Sciences,* 51:1251–1265.

Teichner, W. H.
1954. "Recent studies of simple reaction time," *Psychological Bulletin,* 51:127–149.
1958. "Reaction time in the cold," *Journal of Applied Psychology,* 42:54–59.

Thomson, R. M., B. J. Covner, H. H. Jacobs, and J. Orlansky
1963. "Arrangement of groups of men and machines," chap. 8, in *Human Engineering Guide to Equipment Design* (McGraw-Hill Book Company, New York).

Tuttle, W. W., C. D. Janney, and C. W. Thompson
1950. "Relation of maximum grip strength to grip strength endurance," *Journal of Applied Psychology,* 2:663–670.

Tyroler, H. A.
1958. *Preliminary Cross-Sectional Documentation of Physical Measurements of Male Populations at Risk by Geographic Location and Age,* Asheville, N.C., Health Research Foundation.

U.S. Air Force
1953. *B-52 Airplane Mock-up Human Factors Study,*

Technical Memorandum Report WCRD 53–130, Aero Medical Laboratory, Wright-Patterson Air Force Base, Ohio.

U.S. Navy
1949. "Height distribution of recruits," *Statistics of Navy Medicine,* 5:2–3.

Vernon, H. M.
1924. "The influence of rest pauses and changes of posture on the capacity for muscular work," in *The Effects of Posture and Rest in Muscular Work,* Industrial Fatigue Research Board, Medical Research Council (Great Britain).

Virgin, W. J.
1951. "Experimental investigations into the physical properties of the intervertebral disc," *Journal of Bone and Joint Surgery,* 33-B;607–611.

Washburn, C. T.
1932. "Stadium seating," *Architectural Record,* 71:270–272.

Watts, D. T.
1947. *Human Tolerance to Accelerations Applied from Seat to Head During Ejection Seat Tests,* Report TED-NAM-2560-5, Naval Materiel Center, Philadelphia, Pa. (cited in Eiband, 1959).

Weinbach, A. P.
1938. "Contour maps, center of gravity, moment of inertia, and surface area of the human body," *Human Biology,* 10:356–371.

West, C. C.
1945. "Measurement of joint motion," *Archives of Physical Medicine,* 26:414–425.

West, E. S., and W. R. Todd
1961. *Textbook of Biochemistry,* 3d ed. (The Macmillan Company, New York).

White, C. B., P. J. Johnson, and H. T. E. Hertzberg
1952. *Review of Escape Hatch Size for Bailout and Ditching,* Technical Note WCRD 52-81, Aero Medical Laboratory, Wright-Patterson Air Force Base, Ohio.

White, R. M.
1961. *Anthropometry of Army Aviators,* Technical Report EP-150, Quartermaster Research and Engineering Center, Natick, Mass.

Whitney, R. J.
1958. "The strength of the lifting action in man," *Ergonomics,* 1:101–128.

Whitsett, C. E.
1963. *Some Dynamic Response Characteristics of Weightless Man,* Technical Documentary Report No. AMRL-TDR-63-18, Aerospace Medical Research Laboratories, Wright-Patterson Air Force Base, Ohio.

Whittenberger, R. K.
1959. *Improved Seat and Back Cushions,* WADC Technical Report 59-376, Aerospace Medical Laboratory, Wright-Patterson Air Force Base, Ohio.

Wilkie, D. R.
1960. "Man as a source of mechanical power," *Ergonomics,* 3:1–8.

Williams, R. J.
1942. *A Textbook of Biochemistry,* 2d ed. (D. Van Nostrand Co., Inc., Princeton, N.J.).

Woodson, W. E.
1954. *Human Engineering Guide for Equipment Design-*

ers (University of California Press, Berkeley).

Woodworth, R. S., and H. Schlosberg
    1955. *Experimental Psychology,* rev. ed. (Henry Holt and Company, Inc., New York).

Wright, V.
    1959. "Factors influencing diurnal variation of strength of grip," *Research Quarterly,* 38:110–116.

Yochelson, S.
    1930. *Effects of Rest Pauses on Work Decrement,* unpub. diss., Yale University, New Haven, Conn. (cited in Hunsicker, 1955).

Index